Roger Wrubel
627-3506

Risk Assessment in Genetic Engineering

The McGraw-Hill Environmental Biotechnology Series

Series Editors

Published

Ehrlich & Brierley (eds.) • Microbial Mineral Recovery

Labeda (ed.) • Isolation of Biotechnological Organisms from Nature

Levy & Miller (eds.) • Gene Transfer in the Environment

Saunders & Saunders (eds.) • Microbial Genetics Applied to Biotechnology: Principles and Techniques of Gene Transfer and Manipulation

Nakas & Hagedorn (eds.) • Biotechnology of Plant-Microbe Interactions

Edwards (ed.) • Microbiology of Extreme Environments

Forthcoming

Zeikas & Johnson (eds.) • Mixed Cultures in Biotechnology

Risk Assessment in Genetic Engineering

Morris A. Levin
Maryland Biotechnology Institute
The University of Maryland
Baltimore, Maryland

Harlee S. Strauss
H. Strauss Associates, Inc.
Natick, Massachusetts

McGraw-Hill, Inc.
New York St. Louis San Francisco Auckland Bogotá
Caracas Hamburg Lisbon London Madrid
Mexico Milan Montreal New Delhi Paris
San Juan São Paulo Singapore
Sydney Tokyo Toronto

Library of Congress Cataloging-in-Publication Data

Levin, Morris A.
Risk assessment in genetic engineering / Morris A. Levin, Harlee S. Strauss.
p. cm. — (McGraw-Hill environmental biotechnology series)
Includes bibliographical references and index.
ISBN 0-07-037416-3
1. Genetic engineering—Risk assessment. I. Strauss, Harlee S. II. Title. III. Series: Environmental biotechnology.
QH442.L48 1990
660'.65—dc20 90-47782
CIP

1 2 3 4 5 6 7 8 9 0 DOC/DOC 9 8 7 6 5 4 3 2 1 0

The sponsoring editor for this book was Trev Léger, the editing supervisor was Joseph Bertuna, and the production supervisor was Pamela Pelton. It was set in Century Schoolbook by McGraw-Hill's Professional Publishing composition unit.

Printed and bound by R. R. Donnelley & Sons Company.

Contents

Contributors

Judith A. Chambers American Association for the Advancement of Science, U.S. Agency for International Development, Washington, D.C. 20523-1809 (CHAP. 17)

Joel I. Cohen American Association for the Advancement of Science, U.S. Agency for International Development, Washington, D.C. 20523-1809 (CHAP. 17)

Joseph Fiksel Cimflex Teknowledge Corporation, P.O. Box 10119, Palo Alto, California 94303 (CHAP. 15)

Robert J. Frederick U.S. Environmental Protection Agency, RD-682, 401 M Street, S.W., Washington, D.C. 20460 (CHAP. 2)

James R. Fuxa Department of Entomology, Louisiana Agricultural Experiment Station, Louisiana State University Agricultural Center, Baton Rouge, Louisiana 70803 (CHAP. 5)

Daniel D. Jones Office of Agricultural Biotechnology, Department of Agriculture, Room 321-A, Administration Building, 14th Street and Independence Avenue, S.W., Washington, D.C. 20250 (CHAP. 4)

Lori S. Katz Life Sciences Section, Arthur D. Little, Inc., Acorn Park, Cambridge, Massachusetts 02140 (CHAP. 3)

Kathleen H. Keeler School of Biological Sciences, University of Nebraska, Lincoln, 212 Lyman Hall, Lincoln, Nebraska 68588-0343 (CHAP. 9)

Morris A. Levin Maryland Biotechnology Institute, University of Maryland, Fine Arts Building, Room 542, Baltimore, Maryland 21228 (CHAP. 11)

Judith K. Marquis Life Sciences Section, Arthur D. Little, Inc., Acorn Park, Cambridge, Massachusetts 02140 (CHAP. 3)

James H. Maryanski Center for Food Safety and Applied Nutrition, Food and Drug Administration, Switzer Building, Room 1414, 330 C Street, S.W., Washington, D.C. 20204 (CHAP. 4)

Marla S. McIntosh Department of Agronomy, University of Maryland, 0111 H. J. Patterson Hall, College Park, Maryland 20742 (CHAP. 10)

Jonathan S. Naimon Investor Responsibility Research Center, Inc., 1755 Massachusetts Avenue, Suite 600, Washington, D.C. 20036 (CHAP. 14)

O. A. Ogunseitan Program in Social Ecology, University of California at Irvine, Irvine, California 92717 (CHAP. 8)

Betty H. Olson Program in Social Ecology, University of California at Irvine, Irvine California 92717 (CHAP. 8)

Peter Palukaitis Department of Plant Pathology, Cornell University, Ithaca, New York 14853 (CHAP. 7)

Robert W. Pilsucki U. S. Environmental Protection Agency, H7507C, 401 M Street, S.W., Washington, D.C. 20460 (CHAP. 2)

P. A. Rochelle Program in Social Ecology, University of California at Irvine, Irvine, California 92717 (CHAP. 8)

Frances E. Sharples Environmental Sciences Division, Oak Ridge National Laboratory, Oak Ridge, Tennessee 37831-6038 (CHAP. 1)

Harlee S. Strauss H. Strauss Associates, Inc., 21 Bay State Road, Natick, Massachusetts 01760 (CHAP. 11, 13)

C. C. Tebbe Program in Social Ecology, University of California at Irvine, Irvine, California 92717 (CHAP. 8)

Paul S. Teng International Rice Research Institute, P.O. Box 933, Manila, Philippines (CHAP. 12)

Sue A. Tolin Department of Plant Pathology, Physiology, and Weed Science, Virginia Polytechnic Institute and State University, Blacksburg, Virginia 24061 (CHAP. 6)

Charles E. Turner Biological Control of Weeds, U.S. Department of Agriculture, Agricultural Research Service, Western Regional Research Center, 800 Buchanan Street, Albany, California 94710 (CHAP. 9)

Y. L. Tsai Program in Social Ecology, University of California at Irvine, Irvine, California 92717 (CHAP. 8)

Robert Wachbroit Center for Public Issues in Biotechnology, Maryland Biotechnology Institute, University of Maryland, Baltimore County, Catonsville, Maryland 21228 (CHAP. 16)

Jonathan E. Yuen Department of Plant Pathology, University of Hawaii, 3190 Maile Way, Honolulu, Hawaii 96822 (CHAP. 12)

Preface

Environmental applications of genetically altered microorganisms and plants are on the verge of great expansion in the 1990s. Proposed uses for genetically altered microorganisms include controlling pests and weeds in agriculture, cleaning up toxic chemicals at waste sites, microbial leaching of mineral ores, and enhancing oil recovery. Plants are being genetically engineered to enhance many traits: increasing pest and herbicide resistance, tolerating drought or other environmental stresses, decreasing loss of food during storage and transport, and increasing nutritional value of food products.

This book examines the available testing protocols, data, models, and frameworks for the risk assessment of releasing genetically engineered microorganisms and plants into the environment. Chapters discussing hazard end points, fate and transport, horizontal genetic transfer, and risk assessment were written by scientists from a variety of academic backgrounds, reflecting the importance of creating an interdisciplinary approach to risk assessment. Each chapter provides the reader with an overview of the current state of knowledge, an entry point to the literature of the disciplines represented, and the opinion of scientific experts about the risks of environmental release.

Practical help is provided for the design field tests that have to undergo regulatory scrutiny. There are two chapters written by experts from three United States regulatory agencies: the Environmental Protection Agency, the Department of Agriculture, and the Food and Drug Administration. These two chapters focus on the current regulations and testing protocols relevant to environmental release as well as their underlying rationale. The chapter on statistical techniques will aid in the design of scientifically valid field tests.

The final two chapters address some of the broader questions of risk associated with environmental biotechnology, such as public acceptance, social and economic consequences, and the needs of organizations serving less developed countries.

The intended audience for this book includes scientists—developing new organisms for environmental uses, risk assessment professionals, and government regulators and their corporate counterparts concerned with environmental biotechnology. Scientists and engineers with interests in environmental microbiology, bioremediation, toxicology, biotechnology, and agriculture will also find this book of interest.

Morris A. Levin
Harlee S. Strauss

Introduction: Overview of Risk Assessment and Regulation of Environmental Biotechnology

Morris Levin

Maryland Biotechnology Institute
University of Maryland
College Park, Maryland 21228

Harlee S. Strauss

H. Strauss Associates, Inc.
21 Bay State Road
Natick, Massachusetts 01760

Advances in recombinant DNA techniques have extended the possibilities for the production and distribution of biotechnology products. Genetic engineering techniques have moved from the research laboratory to the applied sector. As a result, genetically engineered microorganisms are being used in large-scale industrial settings to produce pharmaceutical and other products. In agriculture, successful research and development has produced transgenic plants, animals, and microorganisms intended for use in the environment. Other potential environmental uses of genetically engineered microorganisms include cleaning up toxic wastes, mining copper and uranium, and improving oil recovery processes.

The recombinant organisms developed for environmental use inevitably have to be tested in the field. However, the prospect of extensive field tests of genetically engineered organisms has given rise to concern about possible risks these activities may pose to humans and the environment. This concern, which at times has resulted in acrimonious debate, has been shared by scientists as well as the general public. One of the foundations of the scientific debate is the quite different viewpoints of scientists trained in different scientific specialties, such as molecular biology and ecology. Another cause of the scientific debate is the lack of appropriate data and validated conceptual or mathematical models for the processes under examination. In addition, there is not yet a commonly agreed upon overarching framework with which to integrate and evaluate the data and models that do exist.

Concern about risks to human health and the environment is not peculiar to biotechnology. Rather, these questions have emerged as an important component in the development, regulation, and promotion of the products of many new and older technologies such as chemicals and pharmaceuticals. Biotechnology is a part of this trend. The new scientific discipline of risk assessment has emerged to better assess the likelihood of hazards and risks associated with materials in the environment. In this volume, we have applied some of the tools and concepts of risk assessment to the evaluation of the risks associated with the environmental release of genetically engineered microorganisms and plants.

Objectives and Organization

Each chapter in this book addresses a different aspect of the scientific and regulatory questions that arise in the risk assessment of genetically engineered microorganisms and plants intended for use in the environment. Individual chapters survey the current status of data, theory, and government regulation associated with the hazard or transport of genetically engineered microorganisms and plants, or of the inserted DNA itself. Other chapters explore means to integrate data from diverse sources, and how safety in environmental biotechnology affects social concerns and international programs.

The book is divided into four major sections:

- Hazards and (more generally) effects identification and assessment
- Exposure assessment
- Integration of hazard and exposure assessments
- Societal impacts

The four chapters in the section on hazards and effects identification and assessment cover a range of hazard end points for microorganisms and plants. Sharples describes the potential hazards posed by microorganisms to ecosystems. Frederick and Pilsucki describe test methods established by the U.S. Environmental Protection Agency (USEPA) to evaluate hazards to nontarget aquatic and terrestrial organisms as well as the underlying rationale of these methods. Katz and Marquis give an overview of the potential toxic and infectious hazards of genetically engineered microorganisms to humans, and how to test for some of them. Jones and Maryanski discuss the U.S. Department of Agriculture (USDA) and Food and Drug Administration (FDA) regulatory perspective on safety considerations in the evaluation of transgenic plants intended for human food use. A key issue examined in each chapter is whether the use of genetically engineered organisms will result in a set of problems completely dissimilar to those encountered in the environmental application of any biological material.

The six chapters in the section on exposure assessment cover the release, fate, and transport of engineered microorganisms and plants, as well as of the inserted DNA itself. The chapters by Fuxa and Tolin include examples from the scientific literature on viruses, bacteria, and fungi. Olson and colleagues describe the available data on horizontal genetic transfer in microorganisms (primarily bacteria) in a variety of environmental media. Palukaitis describes virus-mediated genetic transfer in plants and its consequences, while Keeler and Turner describe potential transfer through outcrossing to compatible plants and the hazards this may pose. Each of these chapters provides data and discusses theoretical considerations regarding the transport and persistence of genetically engineered organisms, and their implication for evaluating and/or reducing the risks associated with field trials. In the final chapter of this section, McIntosh provides a primer on statistical considerations in environmental monitoring. This chapter is intended to help with the planning of field trials and enable the further collection of statistically valid field data.

The five chapters in the section on integration of hazard and exposure data each examine different approaches to combining data from diverse scientific disciplines. The chapter by Strauss and Levin describes the advantages and limitations of using the type of mathematical models known as "fate and transport models," many of which were developed for chemical risk assessment. Teng and Yuen provide examples of the successful use of epidemic models in the area of plant pathology. Strauss reviews several risk assessment frameworks proposed for the evaluation of environmental releases of microorganisms and examines whether the framework used in chemical risk assessment is applicable to microorganisms. Naimon, using the path to the initial field trial of Frostban® as his case study, evaluates the utility of expert and consensus panels in risk-based regulatory decision making. Finally, Fiksel describes how knowledge-based computer systems can facilitate access to and evaluation of data necessary for risk assessment.

The two chapters in the fourth section, exploring societal impacts of risk assessment, serve to begin a discussion of this important topic, which is often neglected by scientists. Wachbroit discusses the types of end points, including social end points, that could (should) be considered in risk assessment, and how their inclusion would affect public acceptance of the evaluation. Cohen and Chambers provide an overview of biosafety from the perspective of an agency funding environmental biotechnology projects in developing nations.

The remainder of this introduction provides an overview of the regulation of biotechnology in the United States and by international organizations. It also includes a synopsis of scientific debate about the safety of environmental releases of genetically engineered microorganisms and the use of risk assessment in regulating environmental biotechnology.

The U.S. Regulatory System

The first federal coverage of the safety of biotechnology, perhaps one could stretch and call it "regulation," and this *is* a stretch of the word since compliance was voluntary, grew out of the Asilomar Conferences and other meetings organized by scientists concerned about the potential biohazards associated with the use of the newly developing recombinant DNA techniques and the possible escape of the newly created microorganisms from the laboratory. The second Asilomar conference in 1975 was immediately followed by the establishment of the National Institutes of Health Recombinant DNA Advisory Committee (NIH RAC) and the development and publication of the well-known RAC guidelines in 1976. It is interesting to note that these guidelines amounted to a regulatory scheme, since NIH-funded researchers had to comply for fear of jeopardizing major funding to their institutions. As other U.S. government funding agencies adopted the guidelines for their grantees, all federally funded molecular biological research was covered. Nonfederally funded researchers were expected to, and did, comply on a voluntary basis. Eventually, guidelines similar to those of the NIH RAC were applied to researchers around the world.

One outcome of this procedure was that the Institutional Biosafety Committees (IBCs) were established to aid the RAC by reviewing some applications at the local level. The IBCs became, and still remain, the reviewers of most of the molecular biology experiments, now called *biotechnology* experiments, requiring review in the United States. Needless to say, the RAC guidelines focused on laboratory worker safety. They functioned with great success via a commitment to containment of the engineered microorganism within the confines of the laboratory. In fact, in the initial version of the RAC guidelines, any type of environmental release was completely prohibited.

With the application of laboratory-based molecular biology research to industrial processes and products, questions regarding the safety of the technology once again emerged. This time, the health and safety of industrial workers, the environment, the general human population, and agricultural products all came under scrutiny. As commercialization began, federal agencies with the relevant regulatory authority became involved with biotechnology processes and products.

The overall policy regarding how biotechnology will be regulated and coordinated in the United States has been set forth in formal notices published in the *Federal Register* (OSTP 1984, 1986). Three agencies share primary responsibility for regulating the organisms and products of recombinant DNA biotechnologies, whether they be designed for closed systems or for environmental uses. They are the USEPA, the FDA, and the USDA. In addition, the Occupational Safety and Health Administration (OSHA) has the responsibility for ensuring the health and safety of biotechnology workers, and the Department of Health and Human Services for ensuring the health of the general public.

In addition to federal regulation, approximately 13 states and several municipalities have enacted biotechnology-related legislation, including provisions related to the environmental release of genetically modified microorganisms. The most comprehensive of these laws was enacted by North Carolina in 1989. State and local laws have been reviewed and analyzed elsewhere (Industrial Biotechnology Association 1990, Fox 1989, Krimsky et al. 1982).

Each of the federal agencies regulating biotechnology is guided in its analysis and decision-making criteria by its specific legislation—i.e., the laws passed by Congress charging each agency with specific responsibilities. These laws differ in their mandate as to what populations to consider with regard to adverse effects (e.g., humans, crops, the environment), as well as in their mandate as to how to strike a balance between risks and benefits. In addition to its specific legislation, each agency must also adhere to the National Environmental Policy Act (NEPA), which is binding on all federal agencies.

All of the legislation used to regulate recombinant DNA includes safety as an important element in decision making. However, none of the laws used to regulate biotechnology products specifically address the safety of biotechnology. Instead, the laws are based on the uses of the products, or the environmental discharge of materials. For example, the environmental protection laws administered by USEPA were written with inorganic and organic chemicals in mind. The laws cover the uses of products (pesticides, chemical intermediates, and many others), the emission of materials into the air and water, and the disposal of solid and hazardous waste. At the USDA, the laws were written with naturally occurring organisms, and especially plant and animal pathogens or toxins, in mind.

The goal of NEPA is to assure that federal actions are environmentally sound. In essence, NEPA requires all federal agencies to evaluate the possible environmental outcomes of their proposed actions and to look for a balance between benefits and possible adverse impacts. Agencies must conduct a thorough review of all pertinent available information and alternatives and seek public comment. At one time, agencies were required to consider worst case scenarios in their NEPA reviews of biotechnology, but this is no longer the case.

U.S. Environmental Protection Agency

The USEPA administers seven environmental statutes. It specifically regulates biotechnology under two of them: the Toxic Substances Control Act (TSCA), administered by the Office of Toxic Substances (OTS), and the Federal Insecticide, Fungicide, and Rodenticide Act (FIFRA), administered by the Office of Pesticide Programs (OPP). These two statutes have been described as "gateway" legislation. They are not abatement-oriented as are other statutes administered by the USEPA (e.g., the Comprehensive Environmental Response, Compensation, and Liability Act, also known as CERCLA or the Superfund law), but instead they are preventative. They are invoked before applications of a new product to the environment. FIFRA covers all pesticidal products and is clearly applicable to biotechnology products. TSCA is invoked through the concept of "newness." New chemicals or new uses for existing chemicals trigger a TSCA review. When asked whether TSCA applied to microbial products or recombinant DNA, the USEPA's General Counsel wrote in 1983 that since "TSCA covers all chemicals and DNA is a chemical, DNA is covered" and added that "a DNA molecule does not have any use except in the life form of which it is part" (U.S. Congress 1984). This position has not been challenged. USEPA reviews under TSCA and FIFRA have been accepted in court as the equivalent of NEPA reviews.

The USEPA regulates wastes resulting from the biotechnology industry under its waste-related laws, such as the Resource Conservation and Recovery Act (RCRA) for hazardous and industrial solid wastes, and the Clean Water Act for wastewater discharges. No special provisions have been made for biotechnology. Regulations and policies affecting the development of genetically engineered microorganisms intended to clean up the environment (bioremediation) at toxic waste sites have been made under CERCLA, but these have been directed at new technologies in general, rather than biotechnology in particular.

In addition to regulating biotechnology, the USEPA in 1983 formally initiated a Biotechnology Risk Assessment research program in its Office of Research and Development to develop methodology for biotechnology risk assessment. The program was intended to develop methods required for regulatory decision making. It covers the engineered microbes, the DNA involved, and the potential for environmental and human health effects. The research includes work on modifying existing or developing new methods to identify, monitor, and track engineered microbes and recombinant DNA in field situations, as well as work geared to developing methods to identify and quantify environmental and health effects (Levin et al. 1987).

Between 1983 and 1989, the OPP and OTS offices of the USEPA reviewed or were in the process of reviewing approximately 56 biotechnology products. These are summarized in Table I.1. As can be seen, 55% of the reviews involved just two bacterial species, *Rhizobium meliloti* and *Bacillus thuringiensis*, either as the host or donor. *R. meliloti* has been reviewed by OTS under TSCA authority as a new chemical, and *B. thuringiensis* by OPP under FIFRA authority as a pesticide. The USEPA's Biotechnology Science Advisory Committee reviews the actions of both OPP and OTS and attempts to coordinate and unify the procedures of both offices.

U.S. Department of Agriculture

The USDA has responsibility for food and fiber products. It is in one sense outside the circle of ecological safety, in that its primary concerns are the safety of crop plants and food animals, and the safety and wholesomeness of food products. The USDA has three arms that deal with biotechnology: the Agricultural Research Service (ARS), the Food Safety and Inspection Service (FSIS), and the Animal and Plant Health Inspection Service (APHIS). The ARS focuses on research issues and has formed a group equivalent to the NIH RAC, called the Agricultural Biotechnology Recombinant DNA Advisory Committee (ABRAC), to review proposals. The ARS is also instituting an information service as part of its National Biological Impact Assessment Program. The FSIS functions to assure the safety and the wholesome characteristics of food products. Via APHIS, the USDA meets its responsibilities for licensing veterinary biological material and issuing permits for transport of biological material. APHIS has formed the Biotechnology, Biologics, and Environmental Protection Division (BBEPD) with responsibility for all biotechnology products. Under NEPA, the USDA has a responsibility for en-

TABLE I.1 Recombinant Microbe Tests

Microbe (no. of tests)	Trait (no. of tests)	Source of genetic material
Rhizobium meliloti (20)	Nitrogen fixation (11) Marker gene (8) Nitrogen and marker (1)	Bacteria
Bacillus thuringiensis (11)	Insecticide ability (11)	Other *B. thuringiensis*
Bradyrhizobium japonicum (6)	Marker gene (4) Nitrogen fixation (2)	Bacteria
Pseudomonas fluorescens (4)	Frost protection (2) Insect resistance (2)	Gene deletion, *B. thuringiensis*
Trichoderma harzianum (4)	Host range (4)	Other trichoderma, undirected mutagenesis
Clavibacter xyli (2)	Insect resistance (2)	*B. thuringiensis*
P. syringae (2)	Frost protection (1) Insecticide ability (1)	Gene deletion, other *P. syringae*
Colletotrichum gloeosporiedes (2)	Host range (2)	None: chemical mutagenesis
P. aureofaciens (1)	Marker gene (1)	*Escherichia coli*
Vaccinia virus (1)	Immunization potential (1)	Virus (rabies)
Sclerotinia sclerotivorum (1)	Host range (1)	None: mutagenesis
T. viride (1)	Host range (1)	None: undirected mutagenesis
Autographa californica (1)	Persistence (1)	Gene deletion

suring safe ecological utilization of crops, livestock, and veterinary products, whether they are produced by genetic engineering or traditional means.

Table I.2 provides a summary of the 56 field tests of recombinant plants that entered into the USDA review process between 1983 and 1989. The USEPA and USDA have collaborated on the review of 23 of the engineered plants, mostly tobacco and tomato. The table shows that the source of genetic material is highly varied, and includes other plant genera, bacteria, viruses, and mammals. Field releases of transgenic plants have also been reviewed and approved in several countries outside the United States. For example, the Canadian government authorized 39 field trials of engineered plants during this time period, mostly with canola and flax.

Food and Drug Administration

The FDA regulates biotechnology under the authority of the Food, Drug, and Cosmetic Act (FDCA) and the Public Health Services Act. The agency has a mandate to ensure efficacy and safety of food and pharmaceutical products which

TABLE I.2 Recombinant Plant Tests

Plant (no. of tests)	Trait (no. of tests)	Source of genetic material
Tomato (22)	Herbicide resistance (8)	Bacteria, virus, tobacco, petunia, soybean, pea
	Insect resistance (8)	*B. thuringiensis*
	Ripening delay (3)	Bacteria, tomato, virus
	Resistance to mosaic virus (3)	Virus, bacteria, pea
Tobacco (12)	Herbicide resistance (4)	Bacteria, virus, tobacco
	Wound enzyme (2)	Potato, bacteria
	Insect resistance (2)	Cowpea, virus, *B. thuringiensis*
	Resistance to mosaic virus (1)	Virus, bacteria
	Resistance to fungi (1)	Bacteria, bean
	Reduced heavy metals (1)	Mouse, bacteria, virus, pea
	Increased lysine synthesis (1)	Bacteria, virus, pea
Cotton (6)	Herbicide resistance (5)	Petunia, bacteria, virus, soybean, tobacco
	Insect resistance (1)	*B. thuringiensis*
Soybean (3)	Herbicide resistance (3)	Petunia, virus, bacteria
Potato (2)	Resistance to potato virus (2)	Virus, bacteria
Alfalfa (3)	Herbicide resistance (2)	Bacteria, virus
	Resistance to mosaic virus (1)	Bacteria, virus
Walnut (1)	Marker gene (1)	Bacteria
Poplar (1)	Wound enzyme (1)	Bacteria, potato
Cucumber (1)	Resistance to mosaic virus (1)	Bacteria, virus

are segregated into a number of classes. It has been estimated that the FDA has already ruled on thousands of biotechnology products, although not all agree on what is required to label a product "biotechnology." FDA's criteria for product evaluation, as set forth in the FDCA, revolve around purity, lack of adverse effects, and efficacy. FDA's population of concern regarding adverse effects is humans. FDA is also responsible for administering the tolerance limits for pesticides used on crops that are set by other agencies (e.g., USEPA).

The FDA has a major responsibility in biotechnology in that 60% of the current market share of biotechnology products passes through the agency for review. However, the FDA's environmental concerns and responsibilities are only in regard to NEPA. The USEPA is responsible for ecological and related public health issues.

Occupational Safety and Health Administration

OSHA has broad responsibility for worker safety. However, OSHA's active involvement in biotechnology regulation is minor. Existing guidelines and practices are seen as adequate to the need.

A committee to monitor developments in biotechnology that could affect worker health and safety has been established within the National Institute for Occupational Safety and Health (NIOSH). However, NIOSH serves OSHA only in an advisory capacity. Administratively, NIOSH and OSHA are located in separate departments (NIOSH is in the Public Health Service in the Department of Health and Human Services, while OSHA is in the Department of Labor).

Coordinated federal frameworks

There are two important dates to remember in connection with coordination of biotechnology regulation among U.S. agencies: 1984 and 1986. In 1984, the Office of Science and Technology Policy (OSTP) published a proposed coordinated framework for biotechnology that resulted in the formation of the Biotechnology Science Coordinating Committee (BSCC) (OSTP 1984). In 1986, the BSCC published in the *Federal Register* many elements of an altered proposed framework, a matrix of applicable legislation, and responsibilities for oversight of biotechnology products (OSTP 1986). The revisions were based, in part, on comments received on the 1984 proposal.

Table I.3, taken from the 1986 *Federal Register* notice, summarizes the scheme of regulatory authority that is currently operative in the United States. The table indicates the lead and cooperating agencies for the various uses of biotechnology products. Note that there is overlap between agencies in the foods and food additives area, plants and animals area, and in pesticides. In each of these areas of overlap, a lead agency has been designated. A case-by-case approach has been used to determine responsibility of oversight of other uses of microorganisms. The BSCC has been involved in resolving differences in definitions and in establishing communication between agencies, as well as dealing with international issues.

Growth of Biotechnology Products Requiring Regulation

The increase in the number of biotechnology products subject to regulation requires growth in staff and capabilities in the relevant regulatory agencies. Several measures, including the number of patent applications, have been used to predict the increases in personnel that will be necessary to ensure the timely processing of permit applications and premanufacturing notices. Sufficient data are already available to demonstrate that growth in permit applications has been rapid, indeed exponential, throughout the 1980s, and to point out where problems lie.

Table I.4 is a summary of a survey of biotechnology patents issued between 1980 and 1984 that could result in permit applications or premanufacturing notices to the USEPA. More than 3000 patents were issued during the 5-year period, and the patent office had a backlog of 7000 patent applications that have not been included in the survey. Thus, a total of 10,000 patents were applied for or granted between 1980 and 1984. The impact of these patents on the USEPA and other federal agencies is just beginning to be evident.

TABLE I.3 Coordinated Framework—Approval of Commercial Biotech. Products

Subject	Responsible agency (or agencies)
Foods and food additives	FDA [a], FSIS [b]
Human drugs, medical devices, and biologics	FDA
Animal drugs	FDA
Animal biologics	APHIS
Other contained uses	USEPA
Plants and animals	APHIS [a], FSIS [b], FDA [c]
Pesticide microorganisms released in the environment	USEPA [a], APHIS [d]
Other uses (microorganisms)	
Intergeneric combination	USEPA [a], APHIS [d]
Intrageneric combination	
Pathogenic source organism:	
Agricultural use	APHIS
Nonagricultural use	USEPA [a,e], APHIS [d]
No pathogenic source organisms	USEPA report
Nonengineered pathogens	
Agricultural use	APHIS
Nonagricultural use	USEPA [a], APHIS [d]
Nonengineered nonpathogens	USEPA report

[a]Lead agency.
[b]FSIS, Food Safety and Inspection Service, under the Assistant Secretary of Agriculture for Marketing and Inspection Services, is responsible for food use.
[c]FDA is involved when in relation to a food use.
[d]APHIS, Animal and Plant Health Inspection Service, is involved when the microorganism is plant pest, animal pathogen, or regulated article requiring a permit.
[e]USEPA requirements will only apply to environmental release under a "significant new use rule" that EPA intends to propose.
SOURCE: OSTP 1986

TABLE I.4 Microbial Patents Issued, 1980–1984

Area	No. of genera	No. of patents
Agricultural chemistry	8	37
Conversion of biomass	14	809
Industrial chemical production	14	528
Monitoring, measurement, biosensor	7	36
Energy	4	217
Polymer or macromolecule production	16	787
Enhanced oil recovery	9	25
Waste and/or pollutant degradation	17	541
Mining and metals recovery	10	80

TABLE I.5 Number of Reviews by USEPA and USDA

	Permit applications	
Year	USEPA	USDA
1984	4	
1985	7	
1986	4	
1987	16 [a]	5
1988	16	16
1989	25 [b]	30

[a]First generic application (more than one strain or site per application).
[b]Estimate.

Table I.5 demonstrates the growth in the number of biotechnology reviews conducted by the USEPA and USDA from 1983 through the first half of 1989. The rate at which applications are being received is increasing exponentially. Actual recombinant DNA technology, as defined by the NIH RAC, was involved in relatively few cases. In general, the source of the recombinant DNA is a closely related species, and the use of traditional methods to produce the recombinant organism is the rule rather than the exception.

Table I.6 identifies the major ecological issues which were highlighted in the agency reviews. Between USEPA and USDA, approximately 200 field trials had been authorized before the end of 1989. A small number have been controversial, largely because they were not properly authorized. These releases are summarized in Table I.7.

TABLE I.6 Summary of Releases of Engineered Organisms

	No. of genera	No. of tests	Major ecological issues
Plants	6	52	Gene transfer, presence of related plants, exposure of endangered species, possibility of enhanced herbicide use, insect resistance to pesticide, weed resistance to herbicides, use of antibiotic resistance genes as markers
Microbes	14	56	Gene transfer and impact on microbial community structure, exposure of endangered species, alteration of host range, competitive ability of recombinant, use of antibiotic resistance genes as markers, mechanism of dispersal (or transmission) of microbe

TABLE I.7 Controversial Releases

Unauthorized Releases
Bacterial
1. Advanced Genetic Sciences, Oakland, California Release of engineered *P. syringae* prior to issuance of permit; fined $20,000.
2. Montana State University, Bozeman Treatment of Dutch Elm trees with recombinant bacteria prior to issuance of permit; reprimand to university.
Viral
1. Wistar Institute, Philadelphia, Pennsylvania Field test of vaccinia/rabies vaccine in Argentina in collaboration with the Pan American Health Organization: Argentine government alleged that the test was conducted without permission.
2. Oregon State University, Corvallis, Oregon Vaccine tested in New Zealand with the approval of the New Zealand Government but not of the USDA, which claimed prior jurisdiction.
Other Controversial Releases
1. Biologics, Inc., Omaha, Nebraska Permission to field-test a swine vaccine was granted by USDA. However, the agency was challenged in court because it had not followed proper procedure. NIH rebuked Baylor University for the part one of its researchers played in the trial, Baylor and Texas A&M rebuked individual researchers; USDA reissued the permit.

International Biotechnology Guidelines and Regulations

Many industrialized countries have instituted mechanisms for the regulation of biotechnology, including the environmental release of genetically engineered organisms. Thus far, only a few of the more advanced developing countries have biosafety guidelines and regulations, and those appear to be only for contained uses. However, developing countries are also being increasingly faced with requests to conduct field trials of genetically engineered plants and microorganisms.

Several international organizations are developing, or have completed, guidelines related to biotechnology safety that have gained some measure of international consensus. These guidelines cover laboratory safety, contained industrial uses, and more recently, field trials of genetically engineered plants and microorganisms.

The Organization of Economic Cooperation and Development (OECD) has been the major forum thus far in the effort toward international harmonization of biotechnology regulations for both contained uses and environmental releases. The OECD's 1986 monograph, *Recombinant DNA Safety Considerations*, although not binding on member countries, has had a major influence in the development of regulations in developed countries. A revision of these guidelines has been in the works for several years. In addition, a new set of OECD guidelines, "Good Developmental Practices," which covers protocols for field releases, was released in May 1990.

The European Community (EC) issued draft directives in the spring of 1988 on contained use of genetically modified microorganisms, and on the deliberate release to the environment of genetically modified microorganisms. These are still moving toward final form and passage in 1990. The final form of these directives will be binding on all EC countries, and will set minimum standards of regulation. However, each country will have to implement the directives individually, and may impose more stringent rules (Newmark 1988, Dixon 1989). Thus, environmental release regulations may differ substantially in these countries. Currently, regulations governing environmental release of microorganisms in EC countries range from the lack of any regulation in Italy and Spain to a nearly complete prohibition of environmental releases in the Federal Republic of Germany. In the United Kingdom, only notification of deliberate release experiments is required, although detailed guidelines are available to indicate how to balance the level of risk and the cost of taking precautions to avoid the risk. In Denmark, a pre-1989 near prohibition of environmental releases has ended, a field trial with genetically engineered sugar beets was approved in late 1989, and more permissive legislation may be enacted (Newmark 1988, Simpson 1990).

Several United Nations agencies, notably the World Health Organization (WHO), the United Nations Environmental Program (UNEP), the United Nations Industrial Development Organization (UNIDO), and the Food and Agriculture Organization (FAO) have keen interests in biotechnology, including the biosafety and regulatory aspects of environmental biotechnology.

UNIDO, WHO, and UNEP have participated in a Working Group on Biotechnology Safety for several years. A document, *Safety Guidelines and Procedures for Bioscience-Based Industry and Other Applied Microbiology*, describes the history and possible problems the Working Group would like to examine, and reviews the types of data available for assessing health risks, primarily to workers, in the biotechnology industry (UNIDO 1986). In 1989, the Working Group launched a program to develop strict new safety guidelines to minimize medical and other hazards that might arise from the environmental release of genetically engineered microorganisms (UNIDO 1989).

Brief History of the Risk Assessment Debate in the Scientific Community

The debate over the "safety" of using genetically engineered organisms in the environment has been marked by public controversy and a large degree of polarization. The initial focus of the debate was agricultural uses of plants and microorganisms; however, the debate was really of importance to all end uses of environmental biotechnology, including bioremediation, oil recovery, coal desulfurization, mining, and others.

In the scientific literature, the safety debate was highly visible in the pages of *Science* magazine. Winston Brill, a microbiologist with both an academic and industrial biotechnology background, published an article on safety concerns and genetic engineering in agriculture (Brill 1985) in which he argued that plants and animals have been altered by traditional breeding by humans for centuries without any serious problems. In addition, bacteria and fungi, including pathogenic organisms, have been added to soils and plants in an effort to find beneficial uses of these organisms. Brill concluded that these observations comprised an adequate and appropriate data base from which to draw risk assessment conclusions, including for genetically engineered organisms where only a few genes have been added.

Brill's article generated a response from ecologists Robert Colwell, Elliott Norse, David Pimentel, Frances Sharples, and Daniel Simberloff (Colwell et al. 1985), who presented a very different view of the risks. Colwell et al. took Brill to task for his qualitative risk views of "seems very small" and "is extremely unlikely" and suggested that quantitative risk assessment is both necessary and appropriate. The ecologists argued that mutations intended to increase the range of conditions under which an organism is intended to survive are not small from an ecological point of view. They also argued that the phenotype of a microorganism, especially its ecological traits and population dynamics, is not fully predictable from genotype alone. Finally, Colwell et al. (1985) advocated assessing the risk of releasing genetically engineered organisms into the environment on a case-by-case basis.

In a policy forum set up by *Science* magazine a few years later, side-by-side articles by ecologist Frances Sharples (1987) and bacteriologist and molecular biologist Bernard Davis (1987) were published.

Sharples (1987) continued the ecologists' argument that risks associated with the environmental release of engineered organisms must be evaluated on a case-by-case basis. She noted that while the "absence of human health effects of biotechnology workers in contained environments is admirable, it is not particularly relevant to the question of whether uncontained uses of modified organisms will be equally harmless." Sharples pointed out that "shifts in environmental contexts may be as important as genetic modifications in determining whether the ecological relationships of an engineered organism will be unique relative to those of a parental form."

Davis (1987) reiterated the molecular biologists' arguments that the use of modified microbes is not really novel, "but is an extension of the old process of domestication of wild organisms—including the selection of microbial variants to make bread, wine, antibiotics, or vaccines." Davis also argued that the "experience of ecologists with transplanted higher organisms is less pertinent than are the insights of fields closer to the specific properties of engineered microorganisms: population genetics, bacterial physiology, epidemiology, and the study of pathogenesis."

Discussions and debates between molecular biologists and ecologists have also proceeded via symposia sponsored by professional societies. For example, in 1985, the American Society for Microbiology sponsored a cross-disciplinary symposium entitled Engineered Organisms in the Environment: Scientific Issues. The published proceedings of that conference (Halvorson et al. 1985) remain a frequently cited reference regarding scientific considerations and uncertainties in risk assessment. An international conference on the Release of Genetically Engineered Microorganisms (REGEM I) was held in South Wales in 1988. The conference proceedings (Sussman et al. 1988) provide many viewpoints regarding what is necessary to ensure the safety of environmental release experiments.

The U.S. National Academy of Sciences (NAS) published a pamphlet intended for a general scientific audience in 1987 entitled *Introduction of Recombinant DNA-Engineered Organisms into the Environment: Key Issues* (NAS 1987). The NAS made some general conclusions regarding risk, including:

- "There is no evidence that unique hazards exist either from the use of recombinant DNA techniques or from the movement of genes between unrelated organisms."
- "The risks associated with the introduction of genetically engineered organisms carrying recombinant DNA are the same in kind as those associated with the introduction of unmodified organisms and organisms modified by other methods."
- "Assessment of the risks of introducing genetically engineered organisms carrying recombinant DNA into the environment should be based on the nature of the organism and the environment into which it is introduced, not on the method by which it was produced."

While these points, especially the third one, were common in the mainstream discussion of potential risks and regulation, the pamphlet drew protests from scientists (and others) with environmental concerns because of the lack of documentation of the assertions in the report.

In 1989, the NAS published a second report on environmental release, entitled *Field Testing Genetically Modified Organisms: Framework for Decisions*, (NAS 1989). This report was more detailed than the 1987 pamphlet, and was intended for more technical audiences. In a view of risk assessment that differs quite a bit from that used for chemical risk assessment, the NAS posed three questions to use in making initial judgments of risk, and thus suggested the extent of the informational requirements for risk assessment (NAS 1989):

- Are we familiar with the properties of the organism and the environment into which it may be introduced?
- Can we confine or control the organism effectively?
- What are the probable effects on the environment should the introduced organisms or a genetic trait persist longer than intended or spread to nontarget environments?

At approximately the same time as the 1989 NAS report appeared, the Ecological Society of America (ESA) published a special feature entitled "The Planned Introduction of Genetically Engineered Organisms: Ecological Considerations and Recommendations" (Tiedje et al. 1989). It focused on the ecological and evolutionary aspects of planned introductions of genetically engineered organisms, and included microorganisms, plants, and animals.

The ESA report set forth a preliminary set of specific criteria for the scaling of regulatory oversight. Attributes of organisms and environments were divided into four categories: (1) genetic alteration, (2) parent (wild type) organism, (3) phenotypic attributes of engineered organism in comparison with parent organism, and (4) environment. For each category of attribute, the authors defined numerous specific attributes, and a scale on which to base the level of scientific consideration necessary for a risk assessment. The authors suggested that the level of regulatory scrutiny be commensurate with the level suggested by the scientific attribute scales.

From the preceding discussion, it is clear that the scientific debate over the safety of the environmental release of genetically engineered organisms has evolved greatly since its initial stages. In the mid-1980s, people argued that their view of the world was absolutely correct, and cited historical examples that supported their points. In the late 1980s, more thoughtful discussions took place about what considerations are important for risk assessment, how to develop overarching risk assessment frameworks, and how to scale regulatory scrutiny on the basis of risk. It should be noted that throughout this period, governmental, academic, and other groups were examining the scientific needs for risk assessment in less public fora (some of these efforts are briefly described by Strauss in Chap. 13).

Risk assessment can play a key role in the regulation of environmental biotechnology. Work on scaling of scientific scrutiny can assist in the determination of what organisms should be regulated, and with what level of attention. The choice of overarching framework for risk assessment will have a profound impact on the type of data required for the risk evaluation, and thus what data will be required for permit applications. However, these risk assessment discussions may not solve the problems of public acceptance of biotechnology; this is an issue that must be handled within the larger social and economic context.

In putting together this volume, it has been our intention to extend the discussion of what we know and what we don't know about how to predict the safety of environmental releases of genetically engineered plants and microorganisms. We hope that this book will stimulate further advances in thinking about risk assessment as it relates to this important field. Although most of the chapters focus on purely scientific or regulatory needs in risk assessment, we have also included two chapters on the social and international aspects of environmental biotechnology. These topics are likely to be increasingly interwoven into discussions of the risks associated with the commercialization of environmental biotechnology.

References

Brill, W. J. 1985. Safety concerns and genetic engineering in agriculture. *Science* 227: 381–384.

Colwell, R. K., Norse, E. A., Pimentel, D., Sharples, F. E., and Simberloff, D. 1985. Letters, genetic engineering in agriculture. *Science* 229:111–112.

Davis, B. 1987. Bacterial domestication: underlying assumptions. *Science* 235:1329–1335.

Dixon, B. 1989. Inching towards uniformity on Euro-release. *Bio/Technology* 7:742.

Fox, J. 1989. Wide acclaim for North Carolina regulations. *Bio/Technology* 7:1002.

Halvorson, H. O., Pramer, D., and Rogul, M. (eds.). 1985. *Engineered Organisms in the Environment: Scientific Issues*. American Society for Microbiology, Washington, D.C.

Industrial Biotechnology Association. 1990. *Survey of State Government Legislation on Biotechnology*. Industrial Biotechnology Association, 1625 K St., N.W., Washington, D.C.

Krimsky, S., Baeck, A., and Bolduc, J. 1982. *Municipal and State Recombinant DNA Laws: History and Assessment.* Boston Neighborhood Network, Boston.

Levin, M. A., Seidler, R., Bourquin, A., Fowle, J., III, and Barkay, T. 1987. EPA developing methods to assess environmental release. *Bio/Technology* 5:38–45.

National Academy of Sciences (NAS). 1987. *Introduction of Recombinant DNA-Engineered Organisms into the Environment: Key Issues.* National Academy Press, Washington, D.C.

National Academy of Sciences (NAS), National Research Council. 1989. *Field Testing Genetically Modified Organisms: Framework for Decisions.* National Academy Press, Washington, D.C.

Newmark, P. 1988. The view from the continent. *Bio/Technology* 6:1394–1395.

Office of Science and Technology Policy (OSTP). 1984. Proposal for a coordinated framework for regulation of biotechnology; notice. *Fed. Reg.* 49:50856–50907, Dec. 31.

Office of Science and Technology Policy (OSTP). 1986. Coordinated framework for regulation of biotechnology; announcement of policy and notice for public comment. *Fed. Reg.* 51:23302–23350, June 26.

Organization for Economic Cooperation and Development (OECD). 1986. *Recombinant DNA Safety Considerations.* Paris.

Sharples, F. E. 1987. Regulation of products from biotechnology. *Science* 235:1329–1332.

Simpson, K. 1990. BioEurope. *Genet. Eng. News* 10:7.

Sussman, M., Collins, C. H., Skinner, F. A., and Stewart-Tull, D. E. (eds.). 1988. *The Release of Genetically-Engineered Micro-organisms.* Academic, New York.

Tiedje, J. M., Colwell, R. K., Grossman, Y. L., Hodson, R. E., Lenski, R. E., Mack, R. N., and Regal, P. J. 1989. The planned introduction of genetically engineered organisms: Ecological considerations and recommendations. *Ecology* 70(2):298–315.

United Nations Industrial Development Organization (UNIDO). 1986. *Safety Guidelines and Procedures for Bioscience-Based Industry and Other Applied Microbiology.* ID/WG.463/1. NTIS PB87-140521.

United Nations Industrial Development Organization (UNIDO). 1989. *Genet. Eng. Biotechnol. Monitor* 36:1.

U.S. Congress. 1984. Committee on Science and Technology, Subcommittee on Investigations and Oversight, Serial V. *Environmental Implications of Genetic Engineering.*

Chapter

1

Ecological Aspects of Hazard Identification for Environmental Uses of Genetically Engineered Organisms

Frances E. Sharples

Environmental Sciences Division
Oak Ridge National Laboratory[1]
Oak Ridge, Tennessee 37831-6038

Introduction

The purpose of this chapter is to discuss some of the ecological questions and concerns related to identifying hazards and assessing risks posed by environmental applications of biotechnology. Although biotechnology products offer great potential benefits, particularly in agriculture and as a means of replacing chemicals that themselves produce environmental degradation, there are a number of uncertainties about the risks of such products. These uncertainties arise principally because of gaps in our basic knowledge of ecology in general, of microbial biology and ecology in particular, and of genetics and risk assessment itself. The need to ensure that the benefits of biotechnology prod-

---

[1]Operated by Martin Marietta Energy Systems, Inc., under contract no. DE-AC05-840R21400 with the U.S. Department of Energy. Publication no. 3554, Environmental Sciences Division, Oak Ridge National Laboratory. The submitted manuscript has been authored by a contractor of the U.S. government under contract no. DE-AC05-840R21400. Accordingly, the U.S. government retains a nonexclusive, royalty-free license to publish or reproduce the published form of this contribution, or allow others to do so, for U.S. government purposes.

ucts are safely realized and that their environmental applications do not result in new sources of environmental degradation provides the fundamental rationale for systematic assessments of risk before field testing or commercial use.

It must be stated at the outset that the point of view expressed in this paper is that of an ecologist. Ecologists are the practitioners of a scientific discipline that concerns itself with, among other things, understanding the distribution and abundance of life forms. Central to the study of ecology are efforts to answer questions about how communities of organisms are "assembled" and what controls the number of different species found in particular environments. Why do some combinations of species coexist successfully while others cannot? Ecologists have been studying the effects of adding species to or removing them from communities of organisms for decades, and a great deal of information has accumulated. Unfortunately, most of this information deals with higher organisms—plants and animals. Less theory and fewer data are available to explain the workings of microbial communities. Therefore, most of the inferences drawn are based on patterns observed in higher-organism communities. I will return to this point later to discuss whether the differences between microbial and higher-organism communities are putative or real.

Concerns in Risk Assessment

The concerns about environmental uses of genetically "engineered" or altered products are fundamentally different from those for laboratory use. First, it is not primarily humans but rather the vast array of nonhuman species in an ecological community that will be exposed to the released organisms. Second, the spectrum of potential negative effects is not restricted to pathogenicity, although this is certainly a significant concern. Nonindigenous organisms, whether genetically altered or not, can influence the structure (population size and species diversity) and function (energy and material dynamics) of ecological communities through a variety of mechanisms that may displace or destroy indigenous species. Finally, the degree of control afforded by contained experiments is lost in the field. Once released, altered organisms that find suitable habitats may reproduce, spread, and evolve in ways that are beneficial to their own survival and that may subvert imposed biological constraints. For this reason, dealing with organisms obviously differs from dealing with chemical pollutants, which are invariably diluted and usually degraded after a release (Suter 1985). Methodologies for ecological risk assessment have only recently been developed (e.g., Barnthouse et al. 1986) and have focused on chemical effects. The extension of such methods to assessing risks of released organisms remains to be systematically investigated.

Environmental uses of genetically altered organisms therefore give rise to a number of issues that can be grouped under the headings of three general questions. These are:

- Will the genetic alteration modify ecologically or environmentally relevant properties of the organism?

- Can the genetic alteration spread to other organisms, and if it can, what might the consequences be?
- What will be the consequences of adding a new species or genotype to the ecological community in which the product is to be used?

The process of obtaining the answers to these questions can provide a framework for risk assessment. It should be noted that the first and second questions focus specifically on the consequences of altering an organism's genes to modify its properties. The third question, however, is less concerned with modification, per se, than it is with the *ecological interactions* of the released organism, whether its properties result from modification or not.

***Question 1:* Will the genetic alteration modify ecologically or environmentally relevant properties of the organism?** Gaining an understanding of whether a genetic alteration will modify properties that are relevant to an organism's ecological role requires baseline knowledge of the basic biology and ecology of the unmodified parental organism and of the genetic alteration itself—the function of the added or deleted gene(s) and the method by which the modification is created, including the nature of the vector used. The first step in risk assessment should therefore be a screening on the basis of knowledge of the parental organism's properties.

Where information on the basic biology of microbial species is known, it has usually been obtained from single-species culture studies. Conditions for survival and growth—e.g., physical factors, primary nutrient sources, and natural environment—are generally established (Sayler and Stacey 1985) for a wide variety of microorganisms. But other important properties, such as alternate environments, alternate nutrient sources, and whether plasmids are naturally present, may only be well known for limited sets of species. Knowledge of this second group of factors would be useful in predicting whether an altered organism is likely to survive in release environments. Organisms that are known to be unable to survive in particular habitats before modification should not be able to do so afterward *unless* the modification is, in fact, directed at expanding the range of environments in which survival is possible. This kind of information is, however, frequently unavailable.

And unfortunately, even less is typically known about the ecological roles and behaviors of species in mixed populations in natural communities. Information on factors such as the potential for host-range shifts in pathogens, on competitive abilities and the potential of microbial species to dominate one another, on roles in biogeochemical cycles, and on gene-transfer mechanisms is often very limited. The power of biotechnology to manipulate the genetic structure of microorganisms far outstrips our knowledge of their roles in the environment. In fact, newly described species may find their way into the laboratory for genetic alteration before virtually anything is known of their basic biology or ecology. The first step in evaluating the need for more or less detailed risk assessment should, therefore, be guided by how much is known about the parental organism. Risk can be minimized by the preferential selection of parental organisms that are generally recognized as safe because of a long history of use. *Vaccinia* virus for vaccines and *Rhizobium* species in agricultural use provide examples of such organisms. In general, the better the biological and ecological properties of the parental organism are understood, the less will be the necessity to collect new information in the risk assessment process.

The nature of the genetic modification is also important, and is, in fact, much more relevant to determining the need for and level of risk assessment than is the *number* of genes affected by the alteration process. Many examples of important changes in phenotype involving only one gene are known. For example, Nester et al. (1985) reported an interesting case of a potential "host-range shift" related to a single gene in the pathogen *Agrobacterium tumefaciens*, the etiologic agent of crown gall disease. A wide-host-range strain of this bacterium possesses one gene to which resistant grapevines respond with a hypersensitivity reaction. This response prevents tumor formation and, therefore, the development of the disease. A deletion of this single bacterial gene would result in a loss of the plant's ability to recognize the bacterial pathogen and thereby allow infection. Another example of a single gene shift with potential ecological significance concerns the *Pseudomonas* species engineered by Monsanto (described, for example, in Watrud et al. 1985) to produce the δ-endotoxin of *Bacillus thuringiensis*. Although *B. thuringiensis* has been used safely to combat leaf-eating lepidoptera for many years, the new product is designed to affect soil organisms. The shift of a gene from a leaf-dwelling to a soil-dwelling bacterium will expose many other soil organisms to a toxin that they have quite likely not encountered before in significant quantities in their own microenvironment. Both new target organisms and beneficial nontarget species could be affected.

Slight genetic alterations could also be used to produce significant modifications in the rates at which important environmental substrates (e.g., cellulose, lignin) are produced or transformed, or to expand the range of environments in which an organism may survive. Genetic changes that involve deliberate expansions of an organism's tolerances to environmental conditions, such as temperature or salinity, may produce profound ecological shifts with important implications. For example, such an alteration may promote the survival of an organism by providing it access to another microhabitat or nutrient source, thereby removing the necessity for it to compete with its parental type. In short, small genetic changes do not *invariably* translate into small changes in ecologically relevant aspects of phenotype. There is not necessarily a proportionality between the two, and there is no equivalent of "structure-activity" relationships for predicting the effects of a genetic modification on phenotype. From the ecological viewpoint, the number of genes added, or subtracted, is not an appropriate standard by which to judge the necessity for risk analysis. How does the genetic modification affect the distribution of the altered organism in the environment relative to its parental type? Does the genetic alteration affect the organism's relationships with other members of the ecological community? Will the organism's modification cause it to influence the biochemical pathways for production or degradation of important organic (e.g., cellulose, lignin) and inorganic (e.g., nitrate, phosphate, trace elements) materials? These are some of the important environmental questions that warrant evaluation.

Last, it should not be assumed that a genetic alteration will inevitably decrease an organism's fitness. It is frequently argued that altered organisms cannot survive after release because the addition of "engineered" genetic material is likely to disrupt the coadaptation of the natural genome or because it poses a physiological burden. Although it is true that many kinds of alterations are more likely than not to lower the fitness of an organism, the study of evolutionary change clearly indicates that not all genetic novelties are disadvantageous. Furthermore, the extent to which a genetic alteration truly functions as a handicap and reduces fitness may be dependent on the *environmental context* in which the handicapped organism is placed. Pimm and Levin (1986) provided the following instructive example. Suppose that there is an ecological system with two resources and one consumer organism. This organism uses one resource efficiently, and leaves the other unexploited. Suppose further that a mutant form of the consumer organism arises that is marginally capable of surviving by using the unexploited resource. Although the mutant cannot compete with its efficient parent for the more-favored resource, it has the less-favored resource all to itself. Despite its genetic handicap, the altered form may nevertheless be fit enough to survive in the absence of competition from its more fit parent. Or suppose that the mutant organism finds its way into an environmental setting in which the normal community has been disrupted or which is naturally low in species diversity, as are some of the microbial communities involved in important biogeochemical processes. Again, limited competition in such an ecological context may give even a marginally fit organism the ability to survive. And if it can survive long enough, subsequent natural selection may eventually increase its efficiency and its fitness. Small genetic changes of the magnitude common in genetic engineering may *usually* decrease the fitness of the engineered organism. Nevertheless, a small number of such changes could substantially *increase* fitness, and it should not be assumed, unless empirical observation supports such conclusions, that an engineered organism will invariably be less competitive or that it will always die out after a release.

To summarize thus far, the appropriate starting points for risk assessment are to evaluate the properties of an unmodified organism, the effects that a particular genetic modification may have on its ecologically significant phenotypic traits, and the characteristics of the specific environment that provide the context in which it will be used. The degree of uncertainty about each of these elements can be used to determine the level of detail required in the analysis and to guide the selection of particular risk assessment activities. For example, lower levels of review may be required for microorganisms that are indigenous to the overall environment *and* microenvironment in which use will occur, that have genetic modifications which are well and narrowly defined, and that do not pose any special problems such as pathogenicity, toxin production, or participation in fundamental biogeochemical processes. As uncertainty increases, so should the level of required review.

***Question 2:* Can the genetic alteration be transferred to other organisms, and if it can, what might the consequences be?** The mediating effects of environmental factors (such as temperature, pH, and presence of surfaces) on gene transfer in natural communities are not yet well or comprehensively understood. In vitro studies usually provide optimal conditions for plasmid exchange between microbial species, but evidence collected from such studies is not adequate to predict behavior under complex environmental conditions. Little information on how environmental factors promote or discourage gene transfer has been collected as yet.

Gene transfer clearly occurs in nature in response to powerful selective forces, as is attested by the nearly ubiquitous presence of antibiotic resistance plasmids in diverse microbial taxa. The role of accessory genetic elements in ensuring the survival of microorganisms under many kinds of extreme selection is well understood. Plasmids are, nevertheless, also observed to be maintained and transferred in natural populations even in the absence of identifiable selective factors. [See, for example, the review papers by Stotzky and Babich (1984) and Reanney et al. (1983)]. The explanation for this may be that selection is operating on some cryptic gene of unknown function. The better characterized the functions of the genes in a plasmid vector are, the greater the possibility of having an ability to predict whether it will provide a selective advantage in particular environmental settings. In general, precision of characterization of both the genetic alteration itself and of the vector used for its insertion will be important factors to consider in risk analysis.

In addition to detailed characterization, however, other safety features may reside in the genetic techniques used. For example, selection for recombination of an added gene into the bacterial chromosome may make it less subject to transfer; so can the use of poorly mobilizable plasmids. The use of such mechanisms should be encouraged. And again, the consequences of gene transfer should be assessed in the context of the specifics of the organism and genetic alteration in question.

Finally, although gene transfer may be of more general concern for microorganisms, particularly bacteria, than for plants or animals, there are nevertheless specific concerns for higher organisms that should be evaluated when appropriate. An important example is minimizing the spread of genes for highly advantageous traits, such as herbicide resistance, from crop plants to weed species via hybridization.

***Question 3:* What will be the consequences of adding a new species or genotype to the ecological community in which the product is to be used?** A very large number of ecological data sets indicate that the addition of species or genotypes to communities where they did not occur before *can* have undesirable effects. Such effects include toxicity or pathogenicity to resident species and displacement or destruction of residents through a variety of mechanisms, such as competition and predation. If impacts at the level of one or a few species become sufficiently severe, community-level effects, such as loss of species diversity and disruption of energy and nutrient dynamics, can also result. Although these kinds of serious and widespread effects probably occur in only a small fraction of species-addition events, the problem is being able to predict under what circumstances such effects are more likely. Ecologists have therefore traditionally been interested in determining what kinds of species are more apt to successfully become established in new environments and what kinds of environments are more vulnerable or "invadable."

It is possible to say with certainty only that the outcome of adding a species to a community depends on the specific characteristics of the species being added and the structure and function of the specific receiving community. This, in essence, means that no two species-addition events will have exactly the same features, and that each new situation could present somewhat different aspects for consideration if predictions of effects are wanted. On the other hand, it would appear that certain rules of thumb may define some situations that might be more problematic than others.

Evidence from the ecological literature suggests that certain characteristics facilitate the successful establishment of new organisms in communities. First, organisms with generalized requirements and broad tolerances seem to make better "invading" species than ones with narrow requirements and tolerances. This observation would seem to be eminently logical; the less specific an organism's requirements are, the more apt it is to find an amicable environment that meets those requirements.

Second, if an organism has a unique ability to exploit an unused or underused resource, this can make it more apt to succeed. Such abilities could come from morphological, behavioral, or physiological features. The following examples (all related to higher organisms) are provided only by way of illustration. In the southwest desert of the United States, introduced salt cedar (*Tamarix*) represents a life form that is both a highly successful invader and morphologically unlike any member of the native flora (Vitousek 1985). Its morphological advantage is its unusually deep root system, which provides it access to water that is unavailable to other desert plants. Salt cedar requires wet conditions to germinate and also has an extremely high rate of evapotranspiration. As a consequence of these features, it tends to selectively invade desert watercourses, where, once established, its rapid transpiration may result in complete drying up of natural springs, to the detriment of other desert inhabitants. Its unique morphological capability allows *Tamarix* to affect many other species by altering a basic ecosystem property, the water balance of the desert environment.

Unusual behavior can also confer advantages. This has been observed to be of great importance in cases where the invader is a predatory species. The sea lamprey, for example, is an odd combination of predator and ectoparasite. It kills its prey by attaching itself to another fish and then rasping and sucking its flesh. This unique "style" of predation, to which the large native fish in Lake Michigan were totally unadapted, is believed to have significantly contributed to the sea lamprey's ability to virtually wipe out the prey species there (Moyle 1986). Diamond and Case (1986) also discussed the role of "naïveté" in invaded communities in determining the success of exotic species. They provided a fascinating example of introduced rats on islands. Introduced rats have had negligible effects on the avifaunas of islands that support native rats (e.g., Solomon Islands, Christmas Island, Galapagos), but have quickly decimated bird populations on islands where native rats are lacking. There are, however, also islands with no native rats that have not been adversely affected when introduced rats made their appearance. The explanation for this is that in the last instance instead of native rats, such islands have native tree-climbing crabs that are the invertebrate equivalent of rats. The avifaunas of such islands have apparently evolved adequate antipredator behavior, although the predator is an entirely different one.

Unique physiological features, such as the ability to metabolize unusual compounds, may allow an organism to exploit unused or underused resources. How such a trait may affect a microorganism's fitness was discussed in a previous section (see under "Question 1"). This kind of unique feature could be of particular importance in microbial species that are genetically modified to degrade substrates (e.g., chemical wastes) that are not usable by other species.

The third and last factor that I wish to mention as contributing to the success of an invading species is preadaptation. That is, organisms may be "preadapted" to survive in particular new environments if the environment in which they originally evolved was very similar. Again, in such a case, the organism may start off its existence in the new ecological context with distinct survival or competitive advantages. Examples include European annual grasses, which have succeeded in almost completely replacing the native perennial bunch grasses in California, where the climate is highly similar to that of the Mediterranean source area; and the introduced insect fauna of North America, which is composed of more than 60% European and only 14% Central and South American species, despite the greater proximity of the tropical American source area. It would seem that the greater similarity of European climate, crops, and natural vegetation to those of the United States represent a natural advantage for European insects over western hemispheric exotics (Simberloff 1986). The preadaptation argument may also apply to closely related species, such as species in the same genus, or congeners, which may be very similar to each other. This is one argument against automatically limiting risk review to minimal levels in cases of intrageneric gene transfer. A closely related species may be *more* apt to survive and replace its competing relative if it is ecologically highly similar and is therefore preadapted to survive, given some sort of competitive edge, where it is introduced.

There are also quite a lot of ecological data indicating that certain kinds of environments are more vulnerable to the establishment of new species than others. The most "invadable" environments seem to be generally characterizable as "simple," or low in species diversity. "Species diversity" or species "richness" measure the number of different species present in a community. Species diversity may be low for natural reasons, or because of effects of human activity.

Oceanic islands provide excellent examples of ecological systems that are naturally low in species diversity. The lush appearance of many remote tropical islands is often due to the presence of large numbers of individuals of a few species rather than to high species richness. The "depauperate" nature of species diversities on such islands is due to the fact that it is difficult for organisms to disperse long distances over water to get there. Organisms that have successfully colonized such areas by natural means tend to be those that have dispersal advantages, such as wings. The Hawaiian islands, for example, had no native mammals other than bats until the early Polynesian colonists brought pigs, dogs, and rats there. The degree to which species-addition events in such settings have frequently resulted in establishment of the added species is evident in the following. In over 1400 cases of successful establishment of exotic birds and mammals worldwide, 60% have occurred on oceanic islands while only 20% each have occurred on continents and continental shelf islands (Ebenhard in press). There is also little doubt that some of the most severe impacts of species additions have occurred in these environmental contexts, with introduced species producing total or near total replacement of natural vegetation and extinctions of native animals on many islands. Other kinds of ecological communities with naturally low species diversities also exist. For example, high-temperature springs are inhabited by low numbers of species because of the exigencies of surviving in such environments.

Communities with greater species richness seem to be better able to repel invaders. Diamond and Case (1986) compared the success of invading bird species in becoming established on islands with widely differing numbers of extant native birds. On islands with few native species, invasion success has been very high, while on islands with large numbers of native birds, invaders have failed miserably. In some cases, the relationship between native and introduced species also exhibits habitat-related aspects. On Viti Levu, where there is a relatively large number (48) of native species, avian invaders have been successful only in disturbed habitats that are unsuitable for the natives. On Oahu, however, where only 10 natives are extant, the invaders are also abundant in the remaining natural forests. Similar patterns are observable for other major taxa of organisms. In North America, for example, introduced fish species constitute less than 10% of the fish fauna in the east, where natural diversity is high (Moyle 1986). In the western United States, however, stream drainages are isolated, natural diversity is low, and exotics make up 30 to 60% of the fish fauna in most streams. Similarly, Crawley (1986) observed that introduced plants in Britain made up 50% of the flora of disturbed "waste ground" but only 5% of the flora of unmanaged woodland.

Low species diversity in communities of organisms may also be directly attributable to human activity. Some communities, such as agricultural monocultures, have small numbers of species by design. Others have been disturbed, either physically or with chemicals. Although conventional wisdom has it that most "healthy" natural communities are able to repel invaders, it is important not to take the existence of such natural "defenses" for granted. If natural defenses exist, there may still be a variety of circumstances under which they are inoperative, and this possibility should be taken into account in the risk assessment process.

To sum up the third set of points, ecological data indicate that certain features of organisms may make them more likely to become established in new communities, and certain features of communities may make them more vulnerable to establishment. Such generalizations give us a way to think about how particular combinations of species and communities might, or might not, "work." It might be desirable, for example, to avoid releasing microorganisms with extremely broad environmental tolerances in ecological settings with low species diversities if survival and establishment are *not* sought. Although such generalizations can be used to suggest where greater scrutiny might be judiciously applied, real risk assessments will require empirical study of the specifics. Successful predictions of the outcome of manipulating ecological systems are possible. Ecologist Dan Simberloff (1985) believes that an understanding sufficient to make an accurate prediction of how a new species or genotype is likely to affect others in the community requires field research of about the depth of a doctoral dissertation.

Other Issues

How much of what we know about the dynamics of species interactions among higher organisms is applicable to microbial systems? This question deserves at least brief consideration here. Microorganisms live in ecosystems, just as higher organisms do. Microbes compete, prey on each other, and modify each other's environment chemically and physically. Microbial ecologists (e.g., Alexander 1971) speak of "stable climax communities" of microorganisms that evolve by selection and successional processes as do higher-organism ecosystems. And, as in higher-order systems, microbial ecologists have observed that though microbial communities are normally resistant to penetration by new species, perturbations can result in new species becoming dominant. What, then, are the differences between the dynamics of microbial and higher-organism ecosystems?

One of the more obvious categories of difference seems to reside in the population genetics of microbes. Brock (1985), for example, argues that because of small size and high growth rates in prokaryotes, large numbers of genomes can reside in small amounts of biomass. Prokaryotes may therefore be able to adapt more quickly and dramatically to environmental changes than either eukaryotic microorganisms or higher eukaryotes, assuming that the presence of large numbers of genomes is also directly correlated with genotypic diversity. Brock also points out that microorganisms are readily dispersed, even to distant environments, and appropriate organisms are likely to be present wherever conditions are suitable for them. "Everything is everywhere, the environment selects." His implication seems to be that microbial populations are nearer to being "optimally adapted" to their environments. If this is so, it is in marked contrast with communities of higher organisms, which are rarely saturated with species, are rarely in equilibrium with available resources, and are composed of species that are adequately, but not optimally, adapted (Colwell 1985). It would, however, appear that not all microbial ecologists accept such contrasts. Sayler and Stacey (1985), for example, make the point that an introduced organism in a *microbial* system is more likely to survive where competition with other organisms is limited. "In most natural environments, *it is unlikely that saturation of all niches occurs*. Novel organisms introduced into these environments may not have to compete with indigenous organisms" [*emphasis mine*]. Are the differences between microbial and higher-organism ecosystems real, as some microbiologists maintain? This question bears looking into more closely, perhaps as a cooperative effort between higher-organism and microbial ecologists.

To touch upon one final issue, there has been pressure over the last couple of years in the United States to create categories of organisms and/or genetic modifications that may be exempted from risk review prior to use in the environment. Although review for low-risk products should be expedited and requirements for more detailed scrutiny should be reserved for higher-risk products, at this time complete exemption from review of any new product may be premature, especially in light of the fact that little actual experience in risk analysis for genetically altered organisms has yet accumulated. At a minimum, some form of *screening* should be employed for new microbial products for environmental application. The creation of categories of microorganisms completely exempt from review should be postponed until methods for risk analysis are better established and standardized, and experience from empirical testing and monitoring of early biotechnology products has been documented. The participants in a recent workshop convened by the Ecological Society of America (Tiedje et al. 1989) stated that they believed that case-by-case review is currently the most scientifically sound regulatory approach for biotechnology products for environmental applications because of the diversity of products that can be developed and the complexities involved in predicting their ecological fate. These ecologists anticipate that guidelines for establishing review levels, including categories of organisms requiring minimal screening as well as those requiring more intensive review, can be developed after experience is gained from field experiments and from research in ecological risk assessment. They recommended that scaling of regulatory oversight be based on the attributes of organisms and environments, and provided a sample set of scaling factors.

Risk Assessment Needs

The following briefly reviews key areas in which greater knowledge would be beneficial for the successful conduct of risk assessment.

General ecology

- Basic investigations and sound empirical observations on the nature of interactions among coexisting species, the forces that knit ecological communities together, and how the properties of ecosystems are determined by interactions of their structural and functional components could contribute a much stronger theoretical basis for developing predictive capabilities.
- Better and more in-depth analyses of species-addition events could provide greater understanding of the causes of observed ecological patterns.
- Agreement about "regulatory end points" is needed. In other words, what aspects of the environment require protection—structural parameters (such as species diversity), functional parameters (such as nutrient dynamics), or both—to safeguard environmental quality?
- An examination and analysis of the degree to which microbial and higher-order ecosystems actually differ is needed. Considerations include the implications of differences in population dynamics, the role of species richness, degrees of adaptation, and other important community characteristics and organizing principles.

Microbial biology and ecology

- A larger information base on the general properties of microorganisms—including basic taxonomy and its relationship to genotypes and phenotypes, habitats, ability to persist, dispersal mechanisms, pathogenicity, host ranges, immunological responses, and other properties of importance in risk assessment—should be created.
- An extensive data base on the ecologically significant characteristics of microorganisms, including interspecific interactions, population genetics, roles in biogeochemical cycles, and other relevant properties, should be compiled as knowledge grows.
- Greater knowledge of mechanisms of microbial dispersal, including long-range transport, should be developed.
- Information on the efficacy of existing mechanisms for biological containment (including means of debilitating or "disarming" released microorganisms, ensuring the effectiveness of recall or eradication mechanisms, and preventing or reducing unwanted transfer of genetic material to other species) should be compiled and new methods developed where needed.

Genetics

- Understanding of the ecological and evolutionary roles of plasmids could be improved. Considerations include selection factors that promote plasmid maintenance and transfer in natural populations and how environmental factors reduce or enhance gene transfer between microbial species.
- Genetic markers that do not rely on antibiotic resistance traits for detection of modified organisms are needed.
- The reliability and cost-effectiveness of techniques for detecting, quantifying, and assessing persistence and vigor of microorganisms in various environments—including DNA probes, monoclonal antibodies, and immunofluorescent techniques—could be improved.

Risk assessment methods

- Microcosm, mesocosm, and controlled-field-test protocols for use as predictors of responses of natural ecosystems should be developed, standardized, and validated.
- Predictive measures, including predictive models, for estimating field effects from controlled laboratory or other enclosed studies are needed. Also needed are accurate methods of extrapolating results of risk assessment testing from laboratory to field and from one field site type to another.

Conclusion

Ecologists and environmental scientists do not believe that the products of biotechnology will generally be harmful, but only that the many uncertainties merit continued caution for the moment. The most discomforting aspect of these products is not that *some* of them may be harmful, but rather that it may be difficult to determine *which ones* are. The goals of assessing risks in regard to engineered organisms will be to make a reasonably accurate prerelease prediction of the behavior an organism is likely to exhibit in its new ecological context and given its particular genetic modification, and to be able to detect and avert potential problems before they occur. Although ecologists are firm in the belief that these goals can be realized, there is still a great deal of work to do in the development of risk analysis. This work will require the judgment and cooperation of scientists from the many different subdisciplines of biology, from molecular genetics to community ecology.

Ecologists have a natural role in determining product safety because the environmental uses of genetically engineered organisms touch upon many important ecological questions. In addition, the design and production of effective, as well as safe, products could be greatly facilitated by the employment of an ecological viewpoint from the earliest stages of product development. At the close of an interdisciplinary conference on genetically engineered organisms in the environment held in Philadelphia in 1985, the American ecologist Robert Colwell (1985) said, "Ecologists have more to contribute to genetic engineering than worrisome carping. The critical balance between effectiveness and safety that we worry about can only be a challenge if there is effectiveness, and I think ecologists have much to offer in helping to design effective organisms. If genetic engineering is the cutting edge, then maybe ecology is the whetstone."

References

Alexander, M. 1971. *Microbial Ecology*, Wiley, New York.

Barnthouse, L. W., Suter, G. W., Bartell, S. M., Beauchamp, J. J., Gardner, R. H., Linder, E., O'Neill, R. V., and Rosen, A. E. 1986. *User's Manual for Ecological Risk Assessment*. Report ORNL-6251, Oak Ridge National Laboratory, Oak Ridge, Tennessee.

Brock, T. D. 1985. Prokaryotic population ecology, in *Engineered Organisms in the Environment: Scientific Issues*, H. O. Halvorson, D. Pramer, and M. Rogul (eds.), American Society for Microbiology, Washington, D.C., pp. 176–179.

Colwell, R. K. 1985. Session IV. Responses to perturbation: genome to ecosystem, in *Engineered Organisms in the Environment: Scientific Issues*, H. O. Halvorson, D. Pramer, and M. Rogul (eds.), American Society for Microbiology, Washington, D.C.

Crawley, M. J. 1986. What makes a community invasible? in *Colonization, Succession, and Stability*, Symposium of the British Ecological Society, Blackwell Scientific, Oxford, U.K.

Diamond, J., and Case, T. J. 1986. Overview: introductions, extinctions, exterminations, and invasions, in *Community Ecology*, J. Diamond and T. J. Case (eds.), Harper & Row, New York, pp. 65–79.

Ebenhard, T. Introduced birds and mammals and their ecological effects, *Swedish Wildlife* (in press).

Moyle, P. B. 1986. Fish introductions into North America: patterns and ecological impact, in *Ecology of Biological Invasions of North America and Hawaii*, H. A. Mooney and J. A. Drake (eds.), Springer-Verlag, New York.

Nester, E. W., Yanofsky, M. F., and Gordon, M. P. 1985. Molecular analysis of host range of *Agrobacterium tumefaciens*, in *Engineered Organisms in the Environment: Scientific Issues*, H. O. Halvorson, D. Pramer, and M. Rogul (eds.), American Society for Microbiology, Washington, D.C., pp. 191–196.

Pimm, S. L., and Levin, B. R. 1986. Impact on competitive abilities of specific genetic alterations, in *Ecological Issues Relevant to Environmental Applications of Genetically Engineered Organisms*, E. Norse, (ed.). Contract report prepared for the Office of Technology Assessment, U.S. Congress, Washington, D.C.

Reanney, D. C., Gowland, P. C., and Slater, J. H. 1983. Genetic interactions among microbial communities, in *Microbes in Their Natural Environments*, J. H. Slater, R. Whittenbury, and J. W. T. Wimpenny (eds.), Cambridge Univ. Press, Cambridge, U.K., pp. 379–421.

Sayler, G., and Stacey, G. 1985. Methods for evaluation of microorganism properties, in *The Suitability and Applicability of Risk Assessment Methods for Environmental Applications of Biotechnology*, V. T. Covello and J. R. Fiksel (eds.), National Science Foundation, Washington, D.C.

Simberloff, D. 1985. Predicting ecological effects of novel entities: evidence from higher organisms, in *Engineered Organisms in the Environment: Scientific Issues*, H. O. Halvorson, D. Pramer, and M. Rogul (eds.), American Society for Microbiology, Washington, D.C., pp. 152–161.

Simberloff, D. 1986. Introduced insects: a biogeographic and systematic perspective, in *Ecology of Biological Invasions of North America and Hawaii*, Springer-Verlag, New York.

Stotzky, G., and Babich, H. 1984. Fate of genetically-engineered microbes in natural environments, *Recomb. DNA Tech. Bull.* 7(4):163–188.

Suter, G. W. 1985. Application of environmental risk analysis to engineered organisms, in *Engineered Organisms in the Environment: Scientific Issues*, H. O. Halvorson, D. Pramer, and M. Rogul (eds.), American Society for Microbiology, Washington, D.C., pp. 211–219.

Tiedje, J. M., Colwell, R. K., Grossman, Y. L., Hodson, R. E., Lenski, R. E., Mack, R. N., and Regal, P. J. 1989. The planned introduction of genetically engineered organisms: ecological considerations and recommendations, *Ecology* 70(2):297–315.

Vitousek, P. M. 1985. Plant and animal invasions: can they alter ecosystem processes? in *Engineered Organisms in the Environment: Scientific Issues*, H. O. Halvorson, D. Pramer, and M. Rogul (eds.), American Society for Microbiology, Washington, D.C., pp. 169–175.

Watrud, L. S., et al. 1985. Cloning of the *Bacillus thuringiensis* subsp. *kurstaki* delta-endotoxin into *Pseudomonas fluorescens*: molecular biology and ecology of an engineered microbial pesticide, in *Engineered Organisms in the Environment: Scientific Issues*, H. O. Halvorson, D. Pramer, and M. Rogul (eds.), American Society for Microbiology, Washington, D.C., pp. 40–46.

Chapter

2

Nontarget Species Testing of Microbial Products Intended for Use in the Environment

Robert J. Frederick

U.S. Environmental Protection Agency[1]
RD-682
401 M Street, S.W.
Washington, D.C. 20460

Robert W. Pilsucki

U.S. Environmental Protection Agency
H7507C
401 M Street, S.W.
Washington, D.C. 20460

Introduction

Increased interest in, and what may be called a general persuasion toward, the development of biotechnology has been accompanied by caution regarding the widespread use of the new products in the environment. Product testing

[1]This chapter was written by Dr. Robert J. Frederick and Dr. Robert W. Pilsucki in their private capacities. No official support or endorsement of the U.S. Environmental Protection Agency or any other agency of the federal government is intended or should be inferred.

prior to large-scale introductions of microorganisms will be designed not only for determinations of efficacy, but also for untoward effects when the product is in general use. The intent of this chapter is to present the range of considerations that might be addressed in establishing a nontarget testing scheme and, perhaps for some, a starting point in protocol development. Our discussion attempts to address all potential microbial products intended for introduction into the environment, but makes a distinction between microbial pest-control agents (MPCAs) and those products that are not intended for pesticidal use. This has been done in recognition of the availability of guidelines for testing pesticides and the difficulty inherent in choosing a satisfying but not overburdening test or tests for microorganisms that have no known pathogenic effect.

Microbial Products for Environmental Use

Experience with microorganisms deliberately introduced into the environment has primarily been in the agricultural setting and has been generally favorable. Over the last 40 years, since *Bacillus popilliae* was first registered for the control of the Japanese beetle larva, there have been virtually no adverse environmental effects reported that can be attributed to the use of microbial pesticides in agricultural situations. However, the number of acres treated with microbial pesticides has been small when compared with the number on which chemical pesticides have been used.

There are a number of reasons contributing to the small share of the market held by naturally occurring MPCAs. Some of these reasons are (1) limited and costly production methods; (2) relatively short shelf life; (3) slow killing speed, when compared to chemical pesticides; (4) a narrow host range, often restricted to one or a few species; (5) limited persistence in the field after application. (The last two factors, restricted host range and limited persistence, also contribute to the argument that MPCAs are less hazardous to human beings and other nontarget species.) Through the use of genetic engineering, some or all of the above restrictions may be overcome.

In many cases, the host range of naturally occurring microbial insect pathogens is severely limited. There are, for instance, *B. thuringiensis* species active against lepidopteran species only, while others are active against coleopterans only and still others are toxic only to dipteran species. The genes coding for the toxins active against these pest species are generally located on plasmids, which can be passed between different varieties of *B. thuringiensis*, yielding strains that have unique combinations of plasmids and, thus, activity against multiple groups of pests. While intrageneric plasmid transfer results in a widening of the host range for *B. thuringiensis* varieties, the added hosts are still contained within a few orders of insects and the range of insects affected remains small when compared to the broad-spectrum chemical pesticides. In the case of some insect viruses, the host range is even more restricted, often to a single genus and species of insect larva. One notable exception is the nuclear polyhedrosis virus (NPV) of *Autographa californica*, which has been known to infect over 30 insect species (Carbonelli et al. 1985).

The second characteristic of microbial pesticides is that they are rather short-acting when applied to areas where they are available to pests. Two mechanisms contributing to short-lived activity are physical inactivation and movement out of the zone where they can be ingested by the target pest. Environmental conditions, such as ultraviolet (UV) light and desiccation, work to inactivate microorganisms applied to outdoor surfaces. After application, the density of active MPCAs declines so that the number surviving usually falls below some threshold level necessary to maintain effective pest population control. In other cases, such as application of *B. thuringiensis* var. *israelensis* spores to water for mosquito and blackfly control, the spores remain at the water surface where they are available to the target larvae for only a short time before they begin sinking and leave the effective area. In both cases, the active units become unavailable to the target pest.

Lack of persistence coupled with the slowness with which MPCAs begin to achieve control of the target pest, and with their restricted host range, are the major reasons microbiological control of agricultural pests has not penetrated the market to any appreciable extent. What is needed for increased use of microbial control of agricultural pests is the "perfect" microbial pesticide. Efforts are underway to create this "perfect" agent by combining the most effective pesticidal and survival traits using genetic engineering techniques, or by engineering pesticidal traits directly into agricultural crop plants.

The "perfect" MPCA would (1) kill the targeted pest or pests quickly and efficiently, (2) persist from year to year in sufficiently high numbers and be available to the target in order to cause an epizootic before economically significant crop damage occurs, and (3) not affect nontarget organisms.

Other commercial areas that may use microorganisms introduced into the environment include toxic waste cleanup, mining operations, enhanced oil recovery, and snow making. For most of these uses, there have been only a limited number of field trials. Certainly at the field-testing stage, the procedures for application control and site management are made the more manageable because of the ability to define clear boundaries for individual tests. On the other hand, when products are commercially available, there will be larger-scale operations where the spatial and temporal distribution of the microorganisms will be determined by the product's use. The obvious consequence is that many more nontarget species will be exposed when a product is placed on the market.

It may reasonably be expected that the overwhelming majority of these nonpesticide microorganism products will not contain known pathogens and may be intentionally designed to have minimal effect on the environment. For example, the "perfect" microorganism for degrading a particular pollutant would (1) quickly and effectively metabolize or convert the intended substrate completely to a nontoxic product without generating toxic by-products, (2) die off to undetectable levels after the intended substrate was exhausted, and (3) not shift indigenous microbial populations in a detrimental direction or harm the flora and fauna.

Changing characteristics

Over the millennia, microorganisms have evolved efficient and diverse mechanisms for performing their roles in processes such as insect population control or nutrient and geochemical recycling in the environment. However, whether we discuss pesticidal microorganisms or those intended to be used for some other purpose, the existing pool of naturally occurring bacteria, viruses, fungi, and protozoa are relatively inefficient, from the human point of view, at performing these tasks in agricultural settings or in situations where the bioremediation of anthropogenic chemicals in the environment is needed. The goal of genetic engineering of microorganisms is to enhance the desirable traits these organisms carry, thereby increasing their effectiveness in performing a particular environmental task.

With regard to pesticidal microbes, genetic manipulations are aimed at:

1. Improving the "killing speed" by increased toxin production, biosynthesis of the toxin in a more active form, stabilization of toxin-producing genetic elements, and inclusion of "potentiators" which will assist in producing the desired effect
2. Expanding the host range in order to exert control over multiple pests
3. Engineering pesticidal traits into alternate microorganisms which have better survival characteristics in the field
4. Engineering pesticidal traits directly into crop plants.

The introduction of genetically engineered microorganisms has begun with small-scale field tests. Private companies, and university and government researchers, are proceeding slowly and cautiously in the process of bringing modified microorganisms from the laboratory to the field. Since these field tests are small and result in little exposure to nontarget organisms, and since the microorganisms and the introduced genes are well known and not closely related to avian, fish, or beneficial invertebrate and insect pathogens, the potential hazard to nontarget organisms has not been an overriding issue. As the scale of testing increases and as some of these genetically engineered microorganisms begin to be considered for registration and to move into commerce and widespread use, the potential for adverse effects on nontarget organisms and the environment will need to be addressed.

The balance of this chapter will deal with making decisions about testing of genetically modified or naturally occurring microorganisms on nontarget species, including the rationale for nontarget testing (i.e., the answer to the question, Why is nontarget testing necessary?), the philosophy behind the design of nontarget tests, and a description of the current major classes of nontarget single-species testing and testing in a multiple-species, or microcosm, environment. In general, there will be little difference in test design and performance for naturally occurring and genetically modified microorganisms except to take introduced phenotypic traits and characteristics into account when testing the genetically modified species.

The Need for Nontarget Organism Testing

As the efficacy, production, storage, and delivery problems connected with the use of microorganisms for specific tasks in the environment are solved by strain improvement, either by selection or genetic alteration, we should expect to see a significant rise in the use of microorganisms to perform environmental tasks. With this increased usage, our concern for unexpected and possibly adverse effects on the environment, especially nonhuman, nontarget organisms, will increase proportionally. The result of increased use of genetically engineered microorganisms will be artificially elevated levels of these more efficient strains in the environment.

As these new microbial strains, carrying novel recombinant genotypes, are added to particular ecosystems, we may see the introduction of new microbial strains or species carrying new, enhanced, or multiple toxic or pathogenic traits into an ecosystem. Although considerable research effort on the persistence and transfer rates of recombinant DNA in environmental media is currently underway, we have little understanding of the expression of introduced traits within the natural ecosystem. Even though the genetically engineered microorganism may not persist, it is possible that the new trait may move into species already adapted to a particular ecosystem, thereby allowing the trait to persist. At present, we have no method of predicting what effect, if any, the expression of these new traits will have on nontarget organisms inhabiting the receiving ecosystem. Thus, in order to assess the potential effects of a microorganism introduced to an ecosystem in high numbers and carrying new or enhanced toxic or pathogenic traits, empirical testing on representative nontarget organisms is necessary.

Philosophy behind Testing

In the broadest sense, a "nontarget organism" may be defined as any animal or plant species which is not the subject of a control strategy. Traditionally for pesticides, the nontarget organisms of concern include avian species, feral mammals, aquatic animal and plant species, terrestrial plants, and beneficial insects (including predators and parasitic insects). Each of these groups has a different economic and aesthetic value, as perceived by the public, but all have an important role in ecosystem function.

MPCAs are usually highly specific to a single or a small group of target pest(s). In many cases, when microorganisms are being screened for efficacy and host range after being isolated from a diseased insect, species taxonomically close to the host are used. Rarely does one test for effects on organisms outside a host's class because, intuitively, one would not expect to see effects in such distant species.

As mentioned above, one of the goals of genetic engineering is to overcome the narrow-host-range limitation. By increasing the host range, the microorganism becomes a better product—and a more saleable one, because it has a wider market. The consequence of increasing the host range, however, may be an increased potential of the engineered microorganism to cause adverse effects on nontarget, beneficial organisms, especially insects.

The U.S. Environmental Protection Agency (USEPA) under the Federal Insecticide, Fungicide, and Rodenticide Act (FIFRA), as amended (*Federal Insecticide, Fungicide, and Rodenticide Act* 1988), requires that all microbial pesticides, whether naturally occurring or genetically engineered, be tested for safety to nontarget animals and plants prior to registration. A recent revision to the guidelines for testing these pesticidal agents, *Pesticide Assessment Guidelines: Subdivision M* (USEPA 1989), has been published. This document, in addition to delineating recommended testing guidelines, proposes a philosophy for testing the safety of microorganisms toward nontarget organisms.

This philosophy incorporates three basic principles to be followed when performing safety tests: (1) use of a tiered system of testing, in which a set of acute, high-impact, relatively short-term tests are employed initially as a screening tool to detect adverse effects rather than performing a full battery of acute and chronic tests at the outset; (2) performance of tests using a maximum-hazard approach; (3) selection of the appropriate tests and species for a particular MPCA. While this approach is directed toward the registration and use of traditional (nonengineered) and genetically engineered microbial pesticides, the conceptual bases embodied in it are applicable to safety testing of genetically engineered strains released to the environment in high numbers for other purposes.

Maximum-hazard approach

Use of a screening device requires safeguards against false-negative results. Although not guaranteed, one approach to the reduction of false-negative results is to conduct the screening tests under maximum-hazard conditions. These conditions can be achieved through the use of artificially high dosing levels and by using test organisms in a stage of life that is most susceptible to adverse effects from the treatment. The latter condition is relatively easy to fulfill. In almost all cases, the immature stage of a species is the most susceptible. In fact, in some cases an early developmental stage is the only period in an organism's life cycle when it is susceptible to infection by certain microorganisms.

Determination of what constitutes a maximum-hazard dose is more complex. In part, it depends on the concentration of microorganism one expects the nontarget organism to be exposed to in the environment; the desired "safety factor," i.e., some multiple of the expected exposure concentration that one includes to account for species-to-species variation in the level of susceptibility; and anatomical constraints of the test organism, which limit the amount of test material that can be introduced. Thus, for a young bird, the anatomy and maximum volume of dose may be the factor controlling the maximum-hazard dose that can be administered. For aquatic species, some multiple of the expected concentration occurring in water after the microorganism has been used or applied, along with the maximum microbiological loading permissible before water quality in the test system is compromised, will set the upper dosing limit.

Tier progression

Use of a tiered system (Fig. 2.1) is designed to reduce the overall testing burden. Rather than requiring a large number of acute and chronic studies at the

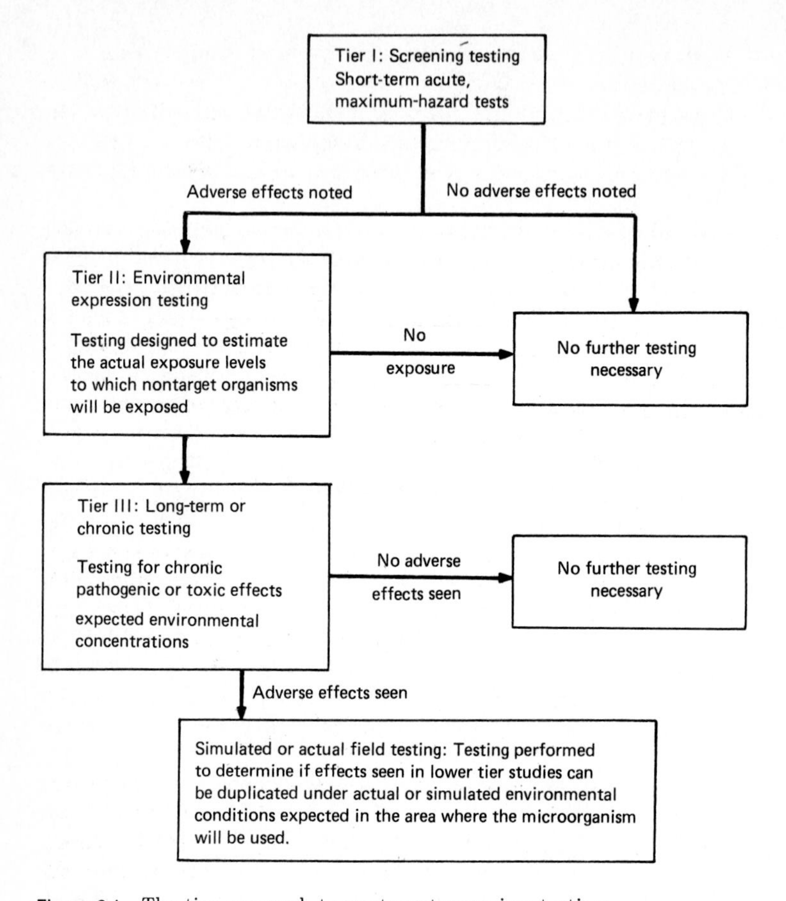

Figure 2.1 The tier approach to nontarget organism testing.

outset, the tier approach uses a few relatively short-term acute studies, performed at relatively high dose levels, as a screening mechanism to identify those microorganisms requiring further study. If no adverse effects attributable to the microorganism under study are seen at the first tier of testing, then no further testing is required. If adverse, microorganism-related effects are seen, then the strain progresses to the next level of testing.

This second tier consists of studies designed to assess the ability of the microorganism to survive and multiply in the environment and result in exposure of nontarget organisms. These studies are intended to estimate the environmental concentration of the microorganism to which nontarget organisms may be exposed under actual use conditions. Survival and persistence studies in Tier II are performed in the media (e.g., soil, water, or sediment media) comprising the particular environment with which the class of nontarget organism for which the effect was seen in Tier I is associated. Thus, if freshwater organisms were affected in Tier I studies, persistence studies in freshwater media would be performed. If the results of the Tier II studies indicated that there would be exposure of nontarget organisms through use of the microorganism, that exposure would be quantified and testing would progress to Tier III. If, on the other hand, definitive data indicated that the organism did not survive in the particular media of interest and that, in consequence, nontarget organisms would not be exposed, further testing would not be required.

With the expected environmental concentration of the microorganism estimated, Tier III testing attempts to further define the extent of the effect seen in Tier I, using dosing concentrations near the expected environmental concentration. The kind of testing varies somewhat with the organism of interest. For vertebrates, this tier of testing would involve exposure of the species to the microorganism at levels expected to occur in the environment over an extended period of time to determine whether or not the effect may occur under typical use and exposure conditions. For invertebrates, the primary focus in testing is to determine the range of nontarget insects and aquatic invertebrates that may be affected through the use of a particular microorganism. Tier III procedures also include, at least for aquatic species, multiple-species testing under laboratory conditions. In multiple-species testing, one takes representatives from all macroscopic trophic levels of an aquatic ecosystem (fish, aquatic insects, invertebrates, and aquatic plants) and exposes them as a unit to the microorganism under study. The purpose of this type of testing is to provide a contained system within which a determination can be made as to whether or not effects seen in single-species testing will occur at multiple trophic levels and whether these effects can disrupt the food chain in aquatic ecosystems.

If an adverse effect is still exhibited at this stage of testing, actual or simulated field testing should be performed to determine if the effect will occur under actual environmental conditions. This fourth tier of testing is comprehensive and can be very expensive. It requires an extensive knowledge of the target ecosystem and the microorganism which will be introduced. Often, a well-designed field test will require several years to complete if one considers all stages: i.e., design, performance, data analysis, and formulation of meaningful conclusions.

Test species selection

The third consideration in designing safety testing is the appropriateness of the species used as test subject. This is perhaps the most difficult area to address. The first question one must ask is whether or not, from a scientific standpoint, testing on a particular class of nontarget organism is warranted for the specific microorganism in question. One approach to answering this question is to examine the nature and characteristics of the microorganism to be released to the environment. Characteristics such as (1) maximum growth temperature, (2) new traits which have been engineered into the microorganism and whether the new traits expand the range of hosts affected, (3) whether or not a toxin is produced and the nature of that toxin, (4) the specific ecosystem to which the microorganism will be released, and (5) the taxonomic relationship of the microorganism under study to known pathogens (whether frank or opportunistic) of a particular class of nontarget species. These characteristics will influence both decisions on which classes of nontarget organism to test and selection of a species within these classes. Microorganisms released to the environment for a particular purpose may accomplish their task either through growth and metabolism (including pathogenesis) or toxic effects. Each of these situations requires a different conceptual approach to safety testing with regard to selection of nontarget organisms.

In cases where elaboration of a toxin or toxins is the chief mechanism of action, the nature of the toxin(s), especially with respect to mode of action and range of species affected, becomes the controlling consideration in test species selection. An illustrative example is *B. thuringiensis*. Some strains of this insecticidal bacterium, most notably *B. thuringiensis* var. *thuringiensis*, produce both a crystalline δ-endotoxin and a soluble β-exotoxin. The endotoxin exerts its toxic effect on certain orders of insects. The structure of this toxin is such that it does not become active toxin until a protoxin is cleaved into smaller units. This splitting most often requires alkaline conditions and, therefore, occurs only in insects with proper gut pH conditions. Toxicity to other orders or classes of organisms with more neutral or acidic gastrointestinal tracts is minimal or nonexistent. Intuitively, one would expect that nontarget organisms not in the insect class with the proper gut pH conditions would not be affected. In general, this hypothesis is supported by experimental evidence.[2]

The β-exotoxin of *B. thuringiensis*, 2-O-(4′-O-5′-deoxyadenosine-5″-yl-β-D-glucopyranosyl)-4-O-phospho-D-allaric acid, has a completely different structure, which closely resembles the structure of adenosine triphosphate (ATP). Intuitively, one would expect that this biomolecule might act as a competitor to ATP and thus cause adverse effects in a wide range of organisms. The toxicity seen is a result of inhibition of ATP incorporation into macromolecules. Indeed, the inhibition by β-exotoxin appears to affect the polymerization of RNA (Bond et al. 1971), presumably replacing the adenosine moiety. Thus, the decision to test all classes of organisms is appropriate because of the ubiquity of ATP in biosynthesis of macromolecules.

A similar problem in species selection exists for testing microorganisms introduced into the environment to degrade anthropogenic chemicals. A concern here may not be with the production of a toxin per se but with the possibility of production of toxic metabolic by-products during the catabolism of the target compound. Again, determination and elucidation both of all metabolites formed during target chemical breakdown and of the final catabolic products, along with a determination of their potential toxicity to nontarget animals and plants, either by structure-activity relationship analysis or some other method, will be a controlling factor in deciding which species should be included in a safety-testing program and which should not.

For microorganisms which do not produce toxins and are introduced into the environment to control pests by an infective process, the task of safety tests is to determine whether the microorganism poses a disease threat to nontarget organisms. Here, the taxonomic relationship of the introduced

---

[2]Cantwell et al. 1966, 1972, 1979, Fisher and Rosner 1959, Flexner et al. 1986, Reichelderfer and Benton 1973, Smirnoff and McLeod 1961.

strain to known pathogens of different classes of nontarget organisms as well as maximum and optimal growth temperatures of these strains are guiding principles for determining whether it is necessary to test a particular class of nontarget organism.

Special Considerations

Avian species

In deciding whether or not it is appropriate to test a particular microorganism in avian species, one must determine (1) whether or not there is a likelihood that avian species will be exposed; (2) whether or not a toxin is produced and, if produced, whether or not the toxin has the potential to affect avian species; (3) whether the released microorganism is related to known avian species or is capable of growth, reproduction, and pathogenicity in birds.

Testing the safety of microorganisms to be released to the environment on avian species may be particularly appropriate for certain insecticidal microorganisms applied to terrestrial sites, especially if the microorganism produces a toxin and/or is capable of growth at, or near, avian body temperatures. Insects supply a substantial portion of the diet of many species of birds. In addition, after death of the insect, the insect becomes easier for birds to find and feed upon. At the same time, because of microorganism growth after insect death, there will probably be a high concentration of the insecticidal microorganism and any toxins produced by it within the insect carcass. Thus, avian species can be expected to be heavily exposed to these microorganisms and toxins.

Another situation in which birds are likely to be exposed is in an aquatic application where applied microorganisms may accumulate in fish or invertebrate species that serve as avian food sources. Use of transgenic plants, in which a gene coding for a toxin is engineered into the plant genome, may lead to exposure of avian species if the toxin is produced in the fruit or other edible portion of the plant. Plant-controlling strains of microorganisms are generally not considered to have a high exposure to birds because they are not generally found in the edible portions of plants that are of interest to birds.

Toxin production by microorganisms presents an exposure to birds if the toxin accumulates in avian foodstuffs. At present, it is difficult to prove that a particular toxin is innocuous to avian species on a theoretical basis alone. Therefore, for toxin-producing microorganisms whose toxin is likely to enter the avian food supply, testing on birds is recommended.

In most cases, pathogenicity depends on infection (entrance into the host body) and the ability of the invader to grow and reproduce, leading to tissue damage. Generally, microorganisms released to the environment will be those that are adapted to the environmental niche of the release site and are expected to have growth characteristics, including growth temperature, optimized for those environmental conditions. For most environmental isolates, the optimal growth temperature is in the range of 20 to 30°C, and many do not grow at or above 37°C. If it were shown that a particular strain failed to grow at 37°C or above and produced no known toxin, it could be argued that this strain would not be able to cause a systemic pathogenic condition in avian species, whose body temperature is approximately 40°C.

Wild mammal species

In general, the same considerations that apply to avian testing apply to wild mammal testing. Common insect pathogens such as Nosema species and other microsporidia have been shown to interact with man and other mammals (Brown et al. 1973, Sprague 1974, and Undeen and Alger 1977). If mammals are going to be exposed and if the strain to be released biosynthesizes a toxin that has not been shown to have no effect on mammalian species, the microorganism probably should be tested for safety to mammals. In terms of pathogenicity, growth at mammalian body temperatures would also argue for testing the strain's pathogenic potential in mammals.

Although there are reports of safety tests for entomopathogens in mammals (Ignoffo 1968, Lamanna and Jones 1963), the paucity of historical data as well as the difficulty of obtaining and maintaining colonies suitable for testing remain as major difficulties to mammalian safety testing. In addition, the choice of test species is complicated by differential sensitivity of species. For these reasons, the mouse and rat models, used for many years in human toxicological studies, are also used to screen for pathogenicity and adverse effects of toxins in wild mammals.

Freshwater fish, aquatic invertebrate, and aquatic insect species

Aquatic species present special concerns for application of both microorganisms and chemicals to the environment. Although chemicals are not the subject of this chapter, they play a role in delineating the concerns about these species. There are three considerations with regard to aquatic species that underpin these concerns. The first is that, unlike other feral species, fish and invertebrates are less able to move away or avoid potential adverse situations, especially in pond and lake situations. Thus, their exposure is somewhat continuous. In addition, the exposures likely to be encountered by freshwater, estuarine, or marine fish are likely to be dual in nature, both by those organisms suspended in the water column and through the oral route from feeding on either infected targets such as mosquito larvae or other invertebrates which may have taken up or concentrated the microorganism. Second, in situations where microorganisms may be applied—agricultural settings, toxic chemical cleanup in aquatic ecosystems, and cleanup of effluents in mining and oil operations—there is a high likelihood that there will be potentially toxic chemicals in the water. Some of these chemicals may put stresses on the nontarget organisms in aquatic systems, leading to a lowered resistance to infection by microbial species that are usually not considered to be frank pathogens of aquatic animals. The third consideration is that aquatic species, unlike birds and mammals, may be exposed at temperatures where the applied microbial strains will be at their most active.

In designing tests for aquatic nontarget organisms, these considerations can be taken into account in a variety of ways. In terms of the dosing, the concentration of the microorganism in the water column should be maximized so that the concentration is as high as possible without compromising water quality or, when there is particulate matter in the test material, causing physical problems (e.g., gill impaction) in test organisms. Additionally, the concentration of microbial species placed in aquaria tends to decrease with time. In order to maximize the availability of viable cells in the testing environment, active cells from freshly prepared stocks should be added periodically to maintain an acceptable level of challenge.

A second exposure route for fish during testing, the oral route, may be employed either by incorporating the test microorganism in food fed to fish or by using infected targets, assuming the fish used will feed on these target organisms. Quantification of the dose using this exposure method is difficult, since there is no reliable way to determine the amount of food or infected pest ingested by the fish during the test. Dose levels would have to be related in some way to either the amount of test microorganism incorporated into the food or the mean concentration found in the infected targets. Oral intubation of fish has the advantage of precise dose quantification, but it also has disadvantages: in cases where very young fish are used, there is a high probability of injury to the fish, and this method unduly stresses any fish if dosing is performed repeatedly.

As discussed above, there may be many cases when microorganisms will be introduced into aquatic environments containing some level of chemical contamination. An example in agriculture is the use of chemical pesticides in conjunction with microbial pesticides. In bioremediation processes done in situ, the chemical(s) being degraded by the introduced microorganism will obviously be present. There is published evidence that heavy metals (Gainer 1973), pesticides (Crocker et al. 1974), and environmental contaminants (Friend and Trainer 1970, 1974) can have synergistic effects with certain viruses. With respect to baculoviruses, chemicals or environmental factors have been shown to enhance viral activity in insects (Himeno et al. 1973, Reichelderfer and Benton 1973, Yamafugi 1964) and shrimp (Couch 1976). In an ideal testing scheme, there should be some way to account for the presence of these chemicals or for the occurrence of environmental events and the stresses they may exert on nontarget organisms. While design of test systems to incorporate this aspect is possible, performing an accurate, meaningful assessment of the hazard to fish in the environment from a combination of anthropogenic chemicals and microorganisms is very difficult. In most cases, the presence of a specific chemical or chemicals, their concentrations in the aquatic ecosystem, or the bioavailability to nontarget organisms will not be known. Thus, there is no practical method to ensure that inclusion of this parameter in nontarget organism testing will reflect environmental hazard and not artifacts of a manipulated test system.

Traditionally, aquatic safety testing of chemical toxicants has been carried out for 48 to 96 h for aquatic invertebrates and fish, respectively. This time frame may also be acceptable for microbial testing where a toxin or toxic metabolite is produced and it has been shown that the microorganism has no infectivity potential in aquatic species. However, in cases where there is the potential for infection of fish and invertebrates, a longer test duration is necessary in order to allow for the establishment of an active clinical infection.

Selection of nontarget aquatic species for test purposes is difficult, as is species selection in all nontarget testing, because of potential differences in species sensitivity and lack of availability of some aquatic species from the commercial sources where many organisms are reared for laboratory studies. Wherever possible, species that are commonly found within the ecosystem to which the microorganism will be released should be used. If there are no clear indications which species of fish are expected to inhabit the receiving ecosystems or if cultures of the selected species are not available, then use of species which have a wide distribution (e.g., the bluegill sunfish) or species that are economically important (e.g., channel catfish and white catfish or similar species) seems to be a reasonable choice.

For the purpose of the present discussion, "aquatic insects" may be defined as members of any insect species which spends a portion of its life cycle in water. During an aquatic application of a naturally occurring or genetically engineered microorganism, whether it be for control of an aquatic pest or some other environmental task, there is a high probability that beneficial aquatic insects, either in the larval or nymph stage, will be exposed if the application is made in the spring or early summer. Knowing this, it is reasonable to include aquatic insect species in a testing program.

Unfortunately, many aquatic insect species, especially those whose habitats include streams, are very difficult to maintain in laboratory culture because of their need for highly oxygenated, flowing water. When testing these organisms, which has been done from time to time, the usual method of obtaining them is to field-collect specimens immediately prior to testing. This method presents two major problems, availability and consistency.

The availability problem is twofold: (1) depending on the species, testing can only be performed during certain times of the year, when the larvae or nymphs are available; (2) even during that time of the year, there may not be a sufficient population in a single stream to supply the number of individuals necessary for testing. Thus, it may be necessary to collect at different streams, which leads to the second problem, consistency.

Field-collected insect larvae and nymphs, especially when collected from different streams where they have been exposed to different environmental conditions and stresses, may show more variability under test conditions than those reared under controlled conditions in the laboratory. This increased variability may mask small but significant differences between treatment and control groups that may have been seen using laboratory-reared insects. Whether small differences between groups when tested under laboratory conditions translate into an ecologically important impact is debatable.

Terrestrial insects

The honeybee, *Apis mellifera*, is an economically important insect and as such has served as a nontarget test organism for a large number of chemicals and naturally occurring MPCAs. Cantwell et al. (1966, 1972, 1979) have collected data on the toxicity and pathogenicity of bacteria, fungi, viruses, and protozoans to honeybees. In general, bees were treated either by feeding adults, where the test material was suspended in some material attractive to bees such as sucrose-water mixtures, or by direct contact (i.e., spraying bees). Mortality was generally under 10% in all cases. The majority of the testing was performed in the laboratory, and the test duration was from 1 to 10 days except for one test using *Nosema locustae*, which was 26 days long.

Because of the economic importance of honeybees and the fact that they visit a variety of agricultural crops, honeybee testing should be included in a safety evaluation program for both naturally occurring and genetically engineered microbial pesticides, as well as in any testing program for nonpesticidal microorganisms used in agricultural settings on plant species where bees are known to forage.

For microorganisms released to the environment for purposes other than agricultural uses, it is more difficult to construct a scenario for including honeybees in a testing program unless the microorganism to be released is closely related to a known pathogen of honeybees. In this case, an evaluation of whether or not honeybees will be exposed through usage of the microorganism must be made.

Over the years, a large number of studies on the toxicity and/or pathogenicity of bacteria, fungi, viruses, and protozoa to insect predators and parasites as well as to other beneficial arthropods has been performed. See Flexner et al. (1986) for a comprehensive review of this subject. Their conclusions, from a survey of the existing literature, were:

1. Indirect effects on predators and parasites as a result of the infection and death of their host are probably more significant than the direct effects of microbial pesticides on the predators and parasites themselves.
2. Long-term mortality or sublethal factors may play a more important role in effects on predators and parasites than direct mortality.
3. Bacterial and protozoan microbial pesticides can cause direct effects on beneficial insects, whereas the same has not been shown for viral pesticides.
4. Direct effects of fungal pesticides have not been well studied and are probably underestimated.
5. Little is known concerning the adaptation and coevolution of parasites, predators, or naturally occurring entomopathogens and their respective hosts.
6. Standard methods for testing need to be developed for evaluating the effect of microorganisms on predators and parasites, including standards for method of exposure, exposure conditions, and test duration.

Interim protocols have been developed by scientists at the USEPA laboratory in Corvallis, Oregon, in conjunction with the Oregon State University (USEPA 1989). These protocols include methods for testing the effects of microbial pathogens on a variety of nontarget beneficial insects, including the common green lacewing (*Chrysoperia carnea*), predatory hemipterans, predaceous coccinellids, parasitic insects of the *Trichogramma* family, aphidophagous syrphids, and predatory mites. These protocols were developed on the basis of current knowledge of testing of the toxicity of chemical pesticides on these various groups of beneficial insects. These protocols have not been performed in the laboratory using microbial pathogens and therefore must not be considered to be a step-by-step approach to testing. Rather, these interim protocols should be used as guides to the development of protocols specific to the microorganism under study.

Three other protocols have been completed and tested using microbial pathogens. The first tests the effect of a bacterial pathogen, *Serratia marcescens*, strain QMB 1466, on a predatory mite, *Metaseiulis occidentalis* (Nesbitt). The second protocol covers testing of a fungal pathogen, *Beauvaria brassiana*, strain RS252cw, on *M. occidentalis* (Nesbitt). The third addresses testing of the fungal pathogen *B. brassiana* (Balsamo) on larvae and adults of the predatory neuropteran *C. carnea* (Stephens). These protocols describe the handling of both the pathogen and test organism, rearing of the test organism, dosing apparatus and methods, and methods for determining effects.

All protocols, interim and validated, are available from the Federal Office of Pesticide Programs (OPP) as an appendix to the *Pesticide Assessment Guidelines, Subdivision M*. As mentioned before, they are intended not as step-by-step testing procedures but as guides for development of detailed protocols for particular testing situations. The protocols contain testing procedures and identify parameters that should be considered when designing tests. It is recognized that modification of these procedures may be necessary, particularly in unique testing circumstances.

Plants

For many years now the USDA has used a procedure described by Wapshere (1974) for testing the host range of nonindigenous species of plant pathogens brought into this country as possible pest-control agents (Bruckart and Dowler 1986). This test begins with plant species closely related to known susceptible hosts and progresses through species increasingly distant taxonomically. The concept of progressive testing has the advantage of experimentally describing the host range of the pathogen and, except for the most virulent, leads to some logical stopping point for testing when no visible signs of pathogenicity are found.

For microorganisms with no known pathogenicity, there will be no indicator of what species to begin with. In such cases indirect tests may suffice. Lelliot et al. (1966) have described the LOPAT series of tests for determining the potential of microorganisms to be pathogenic. The series includes tests for *l*evan formation, *o*xidase production, *p*otato-rotting ability, *a*rginine dihydrolase production, and *t*obacco hypersensitivity. Monsanto used this series of tests to verify that several pseudomonas strains were not plant pathogens (Drahos et al. 1988).

Other forms of testing may substitute for or supplement such indirect tests. For example, representative test plant species may be chosen based on commercial relevance and proximity to some test site or with the intent of providing a survey of representative strains having a range of taxonomic diversity. This has the distinct disadvantage of providing unsatisfying data and providing no clear end point to the number of species that might be challenged. For example, Lindow (1988) tested non-ice-nucleating ("Ice$^-$") *Pseudomonas syringae* on 67 different species of plants in support of his application for an experimental-use permit to field-test the microorganism. Since no adverse effects were anticipated with the strain, all the tests were done in direct comparison with an isogenic "Ice$^+$" strain.

Any microorganism that is a natural plant pathogen or has been genetically engineered to include the ability to in some way affect, be pathogenic to, or control the growth of plants should undergo nontarget plant testing to some degree. In addition, while animal-controlling microorganisms would not be expected to infect or be pathogenic to plants, certain formulations may contain substances which are phytotoxic. Thus, in order to achieve some level of confidence that application of microorganisms or the products which contain them will not cause adverse effects on plant members of the receiving ecosystem, testing on plants is advisable. The degree to which one would test the microorganisms and products just described and the design of the testing protocol, including end points used to measure effects, may vary according to type of deleterious effect expected and the potential of the microorganism to cause it.

Plant pathogens vary with respect to their host selectivity. *Colletotrichum gloeosporioides*, the fungal pathogen of the northern jointvetch, has a very limited range of host-plant species. On the other hand, *Agrobacter tumefaciens*, the causative agent of crown gall disease, has a very wide range of plants which it affects. Selection of plant test species for each of these microorganisms may be very different.

Multispecies testing (microcosms)

The complexity of our environment often confounds attempts to reasonably extrapolate single-species testing results to anticipate the presence or absence of some environmental disturbance in a field-test situation. The use of microcosms is predicated on the premise that "parts of an ecosystem are interconnected," leading to the assumption that "significant effects on one part of the system will eventually lead to effects on other parts" (Giddings 1986). Microcosms are an attempt to reproduce, at least in part, the complex environmental interactions in controlled laboratory testing situations. They range from composites of organisms representing different trophic levels (Armstrong et al. 1987, Shannon et al. 1989, Taub 1989) to an intact piece of the field that is believed to behave ecologically like its counterpart in the actual field (Pritchard and Bourquin 1984, VanVoris 1988). These are discussed more fully by Greenberg et al. (1988) and Fredrickson and Seidler (1989). The use of microcosms for testing has focused primarily on the characterization of the microorganisms in terms of their survival, colonization, or propensity for genetic transfer in a simulation of the environment.

Thorough evaluations of nontarget testing in multispecies systems are now being done. One published demonstration of a system (Fournie et al. 1988) resulted in the conclusion that they have the potential to become standardized procedures for testing. Freshwater and estuarine systems were constructed and successfully maintained for 30 days using combinations of three or four species in a tank. Fournie and his colleagues included histopathological examination as part of the test regimen, and point out that other diagnostic tools for determining infectivity, pathogenicity, or toxicity (e.g., serological testing or electron microscopy) could be utilized. It is clear, however, that additional work will be necessary to establish the general utility of multispecies test systems for nontarget organism testing. With the current research interest and activity in this area, there may be a variety of protocols for such testing available in the very near future.

Conclusion

While we have made some distinction in this discourse between genetically engineered and unmodified (i.e., naturally occurring) microorganisms, it is not clear that these should be tested any differently with regard to their effect on nontarget organisms. As recent National Academy of Sciences (NAS) reports (NAS 1987, 1989) point out, the evaluation of a microorganism should be based on its characteristics and not on its derivation. It may be argued that small changes made through recombinant DNA techniques are not sufficient to cause major differences in the behavior of a particular microorganism (the critical one to this discussion being its pathogenicity) or that the precise nature of the process allows a priori determination of the consequences of the change. However large the majority of cases that would bear this out, it is the rare exception that one hopes would be revealed in some prerelease testing regimen.

References

Armstrong, J. L., Knudsen, G. R., and Seidler, R. J. 1987. Microcosm method to assess survival of recombinant bacteria associated with plants and herbivorous insects. *Curr. Microbiol.* 15:229–232.

Bond, R. P. M., Boyce, C. B. C., Rogoff, M. H., and Shieh, T. R. 1971. The thermostable exotoxin of *Bacillus thuringiensis*. In *Microbial Control of Insects and Mites*, H. D. Burges and N. W. Hussey (eds.), Academic, New York and London.

Brown, R. J., Hinckle, D. K., Trevertan, W. P., Kupper, J. L., and McKee, A. E. 1973. Nosematosis in a squirrel monkey (*Saimiri sciureus*). *J. Med. Primatol.* 2:114–123.

Bruckart, W. L., and Dowler, W. M. 1986. Evaluation of exotic rust fungi in the United States for classical biological control of weeds. *Weed Sci.* 34(suppl. 1):11–14.

Cantwell, G. E., Knox, D. A., Lehnert, T., and Michael, A. S. 1966. Mortality of the honey bee, *Apis mellifera*, in colonies treated with certain biological insecticides. *J. Invert. Pathol.* 8:228–233.

Cantwell, G. E., Lehnert, T., and Fowler, J. 1972. Are biological insecticides harmful to the honey bee? *Am. Bee J.* 112:255–258.

Cantwell, G. E., and Lehnert, T. 1979. Lack of effect of certain microbial insecticides on the honey bee. *J. Invert. Pathol.* 33:381–382.

Carbonelli, L. F., Klowden, M. J., and Miller, L. K. 1985. Baculovirus-mediated expression of bacterial genes in dipteran and mammalian cells. *J. Virol.* 56:153–160.

Conway, K. E. 1978. The effect of *Cercospora rodmanii*, a biological control for water hyacinth, on the fish, *Gambusia affinis*. *Mycopathologia* 66:113.

Couch, J. A. 1976. Attempts to increase *Baculovirus* prevalence in shrimp by chemical exposure. *Prog. Exp. Tumor Res.* 20:304–314.
Crocker, J. F. S., et al. 1974. Insecticide and viral interaction as a cause of fatty visceral changes and encephalology in the mouse. *Lancet* 1974:22–24.
Drahos, D. J., Barry, G. F., Hemming, G. C., Brandt, E. J., Skipper, H. D., Kline, E. L., Kuepfel, D. A., Hughes, T. A., and Gooden, D. T. 1988. Pre-release testing procedures: US field test of a lacZY-engineered soil bacterium. In *The Release of Genetically-Engineered Micro-Organisms*, M. Sussman, C. H. Collins, F. A. Skinner, and D. E. Stewart-Tull (eds.), Academic, New York.
Federal Insecticide, Fungicide, and Rodenticide Act (FIFRA), as amended (7 *U.S.C.* 136 et sec.), 1988.
Fisher, R., and Rosner, L. 1959. Toxicology of the microbial insecticide thuricide. *J. Agric. Food Chem.* 7:686–688.
Flexner, J. L., Lighthart, B., and Croft, B. A. 1986. The effects of microbial pesticides on non-target beneficial arthropods. *Agric. Ecosystems Environ.* 16:203–254.
Fournie, J. W., Foss, S. S., and Couch, J. A. 1988. A multispecies system for evaluation of infectivity and pathogenicity of microbial pest control agents in nontarget aquatic species. *Dis. Aquat. Org.* 5:63–70.
Fredrickson, J. K., and Seidler, R. J. 1989. Evaluation of terrestrial microcosms for detection, fate, and survival analysis of genetically engineered microorganisms and their recombinant genetic material. PNL-6828, National Technical Information Service, Springfield, Virginia.
Friend, M., and Trainer, D. O. 1970. Polychlorinated biphenyl: interaction with duck hepatitis virus. *Science* (N.Y.) 170:1314–1316.
Friend, M., and Trainer, D. O. 1974. Experimental DDT-duck hepatitis virus interaction studies. *J. Wildlife Mgmnt.* 38:887–895.
Gainer, J. H. 1973. Effects of heavy metals on viral infections in mice. *Environ. Health Perspect.* 4:98.
Giddings, J. M. 1986. *A Microcosm Procedure for Determining Safe Levels of Chemical Exposure in Shallow-Water Communities. Community Toxicity Testing*, John Cairns, Jr. (ed.), ASTM STP 920, American Society for Testing and Materials, Philadelphia, p. 122.
Greenberg, E. P., Poole, N. J., Pritchard, H. A. P., Tiedje, J., and Corpet, D. E. 1988. Use of microcosms. In *The Release of Genetically Engineered Micro-Organisms*, M. Sussman, G. H. Collins, F. A. Skinner, and D. E. Stewart-Tall (eds.), Academic, New York, pp. 265–274.
Himeno, M., Matsubana, F., .and Hayashiya, K. 1973. The occult virus of nuclear polyhedrosis of the silkworm larva. *J. Invert. Pathol.* 22:292–295.
Ignoffo, C. M. 1968. Effects of entomopathogens on vertebrates. *Ann. N.Y. Acad. Sci.* 217:141–164.
Lamanna, C., and Jones, L. K. 1963. Lethality for mice of vegetative and spore forms of *Bacillus cereus* and *Bacillus cereus*-like insect pathogens injected intraperitoneally and subcutaneously. *J. Bacteriol.* 85:532–535.
Lelliot, R. A., Billing, E., and Hayward, A. C. 1966. A determinative scheme for the fluorescent plant pathogenic pseudomonads. *J. Appl. Bacteriol.* 29:470–489.
Lindow, S. E., and Panopoulos, N. J. 1988. Field tests of recombinant Ice$^-$ Pseudomonas syringae for biological frost control in potato. In *The Release of Genetically Engineered Micro-Organisms*, M. Sussman, G. H. Collins, F. A. Skinner, and D. E. Stewart-Tall (eds.), Academic, New York, pp. 121–138.
National Academy of Sciences (NAS). 1987. Introduction of recombinant DNA-engineered organisms into the environment: key issues. National Academy Press, Washington, D.C., 24 pp.
National Academy of Sciences (NAS). 1989. *Field Testing Genetically Modified Organisms: Framework for Decisions.* National Academy Press, Washington, D.C., 170 pp.
Pritchard, P. H., and Bourquin, A. W. 1984. The use of microcosms for evaluation of interactions between pollutants and microorganisms. *Adv. Microb. Ecol.* 7:133–215.
Reichelderfer, C. F., and Benton, C. V. 1973. The effect of 3-methylcholanthrene treatment on the virulence of a nuclear polyhedrosis virus of *Spodoptera frugiperda. J. Invert. Pathol.* 22:38–41.

Shannon, L. J., Flum, T. E., Anderson, R. E., and Yount, J. D. 1989. Adaptation of the mixed flask culture microcosm for testing the survival and effects of introduced microorganisms. In *Aquatic Toxicology and Hazard Assessment*, vol. 12, U. M. Cowgill and L. R. Williams (eds.), ASTM STP 1027, American Society for Testing and Materials, Philadelphia, pp. 224–242.

Smirnoff, W. A., and McLeod, C. F. 1961. Study of the survival of *Bacillus thuringiensis* Berliner in the digestive tracts and in feces of a small mammal and birds. *J. Insect Pathol.* 3:266–270.

Sprague, V. 1974. *Nosema connori* n. sp., a microsporidian parasite of man. *Trans. Am. Microsc. Soc.* 93:400–402.

Taub, F. B. 1989. Standardized aquatic microcosm: development and testing. In *Aquatic Ecotoxicology*, vol. II, A. Boudou and Fribeyre (eds.), CRC Press, Boca Raton, Florida, pp. 47–94.

Undeen, A. H., and Alger, N. E. 1977. *Nosema algerae* infection of the white mouse by a mosquito parasite. *Exp. Parasitol.* 40:86–88.

U.S. Environmental Protection Agency (USEPA). 1989. *Pesticide Assessment Guidelines: Subdivision M—Microbial Pest Control Agents and Biochemical Pest Control Agents.* National Technical Information Service, Springfield, Virginia.

VanVoris, P. 1988. Standard guide for conducting a terrestrial soil-core microcosm test. In *Annual Book of ASTM Standards*, vol.11.04, Standard E-1197, American Society for Testing Standards, Philadelphia, pp. 743–755.

Wapshere, A. J. 1974. A strategy for evaluating the safety of organisms for biological weed control. *Ann. Appl. Biol.* 77:201–211.

Yamafugi, K. 1964. Metabolic virogens having mutagenic action and chromosomal previruses. *Entomologia* 27:217–274.

Chapter

3

The Toxicology of Genetically Engineered Microorganisms

Lori S. Katz

Life Sciences Section
Arthur D. Little, Inc.
Acorn Park
Cambridge, Massachusetts 02140

Judith K. Marquis

Life Sciences Section
Arthur D. Little, Inc.
Acorn Park
Cambridge, Massachusetts 02140

Introduction

The innovative technology that is presently utilized to develop products such as new microbial pesticides presents an unusual problem for scientists and regulators alike. The genetic technology that depends on recombinant DNA, or "gene-splicing" as it is popularly known, has already attracted national publicity. It is important to remember, however, that biotechnology has been around for a very long time. Industrial use of biological organisms, e.g., in large-scale fermentation processes, is hardly a new idea. It is the ability to create new genetic strains of biological organisms at will that gives the new technology an aura of mystery and creates suspicion. Of particular interest are genetically engineered microorganisms (GEMs) that are to be released into the open environment.

In determining the possible health hazards of the newly emerging products of biotechnology, however, one need not look beyond the practice of following scientific protocols for testing the potential toxicity of a product. Testing of GEMs for their toxicity and pathogenicity in laboratory animal models, as well as their survival and growth in the environment, is a prerequisite for the registration of these products, just as it is for the registration of conventional chemical products.

The scientific community should perhaps remind itself that failure or inability to acquire adequate data early on in the development of conventional pesticides resulted in the use of millions of pounds of chemicals annually for which only sparse data are available, generally too little to permit risk assessment. As a result, we are now faced with the overwhelming task of reviewing and requiring data for nearly every pesticide on the U.S. market. If anything is to be learned from past experience and applied to the use of GEMs, it is that scientists must assist in and support the development of adequate safety evaluation guidelines, and regulatory agencies must require complete data submission prior to the registration of new products.

The current literature suggests that it is especially difficult to maintain an unbiased approach to this issue, as many who are eager to enter the biotechnology arena insist on the apparent safety of their products. While they may very well be correct in their initial assessment, such conclusions are unreasonable in the absence of real data. Without the help of quantitative data from well-designed studies, conducted in compliance with Good Laboratory Practices, the toxicologist cannot formulate an adequate risk assessment, nor can regulators carry out credible risk management. At present the scientific community, together with the regulatory agencies, has begun to develop a fairly elaborate regulatory structure, portions of which may be waived if it becomes apparent that certain types of organisms really do not pose significant risks (Simonsen and Levin 1988).

What are the potential health risks associated with the release of GEMs into the environment? How can these risks best be evaluated? Risk assessment usually involves two principal processes: (1) identifying health hazards and (2) determining exposure rates. In fact, identifying or predicting the specific type, magnitude, or probability of effects associated with the release of GEMs is really very difficult on the basis of current knowledge. In the discussion that follows, we have first reviewed the basic principles of toxicology as they may be applied to determining exposure to GEMs and the effects of that exposure. Subsequently, the molecular biology of GEMs is reviewed in such a way as to help identify the hazards that may be associated with the environmental release of these organisms.

In one of the many published considerations of the environmental and health impact of the release of GEMs, Colwell (1985) once again called for a reasonable approach to the problem. She and numerous other scientists in this field of study have provided ample evidence that a well-designed evaluation of the properties of the new organisms and the uses to which they are applied can be accomplished by applying the fundamental concepts of toxicology and the basic principles of risk assessment which have been used in all of our efforts to regulate the introduction into the environment of any potentially biologically active materials.

Clearly, the risk assessment process may never be as standardized for GEMs as it is for chemicals. The versatility which is, in fact, a desirable property of GEMs also limits the ability to apply a single set of standardized tests or models to the entire range of products that will be developed.

Overall, the safety of the release of GEMs parallels that of nonengineered microbes, in that it depends on the toxicity, pathogenicity, virulence, and survivability of the individual organism in the natural environment and in humans (Betz et al. 1983). Also, the basic process of the toxicological assessment for GEMs encompasses the same fundamental principles that toxicologists have applied to chemicals (Suter 1985). These include:

- Evaluate the basic data.
- Determine the need for additional information.
- Define the test systems.
- Define the guidelines for interpreting test results.
- Develop models to extrapolate the implications of the test results to actual uses and exposures.

Fundamental Toxicological Considerations

Basic principles

The four major steps of hazard evaluation that are considered in a risk assessment for chemical compounds can also be applied to GEMs:

- Assess overall exposure.
- Measure uptake, or "bioavailability."
- Determine binding or reactivity, i.e., "infectivity."
- Evaluate the biological effects, i.e., "pathogenicity."

Exposure assessment

In evaluating the human health effects of chemicals or xenobiotics, the toxicologist considers the magnitude and frequency of exposure to a potential toxicant. For microbes, there is a special concern with the length of the exposure time, as the exposure level may change over time due to mitosis and doubling of the organisms. The doubling time may also vary, depending on environmental factors, including availability of nutrients, temperature, and light.

Individuals may be exposed to microbes and GEMs, both directly via the usual routes of exposure (inhalation, dermal, oral), and indirectly via food, water, and the transport of GEMs from soil and water into the food chain. In addition, consideration of the possible evolution of derived organisms, analogous to metabolites of chemical compounds, suggests that individuals may be exposed both to the parent organism and to its offspring. Natural immunity to many microorganisms will vary significantly between individuals in a population. Protection by natural immunity to parent microorganisms and the special susceptibility of certain "sensitive" individuals will be discussed later.

As with many chemical toxicants, microorganisms may be nonpathogenic, noninfectious, commensal agents at low densities, and pathogenic at higher doses (i.e., greater numbers of microbes). Furthermore, GEMs at a given tissue site may compete with the natural microbial flora, and they may not find the conditions needed to establish stable colonies at that site. Thus, the actual dose of a GEM may be quite difficult to measure. Nevertheless, if one simply substitutes the number or density of microbes for "dose" and the pathogenicity of the GEM for "toxicity," then most of the fundamental principles of the dose-effect relationship are applicable.

How then does one determine the "dose" of a microorganism in an effort to apply the basic principles of the dose-effect curve to a hazard evaluation? The total dose will depend both on exposure levels and on the transport of the microorganism to biological tissues, as well as the ability of the GEM to successfully compete with natural microbial flora. Organisms with a short doubling time or a higher rate of growth are more likely to establish themselves in a biological host ("infection"). The total dose of toxicant also depends on the products of metabolism and secretions (e.g., exotoxins) of the organisms.

In conventional chemical toxicology, experiments are conducted to measure the kinetics of chemical exposure—i.e., "ADME" studies to measure absorption, distribution, metabolism, and excretion of a compound. In order to address these parameters for GEMs, new techniques may be required to measure the transport of GEMs in the environment and in the potential nontarget host in order to effectively "trace" the microbe.

As for chemical substances, one area of concern for human health effects of GEMs is the generation of toxic intermediary metabolites. Specially designed study protocols may be needed to differentiate the effects of the organism itself from effects of chemical metabolites, and to date no such studies have been developed. Releases of endotoxins, immunotoxic proteins, or other toxic products or components of GEMs are factors to consider.

Just as one determines the half-life of a chemical compound in the environment by residue analysis, and in animals or humans by measuring blood levels, one must know how long a GEM can survive, and how that compares with survival of the parent organism. For example, if the GEM turns out to be more stable than the parent, will it prevail and, in fact, displace the parent from its usual habitat? To date, this has been an unlikely situation, as most genetically engineered plasmids or viruses confer an ecological disadvantage upon the host. It is nonetheless possible that a plasmid may be advantageous to host survival, and that the GEM may exhibit enhanced capacity for multiplication and proliferation.

"Bioavailability" of GEMs

In determining the level of exposure to GEMs, one must also derive a relative absorption factor, analogous to the concept of bioavailability of a chemical. Several considerations are evident. First of all, endogenous microbes abound at all major sites of exposure to GEMs—skin, conjunctiva, upper gastrointestinal (GI) tract, and genitourinary tract. The intestines, for example, are colonized almost immediately after birth with a normal microbial flora that varies with dietary composition, and assumes an important role in normal digestive functions.

Selection pressure against new microbes such as GEMs is especially strong in the GI tract, unless the natural flora is suppressed, e.g., with antibiotics. Thus, patients on antimicrobial therapy or individuals who exhibit altered bacterial profiles in the gut due to disease, diet, or stress, may represent a particularly sensitive subset of the population exposed to GEMs. Given the many possibilities whereby GEMs can alter the indigenous flora of the gut, it seems likely that a sufficient density of a GEM will interfere with normal digestive functions. GEMs may (1) compete with the endogenous microflora, (2) saturate natural nutrients, (3) secrete enzymes and other proteins that alter endothelial and glandular functions, and (4) inhibit the growth of the natural flora. In order to develop data that are needed to approve the environmental release of a GEM, experiments should be designed to provide guideline protocols for measuring the survival and persistence of GEMs in the gut endothelium—e.g., using diffusion chambers or dialysis membranes. Furthermore, similar considerations should be applied to assessing the bioavailability of GEMs via other routes of exposure.

Binding and tissue reactivity

The next step in evaluating tissue-specific or site-specific exposure to GEMs is to measure the microbiological equivalent of binding, e.g., drug-receptor interaction. Specificity and selectivity may be measured for GEMs as for pharmacologically active chemicals: i.e., *selectivity* for target versus nontarget hosts and *specificity* for a particular desired effect versus undesired side effects. These properties are perhaps more important for GEMs than for drugs, as nonspecificity and poor selectivity may be very difficult to control. While pharmacological antagonists are often available, antimicrobial therapy may not be available for an unexpected GEM infection. Furthermore, both specificity and selectivity may be altered in the GEM relative to the parent organism. Genetic manipulation may in some cases expand the potential host range and enable new functions in a given host environment.

Having determined whether a GEM can interact with nontarget tissues, one must then evaluate the potential for colonization or infectivity at that site. GEMs introduce a novel aspect to the end effect of drug-receptor complexation. Colonization, and possibly infectivity, may be defined as toxicity. The parameters that determine the likelihood of colonization include surface-adhesion factors, tissue environment (e.g., pH), competing organisms, and natural barriers such as cell-mediated immunity and cytolytic enzymes. Colonization also depends on the ability of a GEM to invade a tissue site—e.g., penetrate the epithelium—and to survive in the microenvironment of the tissue. Infectivity may occur when microbes can thrive and proliferate in a host tissue. Infectivity is thus enhanced by skin wounds and colonization in well-hydrated areas such as hair follicles or sweat glands. For quantitative evaluation, one can define the virulence of a particular organism as the dose, or number of microbes, required to cause a pathological or disease state. Perhaps an LC_{50} (concentration, or dose, lethal to 50% of the exposed host population) is an appropriate unit of measure of acute toxicity, but it conveys no information about changes in the host's susceptibility to infection by other microbial agents. Thus, more extensive testing is required.

Biological effects (pathogenicity) of GEMs

For chemicals, the biological effects of concern are defined as "toxicity"; for organisms, the effect is "pathogenicity." While it is generally agreed that the release of nonpathogenic organisms seems to be inherently safer than the release of pathogens, authors who make this claim rarely define their particular interpretation of pathogenicity (Keeler 1988).

Rissler (1984), in a discussion of data requirements for defining the environmental effects of GEMs, effectively equated pathogenicity with the ability to cause harm, either by direct infection or by toxin production. She identified several critical points to consider, including whether or not understanding the hazards associated with the parent organism is sufficient to predict the hazards associated with GEMs, whether GEMs have unique toxicological properties, and whether guidelines can be developed for identifying adverse effects. Host cells containing recombinant DNA could demonstrate previously unknown pathogenicity and a relatively novel risk of infection in plants and animals, but it is still not clear just what kind of testing protocol would predict these effects.

It is possible to identify several parameters of pathogenicity. "Virulence" includes the possession of toxins that contribute to infection and injury. With GEMs, host range and relative susceptibility are both important factors, and it may be difficult to predict the likelihood that a particular genetic change will expand the host range of a pathogen. Burges (1981) pointed out that procedures must be developed for testing the range of pathogenicity of GEMs as well as for extrapolating pathogenicity in various test species to other potential hosts. Negative test results may not be as generally applicable for GEMs as they are for chemicals, and it may also be difficult to extrapolate results of pathogenicity testing to any other test species. Clearly, the exchange of genetic material complicates an already difficult application of toxicological principles to the assessment of harmful effects.

The Microbiology and Genetic Toxicology of GEMs

General considerations

The proposed release of GEMs is based on the contention that these organisms are fundamentally similar to their naturally occurring parent strains, but are genetically altered in order to perform specific biological tasks which distinguish them from their wild-type counterparts. Two issues of critical concern to scientists are whether or not the modification of naturally existing microbial genomes could be responsible for environmental and biological hazards, and whether or not our current biotechnological capabilities suffice to predict such hazards prior to release of GEMs.

Techniques of genetic engineering

In order to begin to assess the safety of releasing GEMs into the environment, it is important to consider the techniques used to alter the genetic makeup of the microorganism. Deletions and insertions of genes in the genome are among the simplest modifications proposed. The removal or substitution of small nucleotide sequences may result in simple genotypic changes involving only one or a few nucleotides, but such a change may drastically alter normal functional activities and widen the potential host range of many bacteria and viruses (Keeler 1988).

The transfer of genes into microorganisms which do not naturally express the corresponding trait is one of the most efficient techniques used to modify functional activity. Such transfer can be achieved with the use of free, naked DNA or of vectors such as plasmids or bacteriophages. The genetic modification may involve sequences that code for structural or regulatory proteins which are novel to the recipient strain but not to other microorganisms within the same species or genus. However, this does not insure the safety of the resulting GEM. Such transfer could lead to unexpected changes in the invasive and pathogenic properties of the organism within a given host and alter its stability under certain environmental conditions. For example, modification of regulatory sequences in order to amplify the number of copies of a specific gene may increase or decrease the "fitness" of the organism, and enhancement of particular structural genes may unexpectedly result in the production of large amounts of a product which is harmless in small amounts but toxic if overproduced (Keeler 1985). In order to accurately predict the outcome of the release of GEMs, it is necessary first to elucidate the changes in genetic expression and function in a broad range of hosts (including human microflora) and environmental conditions. In most cases, such information is already available concerning the unaltered parent organisms, but it may not be complete for the genetically altered strains. Therefore, it is not sufficient to rely exclusively on existing data, because, in many cases, simple changes in the genome may result in overwhelming changes in the behavior of the microorganism.

Genetic alteration should be specific to a particular function so that only the targeted property of the microorganism is modified. The degree of specificity of the genetic alteration will determine the extent to which the GEM differs from its parent. Maximizing specificity will result in minimizing these differences. Evidence must be provided that one has not created mutants which are hazardous to nontarget biological systems. Considerable research is necessary to determine the effects of genetic alterations which may result in the expression of "silent" or hazardous genes. Expression of these genes may alter the GEMs' normal metabolic processes, toxin production, ability to compete, and environmental stability.

Determining the characteristics which distinguish GEMs from their naturally occurring parent strains is essential to evaluating their safety as tools in biotechnology. In addition, information about the potential for exchange of the engineered traits between GEMs and other microorganisms is pertinent to the issue of safety. Whether genetic material will be transferred to or from GEMs more readily than within parent strains is dependent on numerous factors. A thorough discussion which addresses all of these factors is virtually impossible because of our inability to predict all of the plausible conditions which would lead to genetic recombination. Furthermore, the effect that genetic recombination of engineered traits between microbial species would have on evolution is beyond the realm of this chapter, and perhaps beyond the scope of current scientific knowledge.

We can begin to assess the potential for spontaneous genetic recombination by looking at the different modes of transfer of genetic material. In general, there are three major modes of transfer: *transformation, transduction*, and *conjugation*.

The process of "transformation" occurs when naked DNA from one cell is taken up by another. Incorporation of the DNA into the genome of the recipient cell may result in the expression of the trait for which the naked DNA coded. Of the three modes of transfer, this is the most primitive.

A second type of transfer is "transduction." Transduction occurs when the DNA from a donor cell is introduced into a recipient cell via infection by a virus or bacteriophage. If the phage vector is virulent, the new DNA will be replicated in the recipient cell and inserted into phage particles which are released subsequent to cell lysis. If a phage is not virulent, then it is considered to be "temperate" and is benign to the host. Specific repressor proteins or degradative enzymes may prevent the recombination of the DNA of the transducing particle with host chromosomal DNA, and therefore may prevent expression of foreign DNA.

"Conjugation" is the most frequent mode of genetic transfer, and is dependent on cell-to-cell contact. This contact can lead to the transfer of extrachromosomal plasmid DNA. Plasmids are autonomously replicating DNA sequences which either exist extrachromosomally or have the ability to integrate into the host genome. Plasmids which are capable of becoming an integral part of chromosomal material are known as "episomes." Some episomes that replicate autonomously contain small segments of chromosomal DNA which can be transferred to a recipient cell during conjugation. The presence of plasmids which are capable of mobilizing portions of chromosomal material between bacteria poses a threat to the security of engineered traits meant to exist only in the bacterial strains in which they are placed.

Ever since the 1959 incident in Japan when the *Shigella flexner* strain was isolated from a case of dysentery and found to be resistant to four different antibiotic drugs, we have been aware that episomes can be responsible for conferring new traits to certain recipient strains of bacteria. The ability of members of the *Enterobacteriaceae* to transfer antibiotic resistance from one organism to another via a plasmid-mediated process has resulted in numerous strains of harmful bacteria which are resistant to drug therapy.

Interhost transfer of genetic material

Adverse human health effects may occur as a result of the transfer of genetic material from GEMs to indigenous strains of bacteria. Some ingestion studies have been done in humans to address the issue of gene transfer between foreign strains of bacteria and the populations which normally exist in the GI tract. Plasmids containing antibiotic resistance genes (R-plasmids) have been used extensively to explore this phenomenon. In most cases, the results of in vivo ingestion studies show no transfer of plasmids in the absence of selection pressures such as antibiotics or chemotherapy (Levin 1981). However, studies by Anderson (1975) indicated that an *Escherichia coli* K-12 strain containing an F-T plasmid colonized and multiplied in the GI tract of human volunteers who ingested it. Four days after ingestion of the donor strain, transfer of the F-T plasmid to a recipient *E. coli* strain occurred at a very low frequency in the absence of any selective pressure. In another study done in dog skin, gentamicin resistance was transferred from nonpathogenic to pathogenic *Staphylococcus* strains without antibiotic treatment (Naidoo et al. 1984). Also, the frequency of in situ transfer exceeded that of in vitro transfer by approximately 10^4-fold because of the skin's ability to facilitate plasmid transfer. Exceptions such as these are of major concern with regard to the safety of releasing GEMs. Clearly, alterations of the genomes of indigenous recipients in humans, as well as in other species, could be deleterious.

The frequency of gene transfer in human organs and tissues will increase parallel to the severity of selective pressures. In order for genetic material to persist in the genome of a recipient cell, there must be a selective advantage for the corresponding traits. Often, the transfer of some plasmids confers a disadvantage to the hosts, and they will not survive unless there is a selective advantage. For example, a plasmid containing genes which increase the survivability of a recipient cell under harsh environmental conditions will have a greater chance of survival if those conditions are present. Bacteria with engineered traits which allow them to consume oil will persist as long as their supply of oil exists. Once the oil source is depleted, the GEMs will be under severe selection pressure to pursue other food sources. Also, selection pressures increase the frequency of random mutations among a population of GEMs, and increase the probability that the organisms will undergo evolutionary changes. This could be hazardous to the finely tuned balance which exists in the environment (Keeler 1988). Therefore, we must be able to identify potential selection forces. In order to accomplish this task, we will need to study the actions of GEMs in field conditions which mimic the natural environment into which they will be released.

The successful transfer of genetic material is also dependent on the nature of the vector which carries it. For example, the pBR322 plasmid is poorly mobilized, and its transfer to normal human intestinal flora subsequent to ingestion occurs at a frequency substantially lower than that for mobile plasmids (Marshall et al. 1981). Genetic transfer can take place under many environmental conditions in a wide range of hosts. However, natural barriers to recombination of exogenous genetic material with host genomic DNA do exist. Many microorganisms produce hydrolytic enzymes which immediately recognize and eliminate alien DNA sequences. In addition, the stability of conjugative plasmids may be compromised in hosts which harbor their own plasmids. Therefore, the predictability of the survival of transferred genetic material involves complex analysis of a number of variables, and it is only the first step in the assessment of potential human health and environmental hazards associated with GEMs.

Special considerations for sensitive populations

As for chemical toxicants, one can identify a number of individuals in a population who for various reasons may exhibit unusual susceptibility to the adverse effects of GEMs. Several different factors, including genetic predisposition, age, gender, nutritional status, preexisting disease, and concomitant exposure to other chemicals, are all capable of modifying an individual's response to a toxicant. These same factors enter into special consideration for sensitive populations and GEMs. Human populations are variable also with respect to diet, occupational and home environment, activity patterns, and other cultural factors that can alter susceptibility to infectious disease (Marquis and Siek 1989).

A number of chemical toxicants, including many organic solvents and pesticides, have been shown to alter immune function and increase susceptibility to infection in laboratory animals (Marquis 1986). Given the possibility of concomitant exposure to such chemicals and GEMs, special attention should be directed to the fact that infectivity may vary with changes in the immune competence of the nontarget host.

It must also be considered that the microorganism, its secreted products, or its cellular components may be directly immunotoxic. Early on in the development of testing guidelines for microbial pesticides, it was suggested that a battery of immunotoxicity tests may be useful for evaluating these effects (USEPA 1982), but the difficulties encountered in interpreting data from these assays appeared to preclude their general applicability and they were eliminated from the more recent guidelines (USEPA 1989).

The effects of GEMs on the environment and on the human population in general may not be a true indicator of how sensitive populations will be influenced. For instance, antibiotic-treated patients and newborns exhibit compromised profiles of intestinal microflora. These populations may be less able to suppress colonization by foreign microbes such as GEMs, and may be more susceptible to invasion. Other sensitive populations include persons with an increased risk of exposure and colonization due to conditions such as skin abrasions, stress, or immune suppression. As with any potentially toxic substance, it is necessary to determine the impact of exposure to GEMs and their by-products on sensitive populations. It is likely that they are at a greater risk than the general population.

Mechanisms for minimizing the hazards of GEMs

It is doubtful that the scientific community will be able to guarantee the safety of GEMs present in the environment, but there are myriad precautions that can be taken in order to minimize the risk of harmful effects. For instance, strict criteria for the selection of recipient microorganisms which will serve to harbor the recombinant genes should be established. Only nonpathogenic microbes which can maintain the engineered trait without becoming harmful should be considered. If plasmids or phages are used as vectors for recombinant genes, then the microorganisms chosen must be able to harbor the foreign genes with minimal chance of transferring the genes to other strains. Also, microorganisms known to be present in the natural microflora of humans and other animals should not be used as recipients because of their ability to colonize and perhaps successfully compete with necessary indigenous strains.

Plasmids are small and autonomous, and they have the potential for copy amplification. All of these characteristics make them convenient tools in the development of GEMs. Plasmids which are unable to transfer genetic material via conjugation to other strains are most suitable as vectors because they eliminate the potential for gene transfer to new hosts. Engineered plasmids are foreign to their hosts and may alter the survivability of the microorganism. In many cases, the plasmids will be lost or will confer an ecological disadvantage to the host in the absence of a selective advantage (Marshall et al. 1981).

Bacteriophages are another means of generating GEMs. Many bacteriophages function in a manner similar to insertion sequences, and are used widely in the development of GEMs (Bukhari 1976). The transfer of nucleotide sequences from one position to another induces mutations within the host ("transposition"). These changes should be specific and affect only the sequences in the host genome which code for target functions.

Regardless of the method used to produce microorganisms with altered genetic makeups, the resultant GEM should be nonpathogenic and easily detectable. Ideally, only a limited number of these organisms—those for which substantial information on microbial function is available—should be released. Also, once the GEMs have performed their designated function in the environment, one must be able to terminate their release in order to prevent them from coevolving with naturally occurring microorganisms.

Numerous biocontrol mechanisms have been proposed to minimize the deleterious effects which could result from the release and persistence of GEMs. "Suicide genes" which are regulated by environmental conditions may be inserted to terminate the microbe in a controlled fashion. The presence or absence of a particular toxin could serve as the trigger for expression of the suicide gene. For example, a GEM designed to biodegrade toxic waste should contain a suicide gene which is activated in the absence of the waste. Other methods, such as partial blockage of particular metabolic pathways to permit the accumulation of toxic metabolites that kill the organism, are feasible alternatives for the biocontrol of toxicity resulting from GEM release. [See Naidoo et al. (1984) for more details on biocontrol.]

Testing Guidelines for Safety Evaluation of GEMs

In March 1989, the U.S. Environmental Protection Agency (USEPA) published revised testing guidelines for microbial pesticides (USEPA 1989). The guidelines were developed to apply many of the basic procedures of toxicity testing, as well as to address the special properties of all microbial pesticides, including both naturally occurring microbes and GEMs. While GEMs were not considered separately, the guidelines recognized that safety evaluation for released microorganisms must take into account their unique characteristics.

As discussed elsewhere in this text (see Frederick and Pilsucki, Chap. 2), the USEPA espoused a tiered approach to testing microbial pesticides, and defined a simple battery of screening assays to evaluate acute toxicity, infectivity, and pathogenicity. In so doing, the guidelines effectively provided definitions for these parameters wherein "toxicity" is equated with "pathogenicity" and "infectivity" with persistence in a particular tissue or biological host. It was also suggested that a microbial agent may be defined as "infective" if it is shown that the microorganism can overcome natural host barriers, or that it can replicate in the host.

Although traditional tests of toxicity and infectivity can provide considerable useful information, GEMs may be sufficiently different from the unmodified or parent organism to require special tests. The hazards that may be unique to GEMs include:

- Possibility of gene transfer and recombination
- Transport of microbes to new environments
- Nontarget infectivity
- Altered dose-response relationships for other microbial agents

Testing guidelines should address the possibility of a worst-case scenario, namely, the large-scale release of a GEM into the environment. Adequate data should be provided to permit the prediction of the risk of hazards to human health that may be associated with such an event. It is not expected that GEMs are likely to survive or thrive after release, and the majority will have no impact beyond their designed effects. Only a very small portion of these "new" microbes may ever have any major impact on a foreign ecosystem. The required tests should address the data needs discussed in this brief review, and provide sufficient data either to justify the elimination of the toxic GEMs or to identify means for further alterations to ensure their safety.

References

Anderson, E. S. 1975. Viability of, and transfer of a plasmid from, *E. coli* K12 in human intestine. *Nature* 255:502–504.

Betz, F., Levin, M., and Rogul, M. 1983. Safety aspects of genetically-engineered microbial pesticides. *Recomb. DNA Tech. Bull.* 6:135–141.

Bukhari, A. I. 1976. Bacteriophage mu as a transposition element. *Annu. Rev. Genet.* 10:389–412.

Burges, H. D. 1981. Safety, safety testing and quality control of microbial pesticides. In *Microbial Control of Pests and Plant Diseases*, H. D. Burges (ed.), London, Academic, pp. 737–767

Colwell, R. 1985. The role of microbial ecology in biotechnology and risk assessment. In *Engineered Organisms in the Environment: Scientific Issues*, O. Halvorson, D. Pramer, and M. Rogul (eds.), American Society for Microbiology, Washington, D.C., pp. 2–3.

Keeler, K. H. 1985. Implications of weed genetics and ecology for the deliberate release of genetically-engineered crop plants. *Recomb. DNA Tech. Bull.* 8:165–172.

Keeler, K. H. 1988. Can we guarantee the safety of genetically-engineered organisms in the environment? *CRC Crit. Rev. Biotechnol.* 8:85–96.

Levin, B. R. 1981. Periodic selection, infectious gene exchange and the genetic structure of *E. coli* populations. *Genetics* 99:1–23.

Marquis, J. K. 1986. *Contemporary Issues in Pesticide Toxicology and Pharmacology*, 2d ed., Karger, Basel, pp. 23–25.

Marquis, J. K., and Siek, G. C. 1989. Sensitive populations and risk assessment in environmental policymaking. In *Hazard Assessment of Chemicals*, vol. 6, J. Saxena, (ed.), Hemisphere, New York, pp. 1–25.

Marshall, B., Schluederberg, S., Tachibana, C., and Levy, S. B. 1981. Survival and transfer in human gut of poorly mobilizable (pBR322) and transferable plasmids from the same carrier *E. coli*. *Gene* 14:145–154.

Naidoo J., and Lloyd, D. H. Transmission of genes between staphylococci on skin. In *Antimicrobials and Agriculture*, M. Woodbine, (ed.), London, Butterworth, pp. 285–292.

Rissler, J. F. 1984. Research needs for biotic environmental effects of genetically-engineered microorganisms. *Recomb. DNA Tech. Bull.* 7:20–30.

Simonsen, L., and Levin, B. R. 1988. Evaluating the risk of releasing genetically engineered organisms. *Tibtech 1988* 6:527–530.

Suter, G. W. 1985. Applications of environmental risk analysis to engineered organisms. In *Engineered Organisms in the Environment: Scientific Issues* O. Halvorson, D. Pramer, and M. Rogul (eds.), American Society for Microbiology, Washington, D.C., pp. 211–219.

U.S. Environmental Protection Agency (USEPA). 1982. *Biorational Pesticides. Subdivision M, Pesticide Testing Guidelines*. Office of Pesticide & Toxic Substances, Washington, D.C.

U.S. Environmental Protection Agency (USEPA). 1989. *Subdivision M, Pesticide Testing Guidelines*. Office of Pesticide & Toxic Substances, Washington, D.C.

Chapter

4

Safety Considerations in the Evaluation of Transgenic Plants for Human Food

Daniel D. Jones

Office of Agricultural Biotechnology
U.S. Department of Agriculture
Room 321A, Administration Building
14th Street and Independence Avenue, S.W.
Washington, D.C. 20250

James H. Maryanski[1]

Center for Food Safety and Applied Nutrition
Food and Drug Administration
Switzer Building, Room 1414
330 C Street, S.W.
Washington, D.C. 20204

Introduction

Gene transfer is becoming an important tool for the improvement of agricultural crops (Sharp et al. 1984, Cocking and Davey 1987, Goodman et al. 1987, Gasser and Fraley 1989). An increasing number of transgenic plants have

[1]Dr. Jones is Deputy Director, Office of Agricultural Biotechnology, U.S. Department of Agriculture, Washington, D.C. Dr. Maryanski is Biotechnology Coordinator, Center for Food Safety and Applied Nutrition, U.S. Food and Drug Administration, Washington, D.C. The views expressed are the authors' and do not necessarily represent the official interpretations of any government agency. The authors acknowledge the helpful comments of Philip Derfler, Esq., Food and Drug Administration, and Dr. David MacKenzie, U.S. Department of Agriculture.

been produced through scientific methods that were not available prior to 1982 (De Block et al. 1984, Horsch et al. 1984, Frey 1988). Examples of transgenic plants in species that are commonly used for human food are shown in Table 4.1. These and other applications have involved various coding genes, regulatory sequences, and methods of gene transfer for a variety of intended effects in crop plants.

Prior to 1982, new genetic traits could be obtained only by conventional plant breeding through sexual crosses and in some cases through random or induced mutation (Frey 1988). In spite of distinct limitations of conventional breeding, important food crops have been modified successfully by this method to exhibit diverse characteristics such as increased yield, improved nutritional content, and resistance to pests and diseases. For example, interspecific hybridization between a cultivated tomato and a weedy variety produced a commercial hybrid that was resistant to the fusarium wilt fungus (Goodman et al. 1987, ref. 5). Traits for salt tolerance and disease resistance from wild grasses have been transferred to cultivated wheat (Goodman et al. 1987, ref. 8). The development of canola oil from rapeseed is an example of the application of plant breeding to improve the safety of a food by reducing the level of a naturally occurring toxicant. Although these examples represent a mere fraction of the useful varieties that have been developed, nearly all food plant varieties produced heretofore have resulted from conventional genetic modification.

Genetic engineering, especially with recombinant DNA techniques, permits

TABLE 4.1 Plants Used for Human Food for Which Transgenic Plants Have Been Produced [a]

Food plant	Reference
Tomato	Fillatti et al. 1987 Fischoff et al. 1987 Nelson et al. 1988
Potato	Jaynes et al. 1986 De Greef et al. 1989 Hoekema et al. 1989
Soybean	Hinchee et al. 1988 McCabe et al. 1988
Maize	Klein et al. 1988 Rhodes et al. 1988
Rye	De la Pena et al. 1987
Rice	Toriyama et al. 1988
Lettuce	Chupeau et al. 1989
Sunflower	Everett et al. 1987
Walnut	McGranahan et al. 1988

[a]This list is meant to be illustrative rather than comprehensive. Gasser and Fraley (1989) report a total of 30 plant species for which transgenic plants have been reported.

scientists to transfer genetic traits between diverse organisms with much greater speed and precision. Investigators have used various coding genes, regulatory sequences, and methods of gene transfer for a variety of purposes. Often the immediate goal of the research is simply the genetic incorporation and expression of a marker gene such as antibiotic resistance in order to demonstrate the feasibility of the method in a particular host plant. In addition to recombinant DNA techniques, genes have been transferred by electroporation, microprojectiles, and cell fusion. To generate stable commercial varieties, these methods of gene transfer must usually be integrated into a breeding program using traditional methods.

As experience with gene-transfer methods has been gained, researchers have often been able to achieve agronomic effects such as herbicide tolerance (Fillatti et al. 1987, Hinchee et al. 1988, De Greef et al. 1989), insect tolerance (Fischoff et al. 1987), and virus tolerance (Nelson et al. 1988, Hoekema et al. 1989). As more experience is gained with gene-transfer methods, it may be possible to engineer plants with more complex agronomic traits such as heat and drought resistance, salt tolerance, and improved nitrogen-fixing ability (Board on Agriculture 1984). These traits and others of economic importance, such as yield and time to maturity, are believed to involve multigene functions and are often referred to as "quantitative trait loci" (QTL) (Helentjaris 1988, Tanksley et al. 1989). The application of restriction fragment length polymorphism (RFLP) mapping is expected to permit scientists to identify and transfer genes that are involved in QTL and to develop improved varieties more easily than with conventional methods alone (Helentjaris 1988, Martin et al. 1989, Tanksley et al. 1989).

Some researchers have also transferred genes not for an agronomic effect in the plant but for an intended technical effect in food prepared from the plant. For example, researchers have transferred synthetic genes into potato in order to improve the amino acid profile of the potato proteins (Jaynes et al. 1986). The shipping properties of tomatoes may also be improved by the incorporation into the plant of an antisense gene that modulates the expression of an enzyme, polygalacturonase, involved in fruit ripening (Comai 1988). As these and other plants are genetically modified for both agronomic and food technical effects, it may be important to address the potential effects of these changes on the safety of food products prepared from transgenic plants. The following discussion provides a brief statutory and regulatory background for food safety.

Food Safety Regulatory Background

In the United States, the federal Food, Drug, and Cosmetic Act (the FDCA) provides authority over the safety of food, including commercial food crops. The FDCA empowers the Secretary of Health and Human Services and, hence, the Food and Drug Administration (FDA) to promulgate regulations to ensure the safety of food and food ingredients that are shipped in interstate commerce, including imported goods. Generally, the FDA's authority extends to food and food ingredients from the farm, manufacturing plant, or import dock until the product reaches the consumer. Under the FDCA, "food" is defined as "(1) articles used for food or drink for man or other animals, (2) chewing gum, and (3) articles used for components of any such article" (U.S. Code, 1982*a*). Thus, the FDA's authority over food is very broad in scope.

The FDCA does not provide authority over research on and development of agricultural crops. However, several other statutes address various aspects of plant health and the environment, and they may apply during research and development of transgenic crops. Notably, the U.S. Department of Agriculture's (USDA's) Animal and Plant Health Inspection Service (APHIS) enforces the Federal Plant Pest Act, the Plant Quarantine Act, the Federal Noxious Weed Act, and other laws that control the movement of plant pests (Jones 1985, OSTP 1986). The U.S. Environmental Protection Agency (USEPA) has authority to establish tolerances or exemptions from tolerances for pesticides in and on raw agricultural crops. Further explanation of these laws may be found elsewhere (OSTP 1986, Gibbs et al. 1987, Korwek 1988).

The FDCA does not contain provisions that are specific to new biotechnology. The FDA believes that the provisions of the FDCA are adequate to ensure the safety of new food products produced through biotechnology (OSTP 1986). The FDA has also stated as policy that issues concerning the regulation of food and food ingredients developed through new technologies will be considered under the relevant statutory or regulatory provisions of the FDCA (OSTP 1986). Under this statute, the burden to establish the safety of new foods and food ingredients rests on the manufacturer or sponsor of the product. Furthermore, most issues that relate to the safety of a food developed through new technologies are expected to involve the application of the adulteration provisions or the food additive provisions. These provisions are described below.

Adulteration of food

The adulteration provisions of the FDCA give the FDA authority to take legal action against a food if it contains a poisonous or deleterious substance that would cause the food to be injurious to health. There are two legal standards that define the burden of proof that the FDA must establish if it is to take action against a food; which standard applies depends on the nature (or source) of the substance in question. Under the first standard, if a substance is a *natural* component of the food (i.e., not an added substance), the FDA must establish that the substance would *ordinarily render* the food injurious to health (*emphasis added*). Except for cases of obvious poisons such as cyanide in food (*United States v. Articles of Food and Drug* 1978), the FDA has applied this standard infrequently.

In most cases of potential adulteration of food, the FDA applies the second standard; this standard requires evidence that a substance that has been *added* to food *may render* the food injurious to health (*emphasis added*). Under this standard, the FDA must show only that there is a reasonable possibility that the food will be harmful if consumed.

Courts have agreed with the FDA's interpretation of this section that any substance that is not a natural component of the food may be treated as an added substance. Added substances include those introduced through human activity (*United States v. Anderson Seafoods, Inc.* 1978) and those substances of environmental or industrial origin that become components of food. For example, mercury in swordfish was considered to be an added substance because it is not produced by the fish (*United States v. An Article of Food* 1975). Contrarily, natural toxicants, such as oxalic acid found in rhubarb, are not viewed as added substances (*United States v. An Article of Food* 1975). Furthermore, the FDA can apply the "may render" standard to an increased level of a natural component of a food to the extent that the increased level is due to some technological change (*United States v. Anderson Seafoods, Inc.* 1980).

The provisions for natural and added substances apply to most of the harmful substances that may occur in food. A food modified by genetic engineering that contains either an added toxicant or an increased level of a natural toxicant that could pose a hazard would be in violation of the law.

The FDA has on occasion been faced with unavoidable substances in food, such as elemental lead (FDA 1974), polychlorinated biphenyls, and aflatoxin. The FDA can issue tolerances by formal rulemaking (*U.S. Code* 1982*b*), as it has done in the case of polychlorinated biphenyls, or it can establish regulatory limits by informal notice and comment rulemaking procedures. Regulatory limits establish the level at which food is adulterated within the meaning of section 402(a)(1) of the FDCA. Alternatively, the agency can set action levels that constitute prosecutorial guidance rather than substantive rules. An action level defines a level of contamination at which a food may be regarded as adulterated based on a scientific evaluation of the health risks posed by the contaminant (FDA 1989, 1990).

In addition to the provisions dealing with the presence of poisonous or deleterious substances in food, other provisions of the FDCA regarding adulteration could apply to food modified by genetic engineering if the food is lacking a valuable constituent, if the food is unfit because of decomposition (e.g., because of increased susceptibility to spoilage), or if the food contains an unapproved food additive as discussed below.

The food additive provisions

The FDCA was amended in 1958 to require premarket approval for new chemicals—i.e., food additives—that would be added to food for technical effects such as sweetness, texture, and preservation (FDA 1986*a*). The definition of a "food additive" includes both artificial and natural substances, and provides, in part, that

> the term food additive means any substance the intended use of which results or may reasonably be expected to result, directly or indirectly, in its becoming a component or otherwise affecting the characteristics of any food.... (*U.S. Code* 1982*c*)

The definition then carves out an exclusion, namely, for any such added substance that is "generally recognized as safe" by qualified experts for its proposed use in food (*U.S. Code* 1982*c*). This exclusion is explained further in a later section.

In addition to the exclusion from the definition for substances "generally recognized as safe" (GRAS), the food additive definition also excludes substances that are covered by other provisions of the FDCA or other statutes, including pesticide chemicals in or on raw agricultural commodities; color additives; substances granted sanctions under the FDCA, the Poultry Products Inspection Act, or the Federal Meat Inspection Act before 1958; and new animal drugs. Thus, a substance that is intended to become a component of food is a food additive and requires premarket approval unless it can be shown to be a GRAS food ingredient or is subject to another exclusion.

A food additive is considered to be unsafe unless it is used in accordance with a food additive regulation. The FDA may not issue a food additive regulation until adequate information is available to establish a reasonable certainty that the intended use of the additive will not adversely affect the health of consumers (OSTP 1986).

GRAS food ingredients

In establishing the premarket approval requirement for food additives, Congress did not intend to require safety testing for the large number of ingredients that were on the market in 1958 and for which there was no evidence of adverse health effects. Congress included a provision for exempting substances introduced after 1958 from the food additive definition if adequate scientific information was available to demonstrate that the safety of the proposed use of the substance in food was widely recognized. Thus, GRAS food ingredients are those generally considered safe by qualified experts based on either (1) a safe history of use in food prior to 1958 or (2) scientific information (*U.S. Code* 1982*c*).

The GRAS exemption is especially important for agricultural crops because the legal distinction between a food and a food additive can become blurred when one food "becomes a component" of another. One author has pointed out the misconception that a raw agricultural commodity is a food and not a food ingredient (Spiher 1975). He interpreted sections of the FDCA as establishing the equivalency of the terms "food" and "food additive" (Spiher 1975). The author cited an example: Foods such as meat and potatoes are used to make stew. Each component eaten alone is a food, but when these foods are added together, they would be "food additives" by definition except that their safe history of use permits them to be considered GRAS food ingredients of the stew. While this example may seem extreme, it illustrates both the broad authority and the discretion that the FDA has in applying its authority in those instances necessary to protect public health.

Ordinarily, substances have been regarded as GRAS without the need for specific regulation if the substance is of natural biological origin, has been widely consumed before 1958 for its nutrient properties, is subject only to conventional processing as practiced prior to 1958, and exhibits no known safety hazard (FDA 1986*b*). The FDA can review the GRAS status of substances of natural biological origin that have undergone significant changes as a result of breeding and selection or a new process introduced into commercial use after 1958 (FDA 1986*c*).

A GRAS substance that has been altered in a manner that would cause it to be no longer generally recognized as safe would be a food additive (OSTP 1986). Adequate published literature is usually required to demonstrate

GRAS status; otherwise the FDA would be inclined to regulate the substance as a food additive. The FDA has relied on a Supreme Court ruling (*Weinberger v. Bentex Pharmaceuticals* 1973) that a pivotal factor in determining whether a substance is generally recognized as safe and effective and thus not a "new drug" hinges on expert knowledge backed by "substantial support in scientific literature" (Spiher 1975). The agency has used a similar criterion of "substantial support in scientific literature" for determinations of GRAS status of food substances (United States v. 45/194 kg. Vegetable Oil).

Agricultural biotechnology and the food safety laws

Food crops modified by genetic engineering have not yet reached the commercial market, and the FDA has not had an opportunity to evaluate and issue an opinion on any of the new applications of biotechnology. Most agricultural commodities are considered GRAS because of an extended history of safe use prior to 1958, and modified crops may be considered in light of whether the new crop is also GRAS (Maryanski 1988).

The regulatory issues associated with living plants are complex. Most food plants, for example, have never been tested for toxicity. Instead, safety has been assumed based largely on experience and history of use. Likewise, little regulatory oversight has been extended to new varieties of agricultural crops because of the extensive safety record.

Many foods contain inherent substances that, if present in sufficient amounts, would be toxic to humans. In other cases, foods such as cassava must be cooked or processed to remove highly toxic substances before the food can be consumed. Examples exist of new varieties that are harmful to human health, even though such examples are rare. One of the most frequently cited examples was a variety of potato, "Lenape," that was developed for better chipping qualities and that inadvertently contained unusually high levels of a toxic glycoalkaloid (Zitnak and Johnston 1970). Concern for the safety of new varieties of crops has been a recurrent theme for government agencies and industry.

The advent of molecular genetic techniques in agricultural research and development has raised questions of food safety both in the scientific community and in public debate. Many of the issues that are being discussed currently are similar to those considered in the 1970s, when concern was raised with respect to potential changes in natural toxicants or nutrient levels of new plant varieties produced by conventional breeding. At that time, the FDA proposed to review new varieties of food crops if the new variety exhibited either a 10% increase in the level of a natural toxicant or a 20% decrease in the level of an important nutrient (Spiher 1975). This proposal met with considerable resistance primarily because it would have established rigid, but arbitrary, levels of regulatory concern (Crop Science Society of America 1974).

It is too early to predict the type and extent of oversight that may be exerted on the products of new biotechnologies, but market introduction is expected early in the 1990s (FDA 1988). The first generation of genetically engineered food plants will most likely be common foods that have been modified to contain one or a few genes of well-defined function(s). Scientific information and experience gained from these "pioneers" should provide a basis not only for evaluations of their safety, but also for evaluating safety of more complex modifications involving QTLs in the future.

A number of factors may be considered in evaluating the regulatory status of genes and gene products such as herbicide resistance, or production of antisense polygalacturonase or potatin, and several approaches to regulation can be envisioned. Conventional plant breeding involves the exchange of genes between varieties of food plants, and these genes and their products have been an integral part of the food supply. Thus, one approach would be to consider genes from common food plants and their close relatives to encode GRAS substances, unless the substance is toxic to humans at the level found in the finished food. This would be consistent with the GRAS status of varieties developed through conventional breeding.

Another approach would be to determine whether the added gene and/or its gene product serves an agronomic function (e.g., growth, yield, photosynthesis, tolerance to salt, weather, chemicals or a food additive function (e.g., preservative, sweetener, thickener, nutrient) in the plant. Substances that have agronomic functions or that could be shown to be GRAS would be exempt from the food additive definition. Again, known toxic substances would not be categorically excluded. This approach would reflect FDAs statutory mandate to regulate substances intended to become components of food, i.e., food additives, unless they are GRAS, and would focus regulatory oversight on substances added by genetic engineering that would be food additives if they were added during processing of the food.

Other approaches are also possible. Jones (1986, 1988) has suggested that a transplanted gene and its gene product, if under adequate genetic control of the host organism, might be considered an integral (natural) part of the host organism rather than an added substance. The suggestion was made in the context of food animals, but it need not be limited to animals. For substances that are not particularly toxic, this approach could simplify some of the questions associated with added substances in food plants. This approach would, of course, have to compete with the broad interpretations of the added substance and food additive provisions of the FDCA established by FDA and the courts (OSTP 1986, Maryanski 1986). The part-of-the-organism approach to gene transfer in plants would probably be uncomplicated only in the case of nonfood tissues such as cotton or wool.

It is likely that the approaches to be followed will only be evident after the regulatory agencies have evaluated the scientific information on the products. The diversity of the applications of new biotechnology to agriculture suggests that different approaches may be appropriate on a case-specific basis. In a recent speech, FDA Commissioner Frank Young called on the scientific community and the industry to develop priority categories based on scientific concern in order to distinguish appropriate scientific questions about and regulatory niches for food products developed through biotechnology (Young 1989). He stated that FDA does not intend to impede new technology, but it will not permit new food products to enter the market unless safety has been established (Young 1989). An industry–academic consortium has prepared a report that addresses many of the food safety issues that may be posed by genetically modified plants and microorganisms (International Food Biotechnology Council 1990).

Potential Food Safety Considerations

For purposes of the following discussion, it may be helpful to clarify several terms that will be used. These are the terms "food additive," "chemical additive," and "genetic additive." The term "food additive" will be used only in its statutory and legal sense with the implications relating to scope, exemptions, burden of proof, etc., described above. The term "chemical additive" will be used to indicate the traditional addition of a chemical substance to raw material, usually during processing for human food use. A chemical additive could be either a food additive in the legal sense or a GRAS or prior sanctioned substance as described above. This use of the term "chemical additive" seems to be in accord with that of legislative hearings on the FDCA (*Food Additives Hearings* 1958).

The term "genetic additive" will be used to indicate a substance that is added to a plant by methods of genetic engineering of the plant genome. In most cases, this will refer to the nuclear genome, but for plants it could also include genetic material in cytoplasmic elements such as chloroplasts or mitochondria. Another possibility is that a genetic additive could be present in somatic cells of the plant but not in germ cells (Howell 1989). Such a "somatic additive" could have an effect on the plant or on food prepared from it, but it would not be passed on to the progeny.

As pointed out above, the legal and regulatory status of genetic additives in food plants may be complex. In many cases, the immediate agent of the intended technical effect in a food product prepared from a plant will be the gene product encoded by the foreign gene. In some cases, it may also be a secondary metabolite produced by an enzyme that is itself the gene product. Whether genetic additives will be defined as food additives in the legal sense or some other category of substance such as GRAS, biological residue, or unavoidable substance is an open question at the present time.

A basic safety question posed by transgenic plants is whether a specific food product prepared from a transgenic plant poses an increment of risk for human food safety. How that risk is assessed may depend to some extent on the intent of the genetic change and on which of the available regulatory categories best fits the genetic change in question. In contrast to the several possible regulatory categories described above, a significant database for quantitative risk assessment does not yet exist for evaluating the safety of food products prepared from transgenic plants. Until such a database becomes available, informed scientific judgment may be the primary means for anticipating potential breaches of food safety from gene transfer, and for resolving them.

The following discussion focuses on a number of points that might be considered in arriving at a scientific judgment of the safety of a food product prepared from a transgenic plant. The discussion is doubtless not as complete as it could be if there were more basic knowledge of gene regulation, developmental molecular biology, and plant genome maps. Information on many of these points is likely to be collected in the normal process of new-plant-variety research and development, including research on the effective expression of foreign genes in transgenic plants. It is in the best interests of sponsors of food products made from transgenic plants to share such safety information with regulatory officials as early as possible rather than allowing safety information to become an emergent issue at the time of commercialization.

The following discussion presents some points to consider in assessing the safety of foods prepared from transgenic plants. Examples and analogies are given where possible, but in many instances of scientific or economic interest, there is a shortage of relevant ones. As experience and safety precedents accumulate, it may be possible to develop more generic criteria for evaluating the safety of transgenic plants and food products prepared from them.

Foreign genes

One primary consideration is the foreign gene itself. Individual genes, even if present in multiple copies, usually constitute a very small part of the biomass of a transgenic plant and the plant tissues that are used for human food. In this respect, it could be argued that a gene is analogous to a chemical additive that is present in very small amounts, and that any risk it may pose at higher levels may be regarded as insignificant at the lower level. While this may be true of a chemical additive, an important difference is that a foreign gene, especially when integrated into a suitable genetic expression system, has the potential to give rise to significant amounts of the gene product for which it codes and any secondary metabolites that the gene product itself might produce. This is an important distinction between genetic additives and chemical additives, and it indicates the importance of the safety review of gene products and secondary metabolites, which is addressed below.

A potentially important aspect of the foreign gene itself is its location within the host genome. At present, most foreign genes are inserted into unknown regions of plant chromosomes in what appears to be a random fashion. It is generally not feasible to know in advance what host-plant sequences will be interrupted by the insertion and what consequences, if any, there will be for the plant (Boyce Thompson Workshop III 1989). In the past, there has been very little experimental control over the location of newly inserted genetic material in recipient genomes, but this situation may be improving with increased proficiency in the use of methods such as homologous recombination (Marx 1988, Capecchi 1989).

Random insertion of new genetic material into a plant genome could have consequences for the plant. Possibilities include changes in the level of expression of host genes near the point of insertion and the expression of novel gene products from regions of DNA at or around newly formed splice junctions (Falco 1989). In addition, some insertional mutagenic events may result from the integration of foreign sequences at unusual sites in the genome, with the associated possibility that gene expression from such sites may not be regulated properly (Friedmann 1989).

Regulatory elements

Another question that may have safety implications relates to promoters, enhancers, and other regulatory elements that may be utilized in transgenic plant expression systems. Regulatory elements that may need to be considered include those in the vector, the host, and any "helper" viruses that may infect the host.

Promoters and enhancers for gene-transfer work are often selected for their ability to maximize the expression of a foreign structural gene. The safety of a gene product or genetic additive in a food product may depend on the amount present. The level of a genetic additive in a plant-derived food product would normally need to surpass a certain minimum level in order to be effective. It may also need to be less than a certain maximum level determined by safety considerations. Do foreign regulatory elements selected for maximization of expression differ in risk potential from other regulatory elements that may not express the foreign gene as profusely? Consideration of this question may be necessary in assessing the safety of some food products made from transgenic plants.

The use of host regulatory elements may help to achieve controlled expression of foreign genes in transgenic plants. In pea plants, for example, regulatory sequences for several pea proteins required for photosynthesis are turned on in response to light. When these regulatory sequences from the pea genes are joined to coding sequences for foreign proteins, the resulting "chimeric" genes behave just like the normal pea genes. That is, the foreign proteins are made only in light-activated leaves that have fully developed chloroplasts (Schell 1988).

In a contrasting hypothetical example, a two-part construct of a foreign structural gene and a high-expression foreign regulatory sequence might be inserted into the plant genome. If the foreign gene is expressed at a level significantly greater than that needed to achieve the intended technical effect in food prepared from the plant, the food could be deemed to be adulterated. This hypothetical example illustrates how overexpression of a foreign gene in a host plant could have an adverse affect on the safety of food products prepared from a transgenic plant. This indicates the importance of adjusting the expression of foreign genes in food plants to appropriate levels, and this subject is considered in more detail in a later section (see under "Gene Expression").

Marker genes

The presence of new genes and gene products in host organisms may not always be easy to assay. This problem can often be resolved by linking the foreign gene with a selectable marker gene which codes for a substance or feature that is easily assayed. Commonly used marker genes include those for antibiotic resistance, herbicide resistance, and various color reactions.

The U.S. Environmental Protection Agency (USEPA) recently held a public meeting on the release of antibiotic resistance markers, particularly in microorganisms (USEPA 1989). Some of the concerns raised about the environmental release of antibiotic resistance genes may also pertain to the use of such marker genes in food and feed crops. One question at issue is the risk to human and animal health of antibiotic resistance genes used for labeling purposes, whether at the molecular, cellular, or tissue level, in food and feed crops. The resolution may have a scientific component consisting of elucidation of the likelihood of horizontal transfer of antibiotic resistance genes in the human or animal gut. There may also be clinical or veterinary concerns related to the chemical identities of the antibiotics encoded by the markers compared to those in current clinical or veterinary use.

There are alternative types of marker genes that may find use in crop plants. Among these are those for herbicide resistance and various color reactions. The primary food safety implications of herbicide resistance marker genes relate to the possibility of novel gene products and associated secondary metabolites (Falco 1989). An example of a marker gene color reaction is the *lacZY* genes incorporated into *Pseudomonas fluorescens* for purposes of monitoring movement of the microbe in the environment (Drahos et al. 1988). The range of available selectable and scoreable transformation marker genes that may find use in crop plants has been summarized elsewhere (Ratner 1989).

Foreign gene product

An assessment of the molecular products, both primary and secondary, of the genetic expression of a foreign gene in a transgenic plant is perhaps the most critical element of the safety evaluation of any resulting food products. Such an evaluation should consider both the chemical and biological properties of the molecular products themselves and the extent of genetic expression of the foreign gene in the host plant. Many of these genetic changes would presumably manifest themselves phenotypically as changes in the agronomic performance of the host plant, sometimes called "pleiotropic effects." If such changes affected the plant's performance adversely, they would in all likelihood be screened out during the process of agronomic evaluation and they would be of no further concern to food safety.

The primary gene product in a transgenic food plant will generally be a protein. One author has called attention to important differences between proteins and industrial chemicals that may affect the ways in which proteins are evaluated for potential toxic effects in foods (Li 1989). These differences include the observation that proteins, with a few exceptions, are usually nontoxic. For example, there appear to be no known proteins that exhibit chronic toxicities such as mutagenesis or carcinogenesis. There are exceptions, including certain cytotoxins, enterotoxins, and neurotoxins. These are highly specialized, acutely toxic proteins that are fairly well characterized as to their source and mode of action.

Another difference is that proteins are usually degraded by digestion in the mammalian alimentary tract. Again, there are some exceptions characterized by fairly well understood chemical interactions which may stabilize parts of protein molecules, thereby enabling them to survive digestion partially intact and enter the portal circulation, where they may elicit immunological reactions.

Some substances in foods are known to cause adverse reactions in some consumers. Adverse reactions to foods can be grouped into those that involve or are suspected of involving immune mechanisms and those that do not. Nonimmunologic adverse reactions to food can result from a wide range of substances such as natural toxins, food contaminants, pharmacologic agents, and secondary metabolites (American Academy 1984).

Adverse reactions thought to involve immune mechanisms have been documented in some consumers of plant products such as nuts, peanuts, chocolate, barley, rice, wheat, citrus, melons, bananas, tomatoes, spinach, corn, potatoes, and soybeans (American Academy 1984). In some cases the immunogenic agent is known to be a protein; in other cases a glycoprotein is involved, and it may not be known whether the protein portion or the carbohydrate portion is the immunogenic agent. Since genetic manipulation of food plants can affect both protein and carbohydrate moieties, developers of foods from transgenic plants should remain cognizant of the possibility of introducing abnormally stable chemical moieties that could lead to adverse immunological reactions.

A final difference between proteins and industrial chemicals is that proteins do not appear to be bioaccumulated as some chemicals are. Li has discussed the implications of these differences for evaluating the human health effects of ingestible proteins (Li 1989).

Concern has been expressed about the possible biological effects of new secondary metabolites because of their greater chemical stability and ease of absorption compared to proteins (Boyce Thompson Work Group B 1988). If a genetic manipulation of a food plant involves an enzyme, for example, that is intended to produce a specific new metabolite, then careful attention may be due to the biological effects of the metabolite. Evaluation of the toxicological properties of secondary metabolites of foreign gene products for their impact on food safety may be similar to that for chemical additives. This would be particularly true if a secondary metabolite of a genetic additive were used for a specific intended technical effect in food. Whether gene transfer with the intent of adding a specific secondary metabolite to a food product would have any advantages over direct chemical addition is not clear at this time.

Review of the biological properties of genetic additives may, however, differ in other respects from that of chemical additives. Genetic additives in a plant may be used for an agronomic effect rather than a technical effect in food. Some gene products, such as the endotoxin from *Bacillus thuringiensis*, for example, are intended to be toxic to species that are pests to the transgenic host. The gene products or secondary metabolites responsible for these agronomic effects in plants will often be carried over into human food products, where they may not serve a technical purpose. They may then be considered as (for example) possible biological residues. Information on possible human toxicity of genetic additives intended for agronomic effects in plants may be important if these substances remain in food made from the plant.

Gene expression

The expression of foreign genes in plants varies with a number of factors such as copy number, and arrangement and position within the genome (Day 1988). The expression of genes in plants also spans a range of susceptibility to human intervention. In nature, for example, genetic variation, expression, and selection proceed independently of human activity. If humans obtain food directly from plants or other organisms in nature, the composition of that food is determined wholly by natural processes and it may include naturally occurring toxicants such as mushroom toxins.

In conventional plant breeding, genetic variation and expression occur more or less naturally, but human intervention enters into the selection process. Conventional plant breeding often involves the deliberate selection and transfer of one or a few genetic traits with the unintended transfer of many other genetic traits unrelated to the trait of interest. Much of this "coattail" genetic material may be replaced by the genetic material of a selected parental strain by repeated back-crossing. Even with back-crossing, however, multigenic transfers under artificial selection sometimes include undesirable genes such as those for susceptibility to Southern corn leaf blight (NAS 1972) or toxic alkaloids in potatoes (Zitnak and Johnston 1970). The blight-susceptible corn variety resulted from a cytoplasmic gene and affected mainly the size of the corn harvest. The high-alkaloid potato, on the other hand, could have had an adverse effect on food safety if it had not been voluntarily withdrawn from commercial development.

The results of genetic modification of food plants can range all the way from no expression of a transferred gene to deliberate overexpression. Between these extremes lie other possible outcomes, including unintended overexpression and various levels of appropriate expression. These different outcomes of gene transfer and expression may have different effects on the safety of food products made from transgenic plants.

One possible outcome of gene transfer in organisms used for food, particularly if the technology has not been fully mastered, is no expression or insufficient expression. This could result either from failure to incorporate the new genetic material or from incorporation without proper integration with appropriate regulatory elements. In either case, the gene transfer would not be successful in achieving an intended agronomic or food technical effect.

In cases where there is no expression or insufficient expression, the genetic manipulation may technically have no effect on the safety of resulting food products. However, some consumers may perceive a heightened food risk from organisms that have been exposed to genetic manipulation even though the organisms were not successfully transformed. The question of the food safety of untransformed animals arising from partially successful gene-transfer experiments in groups of pigs and calves has already been addressed to food safety agencies in the United States.

A second possible outcome of gene transfer in plants may be unintended overexpression. In this scenario the gene is successfully incorporated into the host genome, but the gene product is produced in amounts that may exceed the amount necessary to achieve the intended genetic effect, be it an agronomic effect in the plant or a technical effect in the food. Depending on the identity of the gene product, its overproduction in the host plant could be detrimental to the host plant or to food products made from it. For example, the overproduction of any foreign gene product in a plant could disrupt the energy balance of the plant, thus affecting its health adversely or decreasing the yield of the plant tissues used for food. Overproduction of a foreign substance in the plant tissues used for food could result in amounts of the genetic additive in food that would present an increment of risk. Unintended overexpression of foreign genes in host plants may indicate that greater attention must be given to proper placement of the structural gene with respect to appropriate regulatory elements in the host genome.

The ideal outcome of an artificial gene transfer in a host plant is appropriate expression of the gene product. But "appropriate expression" may not always be easy to define precisely. In general, it would mean sufficient expression to achieve the intended genetic effect in the plant or the food product, but not so much expression that characteristics such as the health of the plant, the yield of the tissues used for food, or the safety of the food itself are compromised.

A final possible outcome of gene transfer in plants is intentional overexpression. This may arise when a plant has been genetically engineered to produce large amounts of an economically valuable substance, such as a pharmaceutical which is harvested from the plant. The plant would thus be serving as an in vivo factory for production of the substance. Many substances produced in this manner would not be suitable constituents of food. The principal impact on food safety would be the necessity to ensure that tissues of spent "in vivo production" plants are not entered into the food supply if the overproduced substance presents a significant food risk.

Specificity of gene expression

The specificity of gene expression in bioengineered plants may affect food safety. Particularly if the gene product or its secondary metabolites are toxic, specificity of gene expression may be one way to separate an intended genetic effect in one part of the plant from the food use of another part of the plant. Such tissue-specific expression could add a margin of safety to the use of pesticidal gene products in plants in which pests and humans eat different tissues. However, it may be of little value for plants in which humans and pests eat the same tissues.

In addition to tissue specificity of gene expression, temporal specificity may also be important. That is, some gene products may need to be expressed only at certain times in the plant's life cycle. Appropriate temporal specificity of gene expression is clearly important for plant-growth substances such as auxins and cytokinins. Appropriate temporal specificity may be important for the expression of other substances as well.

Research Priorities

There are several areas of plant science research that can contribute to the development and safety of food products prepared from transgenic plants. Greater understanding of the basic molecular biology of plant development would be helpful in designing genetic strategies to improve the composition or food characteristics of roots, leaves, stems, seed oils, storage proteins, and other plant components.

Greater knowledge of the organization of plant genomes would be helpful in assessing the positional effects, if any, of new genes in plant genomes. Private research groups have published partial genome maps for maize and tomato (Helentjaris et al. 1986) and have begun using them to identify gene clusters for agronomic traits such as drought resistance (Martin et al. 1989). USDA, in cooperation with the Agricultural Research Institute, recently sponsored a new research initiative in plant genome mapping (USDA 1988). USDA has also established a new office of Plant Genome Mapping Programs within the Agricultural Research Service. Finally, improved methods for toxicological assessment of proteins and/or whole foods would constitute an important advance in the safety review of foods from transgenic plants.

Conclusion

Gene transfer in food plants presents a number of challenging food safety questions. Some of these questions are similar to those addressed in the safety evaluation of chemical additives in food. Others, however, relate to the technical intent, biological mechanisms, and food safety outcomes of genetic expression systems, and they may be unique to biotechnology-modified foods. The interplay of statutory, scientific, and social forces will have important effects on the safety and acceptability of gene transfer in food plants.

References

American Academy of Allergy and Immunology Committee on Adverse Reactions to Foods. 1984. *Adverse Reactions to Foods*, National Institute of Allergy and Infectious Diseases, Bethesda, Maryland.

Board on Agriculture, National Research Council. 1984. *Genetic Engineering of Plants, Agricultural Research Opportunities and Policy Concerns*, National Academy Press, Washington, D.C.

Boyce Thompson Work Group B. 1988. Insect-resistant plants, in *Regulatory Considerations: Genetically-Engineered Plants*, Center for Science Information, summary of a workshop held at Boyce Thompson Institute for Plant Research and Cornell University, Ithaca, New York, Oct. 19–21, 1987, p. 50.

Boyce Thompson Workshop III. 1989. Herbicide resistance, in *Genetically-Engineered Plants: Scientific Issues in Their Regulation for Animal Feed and Human Food Uses*, symposium held at Boyce Thompson Institute for Plant Research, and Cornell University, Ithaca, New York, Apr. 30–May 3, 1989.

Capecchi, M. R. 1989. Altering the genome by homologous recombination, *Science, 244*, 1288.

Chupeau, M. C., Bellini, C., Guerche, P., Maisonneuve, B., Vastra, G., and Chupeau, Y., 1989. Transgenic plants of lettuce (*Lactuca sativa*) obtained through electroporation of protoplasts, *Bio/Technology, 7*, 503.

Cocking, E. C., and Davey, M. R. 1987. Gene transfer in cereals, *Science, 236*, 1259.

Comai, L. 1988. Panel 1: types of transgenic plants, techniques and timing, in *Transgenic Plant Conference Proceedings*, Sept. 7–9, 1988, Keystone Center, Keystone, Colorado.

Crop Science Society of America. 1974. The effect of FDA regulations (GRAS) on plant breeding and processing, Madison, Wisconsin, *Annual Meeting*, Las Vegas, Nevada, November 1973, Hanson, C. H. (ed.), CSSA Special Pub. no. 5, 1974.

Day, P. 1988. Concluding remarks, in *Regulatory Considerations: Genetically-Engineered Plants*, Center for Science Information, San Francisco, summary of a workshop held at Boyce Thompson Institute for Plant Research and Cornell University, Ithaca, New York, Oct. 19–21, 1987.

De Block, M. L., Herrera-Estrella, L., van Montague, M., Schell, J., and Zambryski, P. 1984. Expression of foreign genes in regenerated plants and their progeny, *EMBO J., 3*, 1681.

De Greef, W., Delon, R., De Block, M., Leemans, J., and Botterman, J. 1989. Evaluation of herbicide resistance in transgenic crops under field conditions, *Bio/Technology, 7*, 61.

De la Pena, A., Lorz, H., and Schell, J. 1987. Transgenic rye plants obtained by injecting DNA into young floral tillers, *Nature, 325*, 274.

Drahos, D. J., Barry, G. F., Hemming, B. C., Brandt, E. J., Skipper, H. D., Kline, E. L., Kluepfel, D. A., Hughes, T. A., and Gooden, D. T. 1988. Prerelease testing procedures: U.S. field test of a Lac ZY-engineered soil bacterium, in *The Release of Genetically-Engineered Microorganisms*, Collins, F., Skinner, A., and Stewart-Tull, D. E. (eds.), Academic, New York, p. 181.

Everett, N. P., Robinson, K. E. P., and Mascarenhas, D. 1987. Genetic engineering of sunflower (*Helianthus annuus L.*), *Bio/Technology, 5*, 1201.

Falco, S. C. 1989. Herbicide resistant crops, in *Genetically-Engineered Plants: Scientific Issues in Their Regulation for Animal Feed and Human Food Uses*, workshop at Boyce Thompson Institute for Plant Research and Cornell University, Ithaca, New York, May 1–3, 1989.

Fillatti, J. J, Kiser, J., Rose, R., and Comai, L. 1987. Efficient transfer of a glyphosate tolerance gene into tomato using a binary *Agrobacterium tumefaciens* vector, *Bio/Technology, 5*, 726.

Fischhoff, D. A., Bowdish, K. S., Perlak, F. J., Marrone, P. G., McCormick, S. M., Niedermeyer, J. G., Dean, D. A., Kusano-Kretzmer, K., Mayer, E. J., Rochester, D. E., Rogers, S. G., and Fraley, R. T. 1987. Insect tolerant transgenic tomato plants, *Bio/Technology, 5*, 807.

Food Additives Hearings before a Subcommittee of the Committee on Interstate and Foreign Commerce, 1958, House of Representatives, 85th Cong., on Bills to Amend the Federal Food, Drug, and Cosmetic Act with Respect to Chemical Additives in Food, U.S. Government Printing Office, Washington, D.C.

Food and Drug Administration (FDA). 1974. Poisonous or Deleterious Substances in Food, *Fed. Reg., 39*, 42743.

Food and Drug Administration (FDA). 1986*a*. *Code of Federal Regulations*, Title 21, *Food and Drugs*, Sec. 170.3(o).

Food and Drug Administration (FDA). 1986*b*. *Code of Federal Regulations*, Title 21, *Food and Drugs*, Sec. 170.30(d).

Food and Drug Administration (FDA). 1986*c*. *Code of Federal Regulations*, Title 21, *Food and Drugs*, Sec. 170.30(f).

Food and Drug Administration (FDA). 1988. *Food Biotechnology: Present and Future*, vol. 1, Office of Planning and Evaluation, Rockville, Maryland.

Food and Drug Administration (FDA). 1989. Action levels for added poisonous or deleterious substances in food, *Fed. Reg., 54*, 16128.

Food and Drug Administration, 1990. Action levels for added poisonous or deleterious substances in food, *Fed. Reg. 55*, 20782.

Frey, N. M. 1988. Introduction, in *Genetic Improvements of Agriculturally Important Crops: Progress and Issues*, Fraley, R. T., Frey, N. M., and Schell, J. (eds.), Cold Spring Harbor Laboratory, Cold Spring Harbor, New York, p. 1.

Friedmann, T. 1989. Progress toward human gene therapy, *Science, 244*, 1275.

Gasser, C. S., and Fraley, R. T. 1989. Genetically engineering plants for crop improvement, *Science, 244*, 1293.

Gibbs, J. N., Cooper, I. P., and Mackler, B. F. 1987. *Biotechnology and the Environment: International Regulation*, Stockton Press, New York.

Goodman, R. M., Hauptli, H., Crossway, A., and Knauf, V. C. 1987. Gene transfer in crop improvement, *Science, 236*, 48.

Helentjaris, T., Slocum, M., Wright, S., Schaefer, A., and Nienhuis, J. 1986. Construction of genetic linkage maps in maize and tomato using restriction fragment length polymorphisms, *Theoret. Appl. Genet., 72*, 761.

Helentjaris, T. 1988. Use of RFLP analysis to identify genes involved in complex traits of agronomic importance, in *Genetic Improvements of Agriculturally Important Crops: Progress and Issues*, Fraley, R. T., Frey, N. M., and Schell, J. (eds.), Cold Spring Harbor Laboratory, Cold Spring Harbor, New York, p. 27.

Hinchee, M. A. W., Connor-Ward, D. V., Newell, C. A., McDonnell, R. E., Sato, S. J., Gasser, C. S., Fischoff, D. A., Re, D. B., Fraley, R. T., and Horsch, R. B. 1988. Production of transgenic soybean plants using agrobacterium-mediated DNA transfer, *Bio/Technology, 6*, 915.

Hoekema, A., Huisman, M. J., Molendijk, L., van den Elzen, P. J. M., and Cornelissen, B. J. C. 1989. The genetic engineering of two commercial potato cultivars for resistance to potato virus X, *Bio/Technology, 7*, 273.
Horsch, R. B., Fraley, R. T., Rogers, S. G., Sanders, P. R., Lloyd, A., and Hoffmann, N. 1984. Inheritance of functional foreign genes in plants, *Science, 223*, 496.
Howell, S. H. 1989. Plant genetic engineering: a review of current methods of gene introduction, in *Genetically-Engineered Plants: Scientific Issues in Their Regulation for Animal Feed and Human Food Uses*, Workshop at Boyce Thompson Institute for Plant Research and Cornell University, Ithaca, New York, May 1–3, 1989.
International Food Biotechnology Council 1990. Biotechnologies and food: assuring the safety of foods produced by genetic modification, *Regulatory Toxicology and Pharmacology*, in press.
Jaynes, J. M., Yang, M. S., Espinoza, N., and Dodds, J. H. 1986. Plant protein improvement by genetic engineering: use of synthetic genes, *Trends Biotechnol., 4*, 314.
Jones, D. D. 1985, How the federal government will oversee food products of emerging biotechnologies, *Food Technol., 39*, 59.
Jones, D. D. 1986. Legal and regulatory aspects of genetically engineered animals, in *Genetic Engineering of Animals: An Agricultural Perspective*, Evans, J. W., and Hollaender, A. (eds.), Plenum, New York, p. 273.
Jones, D. D. 1988. Food safety aspects of gene transfer in plants and animals: pigs, potatoes, and pharmaceuticals, *Food Drug Cosmetic Law J., 43*, 351.
Klein, T. M., Gradziel, T., Fromm, M. E., and Sanford, J. C. 1988. Factors influencing gene delivery into *Zea mays* cells by high-velocity microprojectiles, *Bio/Technology, 6*, 559.
Korwek, E. L. 1988. *1988 Biotechnology Regulations Handbook*, Center for Energy and Environmental Management, Fairfax Station, Virginia.
Li, A. P. 1989. Protein toxicology applied to biotechnology products, in *Genetically-Engineered Plants: Scientific Issues in Their Regulation for Animal Feed and Human Food Uses*, Workshop at Boyce Thompson Institute for Plant Research and Cornell University, Ithaca, New York, May 1–3, 1989.
Martin, B., Nienhuis, J., King, G., and Schaefer, A. 1989. Restriction fragment length polymorphisms associated with water use efficiency in tomato, *Science, 243*, 1725.
Marx, J. L. 1988. Gene transfer is coming on target, *Science, 242*, 191.
Maryanski, J. H. 1986. Speech: on Legal aspects of the use of food products from biotechnology. Delivered at the University of Laval, Quebec, Canada, Aug. 21, 1986.
Maryanski, J. H. 1988. Genetically modified agricultural crops: an FDA perspective, in *Genetic Improvements of Agriculturally Important Crops: Progress and Issues*, Fraley, R. T., Frey, N. M., and Schell, J. (eds.), Cold Spring Harbor Laboratory, Cold Spring Harbor, New York, p. 91.
McCabe, D. E., Swain, W. F., Martinell, B. J., and Christou, P. 1988. Stable transformation of soybean (*Glycine max*) by particle acceleration, *Bio/Technology, 6*, 923.
McGranahan, G. H., Leslie, C. A., Uratsu, S. L., Martin, L. A., and Dandekar, A. M. 1988. *Agrobacterium*-mediated transformation of walnut somatic embryos and regeneration of transgenic plants, *Bio/Technology, 6*, 800.
National Academy of Sciences (NAS). 1972. *Genetic Vulnerability of Major Crops*, National Academy Press, Washington, D.C.
Nelson, R. S., McCormick, S. M., Delannay, X., Dube, P., Layton, J., Anderson, E. J., Kaniewska, M., Proksch, R. K., Horsch, R. B., Rogers, S. J., Fraley, R. T., and Beachy, R. N. 1988. Virus tolerance, plant growth, and field performance of transgenic tomato plants expressing coat protein from tobacco mosaic virus, *Bio/Technology, 6*, 403.
Office of Science and Technology Policy (OSTP). 1986. Coordinated framework for regulation of biotechnology, *Fed. Reg., 51*, 23302.
Ratner, M. 1989. "Crop biotech '89: research efforts are market driven, *Bio/Technology, 7*, 337.
Rhodes, C. A., Pierce, D. A., Mettler, I. J., Mascaranhas, D., and Detmer, J. J. 1988. Genetically transformed maize plants from protoplasts, *Science, 240*, 204.
Schell J. 1988. Analysis of gene expression in transgenic plants, in *8th International*

Biotechnology Symposium, Proceedings, vol. II, Durand, G., Bobichon, L., Florent, J. (eds.), Société Française de Microbiologie, Paris.

Sharp, W. R., Evans, D. A., and Ammirato, P. V. 1984. Plant genetic engineering: designing crops to meet food industry specifications, *Food Technol., 38*, 112.

Spiher, A. T. 1975. The growing of GRAS, *HortScience, 10*, 3.

Tanksley, S. D., Young, N. D., Paterson, A. H., and Bonierbale, M. W. 1989. RFLP mapping in plant breeding: new tools for an old science, *Bio/Technology, 7*, 257.

Toriyama, K., Arimoto, Y., Uchimiya, H., and Hinata, K. 1988. Transgenic rice plants after direct gene transfer into protoplasts, *Bio/Technology, 6*, 1072.

United States v. Anderson Seafoods, Inc. 1978. *Fed. Suppl., 447*, 1151.

United States v. Anderson Seafoods, Inc. 1980. *Fed. Reporter, 2d Ser., 622*, 157.

United States v. An Article of Food. 1975. *Fed. Suppl. 395*, 1184.

United States v. Articles of Food and Drug. 1978. *Fed. Suppl., 444*, 266.

U.S. Code. 1982*a*. Title 21, *Food and Drugs*, Sec. 321(f).

U.S. Code. 1982*b*. Title 21, *Food and Drugs*, Sec. 346.

U.S. Code. 1982*c*. Title 21, *Food and Drugs*, Sec. 321(s).

U.S. Department of Agriculture (USDA). 1988. Science and Education 1988. *Plant Genome Research Conference Report*, December 1988.

United States v. 45/194 kg. Vegetable Oil 1989. c.v. 89-0073-MRP, C.D. California, slip op.

U.S. Environmental Protection Agency (USEPA). 1989. Biotechnology Science Advisory Committee; Subcommittee on Use of Antibiotic Resistance Genes, *Fed. Reg. 54*, 269.

Weinberger v. Bentex Pharmaceuticals. 1973. *U.S. Reports 412*, 645.

Young, F. E. 1989. Food biotechnology and the FDA, Remarks to the International Food Biotechnology Council, Washington, D.C., Feb. 23, 1989.

Zitnak, A., and Johnston, G. R. 1970. Glycoalkaloid content of B5141-6 potatoes, *Am. Potato J. 47*, 256.

Chapter

5

Release and Transport of Entomopathogenic Microorganisms[1]

James R. Fuxa

Department of Entomology
Louisiana Agricultural Experiment Station
Louisiana State University Agricultural Center
Baton Rouge, Louisiana 70803

Introduction

Insects, like humans, are susceptible to diseases caused by a wide variety of pathogenic microorganisms including viruses, bacteria, fungi, and protozoa. Researchers have attempted to control insect populations with certain of these entomopathogens since the late 1800s (Kaya 1985). Their advantages as control agents include their environmental safety, persistent action in some cases, versatility in use and application, relative compatability with other control methods and farming practices, and the relatively low cost of development and registration (Falcon 1985, Khachatourians 1986, Fuxa 1989). Four general approaches have been taken in such attempts: introduction and establishment of a nonindigenous microorganism; inoculative augmentation, or a "booster shot," of an indigenous pathogen; environmental manipulation to conserve or enhance the pathogen; and inundative augmentation, or application like a chemical insecticide, for short-term control (Fuxa 1987*a*).

In many ways the research of microbial control of insects has been successful, but the degree to which entomopathogens have actually been used in pest management has been disappointing. For example, 11 entomopathogens have been registered with the U.S. Environmental Protection Agency (USEPA) for

[1]Approved for publication by the Director of the Louisiana Agricultural Experiment Station as manuscript no. 89-17-3524.

insect control (four viruses, five strains or species of bacteria, one fungus, and one protozoan) (Betz et al. 1987), and many others have been demonstrated to be efficacious. However, only two or three of the eleven have been commercially successful. Most of these registered agents were intended for inundative augmentation; there is some indication that entomopathogens have been more successful by introduction and establishment or by inoculative augmentation (Fuxa 1990*a*). The major problems in acceptance of microbial control have related to cost competitiveness and other economic factors, slow activity, harmful environmental effects on the pathogen (particularly sunlight and desiccation), and perhaps inadequate delivery systems (Falcon 1985, Fuxa 1990*a*). In the early 1980s, many microbial control specialists believed that microbial control by inundative augmentation, in spite of its limited success, was already a "mature industry."

The genetic engineering revolution has changed that attitude, and organisms based on entomopathogen recombinants are among the first genetically engineered microorganisms (GEMs) released into the environment in small-plot experiments. Genetic engineering of entomopathogens includes efforts or proposals to increase (or, in one case, decrease) host range, increase virulence, increase pathogen resistance to physical or chemical stresses, overcome insect defenses, increase rapidity of mortality, increase vertical transmission rates, and increase pathogen or toxin production (Miller et al. 1983, Faulkner and Boucias 1985, Kirschbaum 1985). In this chapter, "biotechnology" will refer to "any technique that uses living organisms (or parts of organisms) to make or modify products, to improve plants or animals, or to develop microorganisms for specific uses" (OTA 1989). "Conventional biotechnology" will refer to such techniques exclusive of the "novel" biological methods such as recombinant DNA.

Since entomopathogens are likely to be among the first GEMs to be released, their environmental risk assessment is a topic of current interest. Fuxa (1990*b*) reviewed the environmental risks associated with release of genetically engineered entomopathogens. "Risk" is an estimate of the probability and severity of harm (Rissler 1983), or a prediction of the likelihood of something going wrong in a certain set of circumstances (King 1985). "Risk assessment" is the process of obtaining quantitative measures of risk levels (Fiksel and Covello 1986). One of the essential stages in risk assessment is exposure assessment, or the "measurement or estimation of the intensity, frequency, and duration of human or environmental exposures to the risk agents that are produced by a source of risk" (Fiksel and Covello 1986). One of the elements involved in exposure assessment is the environmental fate of the released organism and its engineered gene (altered genome), including transport outside the target or release site (Fiksel and Covello 1986). Much of the scientific and public concern about the release of GEMs is due to uncertainty about their subsequent movement and dispersal (Jones 1988). There is no single satisfactory method for predicting fate and transport of a GEM until experience is gained in this area; one of the few available methods is a thorough understanding of the population dynamics and transport of parental and related microorganisms in the environment (Tiedje et al. 1989, Fuxa 1989). Thus, knowledge of the transport and fate of released natural strains of entomopathogens is important for evaluation of releases of GEMs. The remainder of this chapter will consist of a review of transport and fate of entomopathogenic viruses, bacteria, and fungi, and consideration of the effect of recombinant DNA techniques on transport and exposure assessment of these agents.

Fate of Released Entomopathogens

Entomopathogens have been released into the environment for experimental or commercial insect control since 1879. Thus there is a substantial body of literature concerning persistence, spread, and population growth or decline of the pathogen populations after these releases, even though such data were recorded in only a small percentage of cases. Basically, the information available for entomopathogens is similar to that for microorganisms in general (Alexander 1988): results of previous releases have ranged from no persistence or dispersal to population growth and spread over wide areas. The previous data are not sufficient in quality or quantity to allow predictions about untested species.

Persistence, spread, and population growth after releases of entomopathogens have been reviewed recently (Fuxa 1987*b*, Fuxa 1989) and will be briefly summarized here. The literature is biased toward reports of successful releases; thus, the summaries of persistence, dispersal, and population growth should be regarded as representative of the capabilities. While the data can be used to demonstrate efficacy of particular entomopathogens, this bias in the data suggests caution in attempts to generalize to predict adverse effects.

Viruses

The persistence of viruses (almost entirely members of Baculoviridae) released into the environment for insect control has varied greatly from species to species, and even within species. The long-term persistence of these viruses is primarily in soil or litter and in the host population. Introduced entomopathogenic viruses have persisted up to 7 years in the host population and more than 5 years in soil, but only 1 to 32 days on foliage or 3 days in bird guts. It is worth noting that certain viruses, for example the *Oryctes* baculovirus and the *Anticarsia gemmatalis* nuclear polyhedrosis virus (NPV), persisted in some introductions but did not become established in others.

Distance of dispersal after introduction also varies widely for viruses. Dispersal after release in a row crop ranged from 0 up to at least 240 m in 45 days; in forests and coconut plantations it ranged from 0 m/yr up to figures such as 700 m/yr, 290 ha/yr, and 3 km/mo. In certain cases, such as the release of the *Gilpinia hercyniae* NPV, the virus can spread throughout the local range of the host populations. Certain viruses, such as the *Lymantria dispar* NPV, spread in some cases but did not spread in others even though they successfully established in the host population.

There have been relatively few estimates of viral population growth or decline after a release. Though there are exceptions, viral population density generally has declined after releases in row crops but increased after releases in forest pests, perhaps due to greater stability of the forest ecosystem. Most releases have resulted in viral population growth or decline within approximately one order of magnitude of the numbers released. In one case the population of *Pseudoplusia includens* NPV had increased more than 800-fold 1 year after release into soybean, but in another it declined after two seasons to 3.6% of the number released. In one interesting example (Thompson and Scott 1979), up to 1.6×10^{15} NPV polyhedral inclusion bodies (PIBs) per acre resulted from a natural viral epizootic in *Orgyia pseudotsugata* in a forest ecosystem, whereas viral population density increased to only 6.9 to 7.4×10^{14} PIB per acre after the virus was artificially released.

Bacteria

Persistence of entomopathogenic bacteria has been studied in *Bacillus thuringiensis* (BT), *B. popilliae*, and *B. sphaericus*. Long-term persistence of BT and *B. popilliae* is in the soil, where spores remain viable for up to 2 years without recycling (i.e., causing disease and reproducing another generation of bacteria). *B. popilliae* has persisted up to 45 years with recycling in the host population. BT spores and δ-endotoxin lose their activity in 20 to 30 days on foliage in certain forest ecosystems, but they may lose activity within hours on row crops. Persistence of *B. sphaericus* in water is variable. The amount of sunlight, amount of water pollution, dosage, and presence or absence of a host population can influence whether this bacterium persists less than 3 hours or more than 60 days. The spores can remain viable more than 9 months in soil.

Entomopathogenic bacteria released into the environment have exhibited more limited dispersal than viruses. *B. thuringiensis* var. *israelensis* (BTI) has been carried by water currents over 7 ha or up to 1250 m. *B. popilliae* has been demonstrated to spread from a release site, but the actual distance was not recorded.

There has been only one estimate of population density of a bacterium after its release. *B. popilliae* can reach levels of 100 billion spores per kilogram of soil in the upper inch. *B. popilliae* is an obligate pathogen to scarabeid beetles; it cannot replicate in soil or water. Laboratory experiments have indicated that, under ideal conditions, BT can grow in soil or increase in insect cadavers by 15- to 100-fold. However, in the environment, it does not grow saprophytically except in very limited niches. BTI can produce 1.4×10^5 spores per mosquito cadaver in the laboratory, but such growth is probably rare in the field. *B. sphaericus*, however, can grow saprophytically in highly polluted water, and it can produce 10^5 to 10^6 spores per mosquito cadaver in the field.

Fungi

The pattern of persistence after release of entomopathogenic fungi is similar to that for the viruses. Different species of fungi are known to persist for at least 1 year in soil and for months or, in one case, up to 22 years in host cadavers or populations. However, they generally persist only a matter of days on plant foliage.

Dispersal of fungi by airborne conidia is probably one of their greatest advantages for spreading in and controlling a population of insects, but it is one of the least studied or understood phenomena in epizootiology of insect diseases. In one case, a fungal spray drifted at least 57 m due to wind currents. Infected hosts have been known to carry fungi up to 400 m after release. Laboratory research has indicated that one of the most commonly released fungi, *Beauveria bassiana*, grows colonies which can spread through as much as 327 cm^3 of soil.

There have been no direct estimates of pathogen population density resulting from field releases of fungi, though in certain cases estimates can be made from data in the literature. Certain fungi can grow saprophytically after release. It has been estimated in laboratory experiments that *B. bassiana* increases 2- to 10-fold in soil, and *Hirsutella thompsonii* has been formulated to grow saprophytically after application to foliage. In certain cases, fungal population density can be estimated from data for disease prevalence and host population density after release. For example, *Nomuraea rileyi*, which is typically released at a rate of approximately 10^{13} conidia per hectare, has been estimated to produce 10^{13} to 10^{16} conidia per 0.4 hectare.

Methods of Dispersal or Transport

Methods of transport of "parental" entomopathogens is one of the most important facets of initial risk assessment efforts for GEMs derived from them. Not only is transport important to exposure assessment, but the initial environmental releases of GEMs are also almost certain to be small-plot field tests. If a problem develops in such tests, the released organism could probably be destroyed, provided it has not dispersed very far from the test site. Thus, it is important to have a thorough understanding of transport of parentals, particularly in the early stages of developing risk assessment schemes. There is a large body of information about transport of entomopathogens, though there also are critical gaps in our understanding of these phenomena.

Viruses

The viruses have probably been the subject of more research of transport mechanisms than any other group of entomopathogens. Their short-range transport, upon which epizootics are probably most dependent, is largely by abiotic agents. Longer-range movement, which is important to risk assessment and to the introduction-and-establishment approach to insect control, is largely by biotic agents.

Abiotic agents Viral transport by abiotic agents has not been well documented, though it likely is critically important to the development of epizootics. The major reservoir of the baculoviruses, the viruses most often researched in ecology and microbial control, and certain other viruses is in the soil; however, epizootics generally result from lepidopterous or hymenopterous insects feeding on foliage above the soil surface. Thus, movement from the soil to the foliage is critical to initiation of epizootics, and such movement is thought to be mediated largely by abiotic agents. Rainfall, air currents, and gravity are the abiotic agents that have been implicated in the movement of entomopathogenic viruses.

Viral polyhedral inclusion bodies (PIBs) are proteinaceous crystals that occlude the virions in certain groups of insect viruses. It is thought that rainfall and gravity carry PIBs down from foliage in the upper parts of plants where the insect dies and release virus to lower plant parts or to the soil. While most experts believe this to be true, there is little experimental evidence. Bird (1961) concluded that rainfall carried NPV of a sawfly, *Neodiprion sertifer*, from the top to lower branches of pine trees, but this was based on circumstantial evidence. However, Young (1990) used sprinkler irrigation to experimentally demonstrate that water flow carried NPVs of two lepidoptera in soybeans from the upper canopies of the plants into the lower canopy. Once the virus reaches the soil, it apparently does not leach very far below the surface. Relatively little NPV of *Trichoplusia ni* was detected below depths of 7.5 cm after application of the virus to the soil surface, even though the soil was monitored for more than 4 years (Jaques 1969). In a similar study, it was also concluded that most granulosis virus (GV) of *Pieris brassicae* stays in the top layers of soil. Some percolates into lower layers of soil and sand after application of large amounts of water (David and Gardiner 1967).

It also has been hypothesized that rainfall splashes viral PIBs from the soil reservoir back up to the foliage to infect insects and initiate epizootics. In pastures, observational research with multiple-regression analysis indicated that the amount of rainfall was one of the most significant factors correlated with prevalence of nuclear polyhedrosis in the fall armyworm, *Spodoptera frugiperda*, presumably due to such transport (Fuxa and Geaghan 1983). This conclusion is supported indirectly by another study of the same NPV and host insect in which there was no such correlation in corn plants, presumably because the latter are much taller than pasture grasses and less likely to be contaminated by splashing transport (Mitchell and Fuxa 1990). In another observational study, the percentage of foliage contaminated with Douglas fir tussock moth NPV increased from 12% before a rain to 100% afterwards (Thompson 1978). Rainfall has similarly been implicated in viral transport in other observational studies (Hostetter and Bell 1985). It is likely that entomopathogenic viruses can be transported by natural waterways such as streams and rivers, but there has been no research of such movement (Hostetter and Bell 1985).

Though entomopathogenic viruses can undoubtedly be transported by air currents, the evidence to date is largely circumstantial. The distribution of fall webworm (*Hyphantria cunea*) NPV in soil was correlated with the distribution of windblown tree leaves, suggesting either that insects killed by virus were transported by wind in the same pattern as leaves or that the virus was transported on the leaves (Hukuhara 1973). A more recent study supported the hypothesis that NPV can be transported by air currents on dust particles

(Olofsson 1988). Patterns of nuclear polyhedrosis in populations of the sawfly *Neodiprion sertifer* were correlated with likely patterns of dust deposition in trees near logging operations or forest roads. Furthermore, road surface materials and air sampled 1 m from the road and 1 m above its surface both contained viable NPV. In experiments with alfalfa grown in flats and a fan creating a 2 to 4 mph wind, viral polyhedra from the alfalfa caterpillar, *Colias philodice eurytheme*, were transported on air currents as far as 2 ft; the virus was transported up to 6 ft after it was mixed with soil and allowed to dry (Thompson and Steinhaus 1950).

Air movement has the potential for a much greater indirect than direct effect on movement of insect viruses. Host insects are an important long-range carrier of these viruses (see next section), and host insects, in turn, can be carried long distances or be strongly influenced by air movements (Rabb and Kennedy 1979).

Biotic agents Evidence to date indicates that biotic agents are primarily responsible for the long-range transport of entomopathogenic viruses. Thus, biotic agents are likely to be of greatest concern in small-plot field releases of GEMs derived from these viruses. The biotic agents known to transport entomopathogenic viruses include host insects; predators, parasites, and saprozoites of the host insects; and grazing mammals.

Host larvae are known to transport viruses. However, since larvae are specialized for feeding rather than movement, they do not carry viruses very far. *Mamestra brassicae* larvae infected with NPV were released into small plots of cabbage; they transported the virus at least 2.5 m, though the authors of the study believed that this was an underestimate of the maximum range of transport (Evans and Allaway 1983). In a similar experiment, NPV-infected larvae of the brown-tail moth, *Euproctis chrysorrhoea*, were released onto bramble bushes. In the spring, larvae apparently carried the virus "many tens of meters" from the release site, and, in the autumn, the virus may have been transported at least 8 m, though most of the virus was detected within 2 m of the release site (Sterling et al. 1988).

Host adults potentially transport virus greater distances than larvae because the adults usually are specialized for movement and mating rather than feeding. Circumstantial evidence indicated that adults of the forest tent caterpillar, *Malacosoma disstria*, carried NPV up to 20 mi from a release site and transmitted it to their progeny (Stairs 1965). Adults of the rhinoceros beetle, *Oryctes rhinoceros*, have consistently carried an introduced baculovirus at least 5 to 20 km from release sites, at rates of 1.0 to 3.9 km/mo (Young 1974, Zelazny 1976, Gorick 1980, Bedford 1986). In some cases, researchers have artificially contaminated external body parts of moths and determined distances that NPV was carried from a release site. In one such study in cotton fields, corn earworm (*Heliothis zea*) adults carried NPV up to 240 m, primarily downwind, and adults of *T. ni* and *Spodoptera exigua* carried their respective NPVs at least 130 m (Gard 1975). In another, females of the fall webworm, *Hyphantria cunea*, transported NPV up to 188 m in a mulberry field (Suzuki and Kunimi 1981). There is a distinct possibility that host insects could carry latent viral infections, including NPV DNA sequences integrated into host genomes (Entwistle 1986); if so, this could be a major factor in transport of virus by host insects, though this has not yet been demonstrated.

Avian and mammalian predators of host insects apparently transport viruses; at least some of the virus from the ingested, infected insects survives passage through the guts of predators. Sixteen species of birds, which represented 40% of the species and 80% of the total birds in a spruce and pine forest, dispersed infective NPV of European spruce sawfly, *G. hercyniae*, in their feces (Entwistle et al. 1977*a*). At one time, in a continuation of this study, 90% of the bird droppings collected from trees contained infectious NPV (Entwistle et al. 1977*b*). Birds generally passed NPV for approximately 2.5 hours after the virus was ingested (Entwistle et al. 1978). Birds probably transported *G. hercyniae* NPV 375 to 750 m within about 1 hour (Entwistle et al. 1983), or up to 7 km/day (Entwistle et al. 1978). In studies of gypsy moth (*Lymantria dispar*) NPV, birds and mammals were identified as possible transporters of virus. In the laboratory, the NPV survived passage through the guts of five mammalian species and three species of birds; birds, shrews, and squirrels passed PIBs in the feces up to 6 hours after ingestion, mice up to 18 hours, and raccoons and opossums up to 70 hours (Lautenschlager and Podgwaite 1979). Among animals captured in a forest, blue jays, towhees, mice, voles, raccoons, and a chipmunk had viable NPV in their guts; robins and shrews did not (Lautenschlager et al. 1980).

Predatory arthropods are also likely transport agents for NPVs. In several laboratory experiments, various species of lepidopterous larvae infected with NPV have been fed to such predators, and then the feces were demonstrated to carry infectious virus. Such experiments have included a predatory carabid beetle, *Calosoma sycophanta* (Capinera and Barbosa 1975); the hemipterans *Oechalia schellenbergii* (Cooper 1981), *Nabis tasmanicus* (Beekman 1980), *Podisus maculiventris* (Abbas and Boucias 1984), and *Nabis roseipennis* (Young and Yearian 1987); and a spider, *Oxyopes salticus* (Kring et al. 1988). In one field experiment, it was demonstrated that one such predator, *N. roseipennis*, transmitted NPV to uninfected hosts (*Anticarsia gemmatalis*) caged on soybean plants (Young and Yearian 1987). There has been no research of dispersal distance by predatory arthropods.

It also is likely that hymenopterous and dipterous parasites of sawflies and lepidoptera transport insect viruses. Several laboratory experiments have indicated that female parasites can mechanically transfer viruses on their ovipositors between infected and uninfected host insects (Thompson and Steinhaus 1950, Laigo and Tamashiro 1966, Irabagon and Brooks 1974, Beegle and Oatman 1975, Raimo et al. 1977, Levin et al. 1979, Hamm et al. 1985). In field observations, the NPV of a heavily parasitized population of the sawfly *Neodiprion lecontei* spread rapidly from tree to tree, but such spread was very slow in a relatively unparasitized population of another sawfly, *N. sertifer* (Bird 1961). An ichneumonid parasite that was artificially surface-contaminated by NPV and released into pine plantations did transmit the virus to uninfected host sawflies (*N. sertifer*), but the dispersal distance was not reported (Mohamed et al. 1981).

Saprozoites can transport entomopathogenic virus by feeding on virus-killed insects; they can become surface-contaminated or can disperse the virus through their feces. Birds might transport virus in this manner since infectious NPV has been detected in bird feces even during winter, when insect cadavers, but not live larvae, of *G. hercyniae* are present (Entwistle et al. 1977*b*). It is likely that the fly *Sarcophaga aldrichi* transports NPV on its proboscis and feet, since it has been observed in the field feeding on forest tent caterpillars (*Malacosoma disstria*) recently killed by the virus (Stairs 1965). Several species of sarcophagid flies captured while feeding on the liquefied remains of cabbage looper (*T. ni*) killed by NPV were contaminated with the virus, which was detected in the feces and probably occurred externally as well (Hostetter 1971).

Finally, grazing mammals can transport NPV, probably by carrying PIBs or contaminated dust on their legs, or perhaps by ingesting virus-contaminated foliage. Prevalence of nuclear polyhedrosis in populations of fall armyworm, *Spodoptera frugiperda*, was greater in pastures with cattle than in pastures without cattle, probably due to their transporting virus within or between fields or from soil to foliage (Fuxa and Geaghan 1983). In a field experiment, the NPV initially infected 53% of the *S. frugiperda* larvae when it was sprayed in a pasture with no cattle; 12 days after the virus was sprayed, there was 22% infection of the insects in the sprayed plot and 0% in insects 27 to 113 m away (J. R. Fuxa and B. H. Wilson, unpublished data). When cattle were present, the percentage figure for initial infection was similar (55%) in the sprayed plot; however, after 11 days, there was 61% infection in the plot, 20% at a distance of 27 to 43 m, 16% at 98 to 113 m, and 10% at 159 to 177 m. Thus, it is likely that the cattle increased the prevalence of disease in the insects and transported the NPV a distance of 159 to 177 m.

Conventional biotechnology Conventional biotechnology can contribute to the dispersal of entomopathogenic viruses being developed for insect control. In some cases, dispersal has been inadvertent outside the target area; in others, virus was released with the intention of allowing it to spread.

Almost all attempts at suppressing insect populations with viruses have involved efforts to increase the pathogen population, usually by isolating virus, growing it in mass culture, and then releasing it at a site where it does not naturally occur in sufficient numbers to control insects. Mass production (Sherman 1985, Shapiro 1986), cell culture (Sherman 1985, Weiss and Vaughn 1986), and formulation and application technology (Young and Yearian 1986) have been reviewed recently. Obviously, human activity can be a major factor in dispersal of these viruses.

Viral formulations for field application can affect their spread after release. The viruses have been formulated for increased persistence by microencapsulation and by mixing with UV-light-protectant adjuvants (Young and Yearian 1986). Presumably, anything that increases persistence will also increase the opportunity for the virus to be transported in the environment. Viruses usually have been applied as aqueous sprays and occasionally as dust formulations (Young and Yearian 1986). Dusts have the potential to be transported by the wind (see "Abiotic Agents," page 87), and a viral spray can drift up to 180 m downwind (Smith et al. 1984).

This group of entomopathogens has been the subject of some particularly novel methods of spreading disease within an insect population. Autodissemination has been attempted several times, whereby infected or contaminated host insects themselves are used to carry the virus into the environment (Young 1974, Gard 1975, Ignoffo et al. 1978, Suzuki and Kunimi 1981). Similarly, a hymenopterous parasite contaminated with NPV was able to transport the virus into a pine plantation to a degree sufficient to infect sawflies (*Neodiprion sertifer*) (Mohamed et al. 1981). A trench mortar has even been used to blow NPV-contaminated forest litter up into the trees in order to contaminate foliage and initiate a viral epizootic in larvae of the nun moth, *Lymantria dispar* (Podgwaite 1985).

Bacteria

The knowledge of transport mechanisms for entomopathogenic bacteria is very sketchy, perhaps because relatively few species have been researched for microbial control, and the best known of them, BT, has poor persistence and spreading capability. For example, one experiment indicated that there had been little movement of BT 1 year after its application to terrestrial test plots (DeLucca et al. 1981). This bacterium apparently has no satisfactory method for dispersal in the natural environment, which is one explanation for the rarity of its natural epizootics in insect populations (Dulmage and Aizawa 1982).

Abiotic agents The few experiments to date would seem to indicate that, if care is taken not to release bacteria into wind or water currents, they are not likely to disperse very far.

Wind and rainfall are known to disperse bacteria (e.g., Franc et al. 1985), but there is very little such information about entomopathogens. Ground sprays sometimes do not drift (Dulmage and Aizawa 1982), but spray drift can transport BT spores in certain circumstances. In one release at high altitude, wind currents carried BT spores 15 km from the point of release, and spores were detected in air for a few months after application (Grison et al. 1976). The only other research of dispersal by abiotic agents has involved transport of BTI by water currents. BTI has been released into flooded rice fields for control of mosquito larvae. When rice plants were small and did not restrict wind currents at the water's surface, the currents distributed BTI over 4.9 to 7.3 ha (McLaughlin and Vidrine 1985). In similar experiments, topography and levee configuration influenced dispersal of BTI by point-source introduction in rice; water movement in this study distributed BTI over at least 25 acres (Finch et al. 1986). BTI has also been released in streams for control of larval blackflies; water current carried viable spores at least 1250 m downstream within 15 min (Gibbs et al. 1986).

Biotic agents As with the viruses, biotic agents are the ones that have the most potential for long-range transport of bacteria and hence for promoting their escape from experimental plots. However, there is relatively little information compared with that for viruses.

The host insects apparently can transport bacteria, but actual field data are rare. *Pseudomonas aeruginosa, Serratia marcescens,* and *Bacillus cereus,* among others, have been found in field-collected grasshoppers to the extent that they probably are transported by these insects (Bucher and Stephens 1957, Bucher 1959). Based on experimental results and field observations, adult Japanese beetles, *Popillia japonica,* are likely a major factor in the dissemination of Type A milky disease caused by *Bacillus popilliae,* even though only 0.26% of adults were infected (Langford et al. 1942). Adults, as well as other stages of storage moths, *Plodia interpunctella,* became surface-contaminated with BT, apparently from their contaminated food source; but the importance of this phenomenon to dispersal was unknown (Burges and Hurst 1977).

Other biotic agents can also act as carriers. Parasites, ants, birds, moles, skunks, and mice apparently can disperse the milky disease bacteria (White and Dutky 1940, White 1943). Spores of milky disease bacteria have been detected in bird droppings and have survived passage in the guts of chickens and starlings. Ants carry the diseased grubs at least 10 ft, and skunks, moles, and mice have been implicated in spread of the bacteria since they feed on the host larvae (White and Dutky 1940). BT spores can remain viable after passage through the guts of three birds and one mammal (Smirnoff and MacLeod 1961), but such survival probably is unimportant in nature since dispersal of this organism is so minimal. Similarly, it has been demonstrated in the laboratory that a parasite of the Mediterranean flour moth, *Anagasta kuhniella,* can vector BT (Flanders and Hall 1965), but this probably is unimportant in natural movement of the bacterium. Undoubtedly, humans have been the most effective biotic agent for transport of bacteria. Certain varieties of BT that are common in stored-product insect pests (Burges and Hurst 1977, Krieg 1987) or that have been used extensively in microbial control (Krieg 1987) have been distributed worldwide.

Conventional biotechnology As with the viruses, conventional biotechnology for bacteria is aimed primarily at increasing the pathogen population density in the host insect's habitat, which necessitates growth of large numbers of bacteria. Mass production, formulation, and application technology have been reviewed previously for the three major species that have been subjects of microbial control research: BT, *Bacillus popilliae,* and *B. sphaericus* (Angus and Lüthy 1971, Klein 1981, Singer 1981, Lüthy et al. 1982, Morris 1982). BT has been formulated to be used in aqueous sprays, dusts, and granules (Lüthy et al. 1982), but it is not known whether any of these is more likely than the others to be transported by wind currents. As with viruses, formulations sometimes include adjuvants designed to increase persistence (Morris 1982), which normally might increase the opportunity for movement.

However, there is no evidence that they have done so. A wide variety of ground and aerial spray equipment has been used for BT releases; it is likely that the higher the altitude of the spray equipment above the target (e.g., host-plant foliage), the greater the likelihood and distance of spray drift on wind currents (Grison et al. 1976, Dulmage and Aizawa 1982). The only other application technique that has seemed to affect transport of entomopathogenic bacteria is the intentional use of water currents in rice fields and streams, which was discussed above.

Fungi

Movement of fungi is one of the major unknowns concerning releases of entomopathogens. It is possible that conidia or other spores that are released, or in some cases forcibly ejected, into the air can be transported great distances on air currents; yet there have been no studies of such dispersal after release. In addition to air currents, fungi can be transported by a variety of abiotic and biotic agents. The fungi are another group that has the potential for widespread dissemination after a release.

Autodispersal Of the entomopathogenic viruses, bacteria, and fungi, only the latter are known to have candidates for microbial control that can transport themselves in the environment to any significant degree. Members of the aquatic genus *Coelomomyces*, in the class Chytridiomycetes and order Blastocladiales, have actively swimming zygotes that infect mosquito larvae and free-swimming meiospores that infect copepods, the alternate hosts (Whisler 1985). Similarly, the genera *Lagenidium* (class Oomycetes, order Lagenidiales) and *Leptolegnia* (class Oomycetes, order Saprolegniales) have free-swimming zoospores that infect mosquito hosts (Domnas 1981, Sweeney 1981). In terrestrial habitats, most members of the Entomophthorales (class Zygomycetes) forcibly eject primary, secondary, or even tertiary conidia (King and Humber 1981), which aids their transport for short distances away from the dead host insect.

Abiotic agents In contrast to the viruses and bacteria, abiotic agents, particularly air currents, are quite possibly long-distance carriers of certain fungal stages according to what little information is available. This is a subject that clearly requires further research and that will have to be considered when microorganisms based on these fungi are genetically engineered for small-plot releases into the environment.

In spite of its likely importance, there is little information about wind dispersal of entomopathogenic fungi. Conidia of *Nomuraea rileyi*, a pathogen of lepidopterous pests, were aerially sampled in a soybean field; the density of conidia in the air peaked at 70,000 conidia sampled in 1 h (Kish and Allen 1978). Based on the settling rate of the conidia in air, the authors estimated that conidia produced on a dead insect on a soybean plant would take 1 min to fall to the ground, during which time they would be displaced laterally by any air currents. Wilding (1975) aerially sampled conidia of members of the

Entomophthorales. The concentration of conidia 45 cm above the ground was correlated with disease prevalence in the host insect. In another study, five species of Entomophthorales conidia were aerially sampled, with peak numbers reaching almost 2000 conidia per cubic meter in 1 hour of sampling (Wilding 1970). Neither study included any estimate of dispersal distances. Studies of plant-pathogenic fungi have indicated that the vast majority of conidia are deposited within a few centimeters of their source, but also that the numbers produced are so large that some escape into higher airflow layers and are transported great distances: 300 to 700 mi and, quite possibly, further (Jeger 1985).

There has not been much more research on movement of fungi by water currents than there has been on their movement by air. Flooding dispersed *Lagenidium giganteum*, which infects mosquito larvae, throughout an 800 m^2 site from a single release point (Jaronski and Axtell 1983). The only other research has been on terrestrial species in soil. Water percolating through soil probably carried spores of *Metarhizium anisopliae* from the top 10 cm of soil to a depth of 20 to 30 cm (Latch and Falloon 1976). Similarly, water carried conidia of *Beauveria bassiana* at least to a depth of 31 cm, but more than 90% of the viable conidia remained in the top 15 cm of soil (Storey and Gardner 1987). Presumably, fungal conidia of terrestrial species can be dispersed from the soil reservoir by the splashing of rainfall just as with the viruses, but there has been no research on this.

Biotic agents As usual, biotic agents provide opportunity for long-range transport of entomopathogenic fungi.

Host insects are important carriers of the fungi in two ways. First, in certain fungus-insect systems, the insect tends to die attached to the top of the host plant, aiding the dissemination of spores by air currents. Examples include *Empusa aphidis* in the pea aphid *Macrosiphum pisi* (Rockwood 1950) and *Entomophaga grylli* in the grasshopper *Camnula pellucida* (Carruthers and Soper 1987). In the pea aphid example, whereas the apterous, dying aphids crawl to the tops of host plants, the mature, alate aphids tend to migrate in early stages of the disease, thus transporting the fungus (Rockwood 1950). Second, the host insects themselves can transport fungi horizontally. The best-known example is the fungus *Massospora cicadina*, a pathogen of periodical cicadas. Infected adult cicadas live long enough to slough off abdominal segments containing masses of conidia or resting spores while they fly; the adults have been known to transport the fungus at least 132 m (Lloyd et al. 1982). Circumstantial evidence has indicated that *Metarhizium anisopliae* was transported up to 300 to 400 m by infected adult beetles, *Oryctes rhinoceros* (Latch and Falloon 1976). Larvae of *T. ni* infected with *Nomuraea rileyi* and released onto soybean plants transported the fungus throughout the plant on which they were released (Ignoffo et al. 1977). Subterranean termites infected with *M. anisopliae* have been demonstrated to carry the disease and infect other termites within the colony (Kramm et al. 1982).

Other biotic agents also are known to be able to transport entomopathogenic fungi. *M. anisopliae* was transported several centimeters by Acari, Collembola, and small dipterous and coleopterous larvae when these soil organisms were forced to migrate through soil contaminated with the fungus (Zimmerman and Bode 1983). Phoretic mites transported *M. anisopliae* from dead pales weevils, *Hylobius pales*, to live weevils a few centimeters away in small-cage tests (Schabel 1982). Schabel cited several other studies in which such mites apparently were involved in transport of entomopathogenic fungi. A crab, *Sesarma gardineri*, was identified as a probable agent of spread for the aquatic fungus *Coelomomyces stegomyiae*; the crab was observed darting from one small source of water to another in an area where the fungus had been introduced and had spread (Laird 1985).

Conventional biotechnology Conventional biotechnology for the fungi again has been aimed largely at production of large amounts of fungus and delivery of the inoculum such that field populations of pest insects develop high prevalence of disease. Production and application have been reviewed previously for *Beauveria bassiana, Metarhizium anisopliae, Verticillium lecanii, Hirsutella thompsonii, Nomuraea rileyi, Culicinomyces, Lagenidium, Coelomomyces*, and various Entomophthorales (Burges 1981, Couch and Bland 1985). Other than this physical transport by humans of fungal inoculum, the literature concerning the effect of biotechnology on transport of fungi is scanty. Spray drift carried hyphal bodies of the fungus *Paecilomyces farinosus* distances as great as 57 to 76 m from a site of application (Agudelo and Falcon 1983).

Effects of Genetic Engineering on Transport

At this time, one can only speculate about the potential effects of genetic engineering on transport of microorganisms. Also, such speculation will be more appropriate for specific releases rather than for generalizations about the effects of manipulations on groups of microorganisms. Releases of GEMs have been minimal to date, and questions about transport are almost impossible to answer definitively in laboratory or microcosm experiments, because of the wide range of environmental variables that can affect transport. Speculation presently must be based on knowledge of the parental organisms and their transport through the environment and on the nature of the changes in the end-product organism as well as the environment into which it will be released.

Transport of genetically engineered entomopathogenic viruses

The genetic manipulations of entomopathogenic viruses that are being researched or planned for the near future are aimed at increasing virulence, speed of kill, host range, environmental stability, or vertical transmission. Conventional selection techniques have been used in attempts to increase viral virulence (Veber 1964), resistance to UV radiation (Brassel and Benz

1979), and rate of vertical transmission (J. R. Fuxa, A. R. Richter, and E. Weidner, unpublished data). Genetic recombination among baculoviruses is being researched in order to increase the virulence and host range of these usually very host-specific pathogens (Maruniak 1988). Recombinant DNA techniques are being researched to increase virulence and speed of kill by introducing genes for insect-specific peptide toxins or peptide hormones (Maeda 1988). Recombinant DNA research has been proposed to increase the effective viral host range by insertion of insect-specific neurotoxin or behavior-modifying genes, without increasing the host range in which the virus would replicate (Miller 1987). Improvement of environmental stability is likely to be a goal of recombinant DNA research at some time in the future (Summers and Smith 1985).

Increases in virulence or speed of kill would not be expected to increase transport capability of these viruses. Rather, increasing virulence or speed of kill should decrease transport because the infected insect would not travel as far as normal. It should die more quickly and at a smaller size than an insect infected with wild virus. Additionally, viral replication in the environment would be reduced as a result of the smaller size of insect cadavers, and the smaller than normal viral population should have less opportunity than a larger population to persist and be transported. Furthermore, though this speculation is the most dangerous of all, the added genes probably would not persist at any high level either in the engineered virus population or in wild baculoviruses if recombination occurs, because modeling studies have indicated that moderate virulence is best for long-term insect control and host-pathogen relationships (Anderson 1982). Even though significant changes occasionally occur in viruses in nature, generally they are so closely attuned to their ecological niches that extreme alterations are lethal, and their long-term survival depends on a critical level of virulence (Kilbourne 1985).

Increases in replicative host range, environmental stability, or rate of vertical transmission are likely to increase the distance that viruses are transported after release. All three changes will increase persistence in the environment. Persistence would be greater in the host populations if host range or transmission were increased; and the viruses would persist longer in the abiotic environment if stability were improved. Increases in persistence will simply increase the opportunity for the virus to be transported by abiotic or biotic agents, or to infect a new host and replicate, thereby increasing the viral population and opportunity for transport. Persistence influences the frequency of transmission and, therefore, transport (Smith 1971); modeling studies have indicated that pathogens with good persistence in their free-living stage can become endemic in a host population even if they induce immunity or are very virulent (Anderson and May 1979). Such studies also have indicated that vertical transmission increases the chances for a successful introduction and establishment of a pathogen (Anderson 1982, Anderson and May 1979, 1980). In addition to increasing persistence, increasing host range or the rate of vertical transmission is likely to increase transport in a more direct manner. Presumably, the more host species a virus has, the more insects will be infected. Similarly, increasing the rate of vertical transmission (parent-to-offspring passage) means that a greater proportion of adults will be carrying the virus. We have already seen that transport by the host insect is one of the major methods by which entomopathogenic viruses are transported. Thus, increasing the numbers of infected hosts is likely to increase viral transport unless the virus is no longer capable of replication.

Finally, researchers currently are releasing "disabled" NPVs, viruses with a polyhedrin deletion mutation. These viruses cannot produce the polyhedrin protein that forms a protective crystal in which virions become embedded. The purpose of this deletion is to decrease environmental stability and thereby decrease regulatory problems by creating viruses that would not persist after release. One factor that has been ignored is that vertical transmission has been detected in many baculoviruses, sometimes at very low rates.[2] Thus these "disabled" viruses could quite possibly persist in the host population, rather than in the abiotic environment, and be available for transport and perhaps recombination with wild strains of baculovirus.

Transport of genetically engineered entomopathogenic bacteria

Genetic manipulations of entomopathogenic bacteria that are being researched or planned are directed at improvement of delivery to the insect pest, increasing virulence, altering the host range, increasing environmental stability of a toxin, or increasing toxin production. Surprisingly, classical genetic techniques have been limited to discovery of natural strains, particularly of BT, with improved qualities or different host ranges (Lüthy et al. 1982, Carlton 1988). Increases in virulence are possible through better understanding of the genes and proteins for the appropriate toxins and by overcoming biochemical host defenses (Kirschbaum 1985). Similarly, efforts toward manipulating host range and toxin production have depended on isolating the protoxin gene for the BT δ-endotoxin and locating the sequence for the active toxin (Kirschbaum 1985). In one case, a BT strain has been produced with activity against both Coleoptera and Lepidoptera (Carlton 1988). The earliest practical successes in manipulating the BT gene for δ-endotoxin, that is, the cases closest to being actually released or having been released, all involve improvement of delivery of the toxin to the insect pest. These efforts include such novel ideas as inserting the gene for δ-endotoxin into another bacterium that is subsequently killed and that protects the toxin from environmental degradation (Valiulis 1985); into various isolates of *Pseudomonas* that inhabit seed coatings, roots, or surface films of fresh water where target insects feed;[3] into host plants of target insects (Adang et al. 1988, Perlak et al. 1988, Van Mellaert et al. 1988, Gasser and Fraley 1989); into endophytic bacteria that colonize host plants (Dimock et al. 1988, Lindow et al. 1989); and into a blue-green alga for the purpose of killing mosquito larvae (de Barjac 1987). Inser-

[2]Payne 1982; Hostetter and Bell 1985; Andreadis 1987; Tanada and Fuxa 1987; J. R. Fuxa, A. R. Richter, and E. Weidner, unpublished data.

[3]de Barjac 1987, Gillett 1987, Marx 1987, Adang et al. 1988, Graham 1988, Lindow et al. 1989.

tion of the δ-endotoxin gene into plants will not be discussed further in the present chapter.

It is doubtful that the manipulations involving increased toxin production, a killed bacterium, or increased virulence will result in increased dispersal of genetic material from these bacteria. The examples of increased toxin production and killed bacteria do not involve living, replicating agents in the environment. The effect of increased virulence on transport of viruses should apply also to the bacteria; in one case involving natural strains of *Bacillus popilliae*, which is used for control of the Japanese beetle, *Popillia japonica*, strains with moderate virulence had better persistence in the host population than highly virulent strains (Krieg 1987). The effects of increasing host range are more dubious. BT already has a wide host range for an entomopathogen, yet is notorious for its lack of persistence and dispersal in host populations. Therefore, even though increased host range might result in greater transport soon after a release, it is unlikely that a transgenic organism would persist in the environment and disperse to any greater degree than a parental, unmodified strain. An increased host range in other species of bacteria would be likely to increase transport for the same reasons as for the viruses.

Insertion of the δ-endotoxin gene into *Pseudomonas* or a blue-green alga is more likely to result in increased transport. In these cases, the recombinant organism is intended to persist at least for a time, and increased persistence simply will increase the opportunity for transport. It is possible that organisms such as *Pseudomonas*, which utilize organic matter, could have an advantage over other microorganisms due to their new capability for obtaining a source of organic matter, namely, dead insects. On the other hand, in laboratory tests involving one such manipulation, the recombinant bacterium did not differ from the parental *P. fluorescens* in its dispersal characteristics (Lindow et al. 1989).

Transport of genetically engineered entomopathogenic fungi

Genetic manipulations of entomopathogenic fungi have not been developed as fully as those for the viruses and bacteria, partly due to the relative genetic complexity and lack of molecular biology of these eukaryotes. The genetic manipulations that have been proposed for these organisms include combining of desirable traits, or increasing virulence, host range, in vivo production, ability to survive exposure to UV radiation, and ability to survive exposure to fungicides. The techniques being researched include parasexual crosses, protoplast fusion, and recombinant DNA techniques (Boucias 1988, Staples et al. 1988). A gene for resistance to benomyl has been integrated into the genome of *Metarhizium anisopliae* (Staples et al. 1988). In some fungal species, certain strains are virulent to insects but are difficult to culture, and others have the exact opposite characteristics; development of hybrid strains with a desirable combination of traits is one objective of current research (Boucias 1988). Other ideas focus on increasing virulence of fungi by improving their germination, changing germ-tube orientation, improving penetration of insect cuticle, and improving vegetative growth and toxicity once inside the insect homocoel (Boucias 1988). Host range of a fungus could be tailored to pest situations by manipulations of conidial attachment capabilities for host cuticle or by manipulations of germ-tube orientation and penetration (Boucias 1988). It is possible that photoreactivation repair genes could be manipulated to enhance resistance to sunlight, or that *Beauvaria bassiana*, a fungus commonly found in soil, could be made more resistant to patulin, a fungal inhibitor also common in soil; such manipulations would improve environmental stability and insect control (Boucias 1988).

The genetic manipulations under consideration for fungi are likely to increase their chances for transport outside the area where they are released. The changes that improve their environmental stability (i.e., resistance to sunlight and fungicides or inhibitors) will improve their persistence and thus indirectly increase the probability that they will be transported. Any increase in host range will simply increase the amount of available nutritional substrate, which will increase the probability of fungal population growth and thus, again, the opportunity for transport. Since the fungi generally "recycle" after release (i.e., replicate and control more than one pest generation), the strategies to increase virulence by improving invasive capability and vegetative growth should result in the production of more conidia than native strains, which in turn will increase the probability of transport over the short term. As with viruses, however, increased virulence probably will not persist in a fungal population over the long term.

For the other proposed manipulations, either the chance for transport should decrease, or the effect on transport would be difficult to predict. Increasing virulence through increased toxicity would probably kill insects more quickly and thus reduce fungal growth and production of conidia available for transport. The probability of transport of recombinants with both virulence and good in vitro growth would be difficult to predict. Increasing virulence, again, could indirectly increase transport over the short term and decrease it over the long term; selection for in vitro growth obviously would indirectly increase transport by increasing population density if it included an increase in numbers of the transmissive stage and if it did not decrease environmental fitness in some other way.

Exposure Assessment

This review of transport mechanisms for entomopathogens provides information relative to the fate of these organisms after release, which, in turn, is a critical element in environmental exposure assessment. Certain generalizations about transport of GEMs can be made, though exceptions undoubtedly will be found to virtually any such generalization, particularly when there are interactions between factors.

Factors affecting transport

Based on this literature review, the environmental factors likely to contribute to long-range transport after release include use of a viral or fungal parental organism that is the recipient of the genetic manipulation, air currents, water currents, insect host adults, predatory birds, parasitoids, saprozoites (especially birds), grazing mammals, and possibly predatory mammals and arthropods. Additionally, release into relatively stable habitats, such as forests or permanent bodies of water, is more likely to result in increased persistence and opportunity for transport than release into unstable habitats, such as row crops (Fuxa 1987*b*, 1990*a*).

Other environmental factors are more likely to contribute only to short-range transport after release: use of a recipient entomopathogenic bacterial parent (exceptions are likely), rainfall (except where it results in runoff to streams or rivers), gravity, water percolation through soil, insect host larvae, autodispersal of fungi, and soil arthropods.

Information about the effects of conventional biotechnology on transport are scanty, but the following factors have the potential to contribute to long-range transport after release: formulation for increased persistence, autodispersal by the host, application of dust or spray under certain (especially windy) conditions, high-altitude release, and product transport by humans.

The genetic manipulations most likely to contribute to long-range transport include increased replicative host range; increased environmental stability; increased rate of vertical transmission; and certain strategies for improved delivery, such as insertion of a gene into *Pseudomonas* or an alga.

Other genetic manipulations are not likely to increase transport, and they could decrease it: increased virulence, increased speed of kill, increased toxin production, increased nonreplicative host range, and insertion of a gene into a plant endophyte.

Keeler (1988) has listed, in a more general sense, classes of genetic modifications that are likely to be inherently safe.

Monitoring entomopathogens in the environment

Methods have been developed to monitor certain entomopathogens in the environment, but this is not a strong area of research in insect pathology. Fuxa (1987*c*) reviewed the methods that have been used to detect entomopathogens. The most frequently used methods involve the host insects: the insects themselves can be sampled and observed for disease; samples from the habitat can be processed and bioassayed; or sentinel host insects can be placed in the field. Pathogen units have been sampled and counted more directly by a wide range of techniques such as flotation, filtration, impression films or slides, aerial trapping, selective media, microscopy of habitat samples, and serology. In direct or indirect comparisons, bioassay of one type or another has generally been the most sensitive, the most biologically meaningful, and sometimes the most accurate method for detection or quantification of entomopathogens (Fuxa 1987*c*). Other researchers are developing monitoring methods for microorganisms in general, including incorporation of certain genes that can be easily detected, DNA probes (Levin et al. 1987, Marx 1987), rRNA sequencing and fingerprinting (Levin et al. 1987), and use of a polymerase chain reaction in conjunction with DNA probes (Chaudhry et al. 1989).

Modeling

Modeling can help predict exposure but should not be relied upon too heavily in risk assessment due to its removal from real-life situations and to the current paucity of research data. Models of entomopathogen population dynamics are not common. They generally have been designed to help understand prevalence of disease in an insect host population within a certain spatial framework. Some of them include components of spread of disease from foci of infection in a population, but these models have not been aimed at understanding transport of pathogens from a release site (Brown 1987, Carruthers and Soper 1987). Other models may be applicable to entomopathogens, such as models of biological invasions (Mollison 1986), of plant disease epidemics (Jeger 1985), and of air pollution (Bhumralker 1985).

Recommendations

Small-plot releases

Certain precautions can be taken to reduce the chance of spread of a released microorganism in a small-plot test or to hamper its spread and thereby perhaps provide an opportunity to reduce the density of the released population. The first consideration in the choice and design of a site for a small-plot release will be the organism to be released and its parental biology and ecology, particularly persistence and dispersal traits of any long-lived survival stages (Strauss et al. 1987). The ability to transfer genetic material to other organisms must be considered; certain genetic manipulations may be possible in some cases to limit transfer of genetic material (Strauss et al. 1987). Tiedje et al. (1989) considered the traits of the organism that could contribute to an increased probability of transport outside a release area. Means of mitigating the released population are a primary concern in case a problem arises. Other preliminary considerations will be the characteristics of the ecosystem in which the release will occur and the likelihood of and manner in which the released microorganism might disrupt it (Domsch et al. 1988). For example, releases into agroecosystems generally will be safer than those into native ecosystems, and releases into closed ecosystems (e.g., ponds) safer than those into open ones (e.g., streams) (Keeler 1988).

The current knowledge of releases of natural strains of entomopathogens and other microorganisms suggests certain arrangements for containment in small-plot field releases of GEMs:

1. If the released microorganism is host-specific, a release site outside the host range would hamper its spread, though it might necessitate artificial infestation with the insect host.
2. If possible, a site should be selected for its lack of candidates for genetic

transfer. For example, baculoviruses can genetically recombine, but a site could be found with no history of NPV epizootics in insect populations, thereby greatly reducing the chance for recombination and subsequent transport of an introduced gene.

3. Climatic-monitoring data from the site should have been recorded for enough years to allow prediction of predominant wind directions and velocities. A relatively calm site should be selected, because direct or indirect transport on air currents will probably be the most difficult transport method to control.
4. Either the natural terrain or artificial dikes should be used to prevent the possibility of water runoff to a stream or other body of water.
5. Though data are scanty, research on plant pathogens indicates that relatively small fields with elongated shapes are detrimental to the spread of crop disease within the field (Fleming et al. 1982). With a lack of any better information, it can be presumed that hindering spread within the field will decrease pathogen population growth and thus the potential inoculum for transport outside the field. Thus, test fields probably should be kept as small as possible, elongated, and the long axis oriented in the same direction as the prevailing wind.
6. Soil composition of the site should be determined. A site with a clay-type soil might be preferable to other types such as sandy soils, because certain entomopathogens, at least, tend to adhere to the finer clay particles and thus would have less tendency to leach downward toward a water table.
7. Buffer zones can be established with barriers outside the perimeter of the release area. Vegetation should be eliminated immediately outside the perimeter, and the vegetation-free zone could be covered with plastic mulch, depending on the agent released. Outside this area, an additional barrier could be comprised of one or more "trap crops," either the same plant species as in the release plot or other species. This would not be an absolute barrier but should, depending on insect species, attract some host and nonhost emigrants from the treated area, thus providing an additional opportunity to destroy arthropods that might be transporting a GEM. Depending on the behavior of the host insect, a trap crop could prove beneficial if the released organism has the capability for vertical transmission.
8. Fencing should be installed around the perimeter of the buffer zones and buried deeply enough to prevent mammals and other large animals from entering the release site.
9. Installation of bird netting above the release site should be considered.
10. The agent should be released under calm conditions. If it must be sprayed or applied as a dust, the droplet or particle size should be as large as possible, and the release should be at the lowest possible height above the ground.

11. Depending on the agent, as many components of the environment as possible should be monitored to detect it: soil, air, potential biotic transport agents, and crop residue.
12. At the experiment's termination, all plant residue from the release site and buffer zones should be destroyed. The soil could be tilled, depending on the GEM's persistence characteristics.

An important disadvantage of these various barriers and precautions is that, the more they are used in a field test, the more that test will resemble a "microcosm" study rather than a true field release. Microcosm studies have their advantages, but they will not permit extrapolation for prediction of the results of field releases any more than will laboratory studies or knowledge of parental organisms (Fuxa 1987*b*). A release into an isolated habitat of some sort, such as irrigated plots in a desert or on an island, would, in a sense, constitute a microcosm study, because the habitat is still isolated from the type of environment for which the GEM is targeted. The basic purposes of a small-plot release will be to evaluate certain parameters of biological or practical interest (e.g., efficacy of crop protection, persistence, spread) and allow as natural as possible an interaction with the rest of the environment (necessary for exposure and risk assessment), yet do so in such a manner to increase the probability of stopping the population growth and spread if a problem arises. In a sense, one can question whether precautions against spread from a small-plot release of microorganisms are necessary; if the escape of some organisms is truly inevitable, as some scientists believe (e.g., Strauss et al. 1987), then sooner or later it will be necessary to "cross the Rubicon" with small-scale field tests in the development of every transgenic organism intended for environmental release. Whether or not the organism spreads from a release site will depend on its capability for persistence, dispersal, and population growth in the ecosystem where it is released, not on containment precautions.

Data requirements

How many data are sufficient before an environmental release? There is no good answer to this question. Tiedje et al. (1989) outlined certain situations that will require more data than others, but they were not able to be specific. Basically, as others have already pointed out, safety data will be negative (e.g., Keeler 1988, Simonsen and Levin 1988), and one will never be able to prove a hypothesis that harm will not occur. However, this is a similar situation, though a much more complex one, to that faced by the USEPA and microbial control researchers in the 1970s in registering natural strains of entomopathogens for insect control (Fuxa 1987*b*). In that case, a conservative phase of registration with strict, changing, and perhaps unreasonable requirements was followed by a burst of case-by-case analyses for registration, and finally by standardized guidelines.

The most important types of data will come from ecology of the parental, recipient organism and from laboratory safety tests with the recombinant organism. The initial cases under consideration for release should be limited to those recombinants that involve the transfer of just one gene. Then the data that will best allow for extrapolation by experts, including at least one ecologist familiar with the recipient parental organism, will be the biology and ecology of that recipient. The data required prior to release of a GEM should include full knowledge of the parental organisms—particularly the recipient—including data on population dynamics after field releases. Laboratory and microcosm studies of the GEM itself can be concentrated on Tier 1 environmental and safety tests developed for releases of entomopathogens by USEPA (Betz 1986), see Chapter 2, as well as on the potential for genetic exchange with several of the most likely recipient organisms that are indigenous at or near release sites. Thus, microcosms will be of limited usefulness except to answer certain specific questions. Their contribution will be to lessen the danger of extrapolations based only on laboratory safety data and biology-ecology of the parental organisms (Fuxa 1987*b*, 1989).

One can only hope that reasonable attitudes will prevail in the controversies surrounding the release of GEMs. As experience is gained with genetically engineered entomopathogens that are not likely to persist or spread in the environment, releases should begin to include organisms that are *intended* to persist and spread, suppressing insect populations permanently rather than through the inundative "firefighting" approach. Zero risk will not be possible nor should this impossibility prevent releases; the potential benefits of release are likely to outweigh the risks (Fuxa 1990*b*). Thus regulatory agencies and the general public probably should recognize that eventually some environmental problem will arise from a release. If such problems can be kept to a minimum and can be used as experience to avoid further mistakes, then society undoubtedly will benefit just as it has from many other introductions during the history of agriculture.

References

Abbas, M. S. T., and Boucias, D. G. 1984. Interaction between nuclear polyhedrosis virus-infected *Anticarsia gemmatalis* (Lepidoptera: Noctuidae) larvae and predator *Podisus maculiventris* (Say) (Hemiptera: Pentatomidae). *Environ. Entomol.* 13:599–602.

Adang, M., DeBoar, D., Endres, J., Firoozabady, E., Klein, J., Merlo, A., Merlo, D., Murray, E., Rashka, K., and Stock, C. 1988. Manipulation of *Bacillus thuringiensis* genes for pest insect control. In D. W. Roberts and R. R. Granados (eds.), *Biotechnology, Biological Pesticides and Novel Plant-Pest Resistance for Insect Pest Management*, Boyce Thompson Institute, Cornell University, Ithaca, New York, pp. 31–37.

Agudelo, F., and Falcon, L. A. 1983. Mass production, infectivity, and field application studies with the entomogenous fungus *Paecilomyces farinosus. J. Invertebr. Pathol.* 42:124–132.

Alexander, M. 1988. A microbial ecologist looks once again at risk analysis. In W. Klingmuller (ed.), *Risk Assessment for Deliberate Releases. The Possible Impact of Genetically Engineered Microorganisms on the Environment*, Springer-Verlag, Berlin, Heidelberg, New York, pp. 1–9.

Anderson, R. M. 1982. Theoretical basis for the use of pathogens as biological control agents of pest species. *Parasitology* 84:3–33.

Anderson, R. M., and May, R. M. 1979. Population biology of infectious diseases: part I. *Nature* 280:361–367.

Anderson, R. M., and May, R. M. 1980. Infectious diseases and population cycles of forest insects. *Science* 210:658–661.

Andreadis, T. G. 1987. Transmission. In J. R. Fuxa and Y. Tanada (eds.), *Epizootiology of Insect Diseases*, Wiley, New York, pp. 159–176.

Angus, T. A., and Lüthy, P. 1971. Formulation of microbial insecticides. In H. D. Burges and N. W. Hussey (eds.), *Microbial Control of Insects and Mites*, Academic, London, pp. 623–638.

Bedford, G. O. 1986. Biological control of the rhinoceros beetle (*Oryctes rhinoceros*) in the South Pacific by baculovirus. *Agric. Ecosystems Environ.* 15:141–147.

Beegle, C. C., and Oatman, E. R. 1975. Effect of a nuclear polyhedrosis virus on the relationship between *Trichoplusia ni* (Lepidoptera: Noctuidae) and the parasite, *Hyposter exiguae* (Hymenoptera: Ichneumonidae). *J. Invertebr. Pathol.* 25:59–71.

Beekman, A. G. 1980. The infectivity of polyhedra of nuclear polyhedrosis virus (N.P.V.) after passage through gut of an insect-predator. *Experientia* 36:858–859.

Betz, F. S. 1986. Registration of baculoviruses as pesticides. In R. R. Granados and B. A. Federici (eds.), *The Biology of Baculoviruses*. vol. 2: *Practical Application for Insect Control*, CRC Press, Boca Raton, Florida, pp. 203–222.

Betz, F., Rispin, A., and Schneider, W. 1987. Biotechnology products related to agriculture. Overview of regulatory decisions at the U.S. Environmental Protection Agency. In H. M. LeBaron, R. O. Mumma, R. C. Honeycutt, and J. H. Duesing (eds.), *Biotechnology in Agricultural Chemistry*, American Chemical Society, Washington, D.C., pp. 316–327.

Bhumralker, C. M. 1985. Possible applications of air pollution models to the movement and dispersal of biotic agents. In D. R. MacKenzie, C. S. Barfield, G. G. Kennedy, R. D. Berger, and D. J. Taranto (eds.), *The Movement and Dispersal of Agriculturally Important Biotic Agents*, Claitor's, Baton Rouge, Louisiana, pp. 289–304.

Bird, F. T. 1961. Transmission of some insect viruses with particular reference to ovarial transmission and its importance in the development of epizootics. *J. Insect Pathol.* 3:352–380.

Boucias, D. G. 1988. Genetic improvement of entomopathogenic fungi. In G. A. Herzog, S. Ramaswamy, G. Lentz, J. L. Goodenough, and J. J. Hamm (eds.), Theory and tactics of Heliothis population management. III. Emerging control tactics and techniques, Southern Coop. Ser. Bull. no. 337, Agricultural Pub., Oklahoma State University, Stillwater, Oklahoma, pp. 66–68.

Brassel, J., and Benz, G. 1979. Selection of a strain of the granulosis virus of the codling moth with improved resistance against artificial ultraviolet radiation and sunlight. *J. Invertebr. Pathol.* 33:358–363.

Brown, G. C. 1987. Modeling. In J. R. Fuxa and Y. Tanada (eds.), *Epizootiology of Insect Diseases*, Wiley, New York, pp. 43–68.

Bucher, G. E. 1959. Bacteria of grasshoppers of Western Canada: III. Frequency of occurrence, pathogenicity. *J. Insect Pathol.* 1:391–405.

Bucher, G. E., and Stephens, J. M. 1957. A disease of grasshoppers caused by the bacterium *Pseudomonas aeruginosa* (Schroeter) Migula. *Can. J. Microbiol.* 3:611–625.

Burges, H. D. (ed.). 1981. *Microbial Control of Pests and Plant Diseases 1970–1980*, Academic, London, 949 pp.

Burges, H. D., and Hurst, J. A. 1977. Ecology of *Bacillus thuringiensis* in storage moths. *J. Invertebr. Pathol.* 30:131–139.

Capinera, J. L., and Barbosa, P. 1975. Transmission of a nuclear-polyhedrosis virus to gypsy moth larvae by *Calosoma sycophanta*. *Ann. Entomol. Soc. Am.* 68:593–594.

Carlton, B. 1988. Genetic improvements of *Bacillus thuringiensis* as a bioinsecticide. In D. W. Roberts and R. R. Granados (eds.), *Biotechnology, Biological Pesticides and Novel Plant-Pest Resistance for Insect Pest Management*, Boyce Thompson Institute, Cornell University, Ithaca, New York, pp. 38–43.

Carruthers, R. I., and Soper, R. S. 1987. Fungal diseases. In J. R. Fuxa and Y. Tanada (eds.), *Epizootiology of Insect Diseases*, Wiley, New York, pp. 357–416.

Chaudhry, G. R., Toranzos, G. A., and Bhatti, A. R. 1989. Novel method for monitoring genetically engineered microorganisms in the environment. *Appl. Environ. Microbiol.* 55:1301–1304.

Cooper, D. J. 1981. The role of predatory Hemiptera in disseminating a nuclear polyhedrosis virus of *Heliothis punctigera*. *J. Aust. Entomol. Soc.* 20:145–150.

Couch, J. N., and Bland, C. E. (eds.). 1985. *The Genus* Coelomomyces, Academic, Orlando, Florida, 399 pp.
David, W. A. L., and Gardiner, B. O. C. 1967. The persistence of a granulosis virus of *Pieris brassicae* in soil and in sand. *J. Invertebr. Pathol.* 9:342–347.
de Barjac, H. 1987. Operational bacterial insecticides and their potential for future improvement. In K. Maramorosch (ed.), *Biotechnology in Invertebrate Pathology and Cell Culture*, Academic, San Diego, pp. 63–73.
DeLucca, A. J., III, Simonson, J. G., and Larson, A. D. 1981. *Bacillus thuringiensis* distribution in soils of the United States. *Can. J. Microbiol.* 27:865–870.
Dimock, M., Beach, R. M., and Carlson, P. S. 1988. Endophytic bacteria for the delivery of crop protection agents. In D. W. Roberts and R. R. Granados (eds.), *Biotechnology, Biological Pesticides and Novel Plant-Pest Resistance for Insect Pest Management*, Boyce Thompson Institute, Cornell University, Ithaca, New York, pp. 88–92.
Domnas, A. J. 1981. Biochemistry of *Lagenidium giganteum* infection in mosquito larvae. In E. W. Davidson (ed.), *Pathogenesis of Invertebrate Microbial Diseases*, Allanheld, Osmun, Totowa, New Jersey, pp. 425–449.
Domsch, K. H., Driesel, A. J., Goebel, W., Andersch, W., Lindenmaier, W., Lotz, W., Reber, H., and Schmidt, F. 1988. Considerations on release of gene-technology engineered microorganisms into the environment. *FEMS Microbiol. Ecol.* 53:261–272.
Dulmage, H. T., and Aizawa, K. 1982. Distribution of *Bacillus thuringiensis* in nature. In E. Kurstak (ed.), *Microbial and Viral Pesticides*, Marcel Dekker, New York, pp. 209–237.
Entwistle, P. F. 1986. Epizootiology and strategies of microbial control. In J. M. Franz and M. Lindaur (eds.), *Biological Plant and Health Protection*, Fischer Verlag, Stuttgart, pp. 257–278.
Entwistle, P. F., Adams, P. H. W., and Evans, H. F. 1977*a*. Epizootiology of a nuclear-polyhedrosis virus in European spruce sawfly (*Gilpinia hercyniae*): the status of birds as dispersal agents of the virus during the larval season. *J. Invertebr. Pathol.* 29:354–360.
Entwistle, P. F., Adams, P. H. W., and Evans, H. F. 1977*b*. Epizootiology of a nuclear-polyhedrosis virus in European spruce sawfly, *Gilpinia hercyniae*: birds as dispersal agents of the virus during winter. *J. Invertebr. Pathol.* 30:15–19.
Entwistle, P. F., Adams, P. H. W., and Evans, H. F. 1978. Epizootiology of a nuclear polyhedrosis virus in European spruce sawfly (*Gilpinia hercyniae*): the rate of passage of infective virus through the gut of birds during cage tests. *J. Invertebr. Pathol.* 31:307–312.
Entwistle, P. F., Adams, P. H. W., Evans, H. F., and Rivers, C. F. 1983. Epizootiology of a nuclear polyhedrosis virus (Baculoviridae) in European spruce sawfly (*Gilpinia hercyniae*): spread of disease from small epicentres in comparison with spread of baculovirus diseases in other hosts. *J. Appl. Ecol.* 20:473–487.
Evans, H. F., and Allaway, G. P. 1983. Dynamics of baculovirus growth and dispersal in *Mamestra brassicae* L. (Lepidoptera: *Noctuidae*) larval populations introduced into small cabbage plots. *Appl. Environ. Microbiol.* 45:493–501.
Falcon, L. A. 1985. Development of microbial pesticides. In M. P. Ferguson and H. G. Alford (eds.), *Microbial/Biorational Pesticide Registration*. University of California (Berkeley), Cooperative Extension Special Publication 3318, pp. 11–17.
Faulkner, P., and Boucias, D. G. 1985. Genetic improvement of insect pathogens: emphasis on the use of baculoviruses. In M. A. Hoy and D. C. Herzog (eds.), *Biological Control in Agricultural IPM Systems*, Academic, Orlando, Florida, pp. 263–281.
Fiksel, J. R., and Covello, V. T. 1986. The suitability and applicability of risk assessment methods for environmental applications of biotechnology. In J. Fiksel and V. T. Covello (eds.), *Biotechnology Risk Assessment*, Pergamon, New York, pp. 1–34.
Finch, M. F., Stark, P. M., and Meisch, M. V. 1986. Point source introduction of *Bti* into ricefields for mosquito control. *Arkansas Farm Res.* July–August:10.
Flanders, S. E., and Hall, I. M. 1965. Manipulated bacterial epizootics in *Anagasta* populations. *J. Invertebr. Pathol.* 7:368–377.
Fleming, R. A., Marsh, L. M., and Tuckwell, H. C. 1982. Effect of field geometry on the spread of crop disease. *Protection Ecol.* 4:81–108.

Franc, G. D., Harrison, M. D., and Powelson, M. L. 1985. The dispersal of phytopathogenic bacteria. In D. R. MacKenzie, C. S. Barfield, G. G. Kennedy, R. D. Berger, and D. J. Taranto (eds.), *The Movement and Dispersal of Agriculturally Important Biotic Agents*, Claitor's, Baton Rouge, Louisiana, pp. 37–49.

Fuxa, J. R. 1987*a*. Ecological considerations for the use of entomopathogens in IPM. *Annu. Rev. Entomol.* 32:225–251.

Fuxa, J. R. 1987*b*. *Risk Assessment of Genetically Engineered Entomopathogens: Effects of Microbial Control Agents on the Environment Including Their Persistence and Dispersal*, American Association for the Advancement of Science/USEPA, Environmental Science and Engineering Fellows Report, Washington, D.C.

Fuxa, J. R. 1987*c*. Ecological methods. In J. R. Fuxa and Y. Tanada (eds.), *Epizootiology of Insect Diseases*, Wiley, New York, pp. 23–41.

Fuxa, J. R. 1989. Fate of released entomopathogens with reference to risk assessment of genetically engineered microorganisms. *Bull. Entomol. Soc. Am.* 35:12–24.

Fuxa, J. R. 1990*a*. New directions for insect control with baculoviruses. In R. Baker and P. E. Dunn (eds.), *New Directions in Biological Control*, Alan R. Liss, New York, pp. 97–113.

Fuxa, J. R. 1990*b*. Environmental risks of genetically engineered entomopathogens. In M. Laird, L. A. Lacey, and E. W. Davidson (eds.), *Safety of Microbial Insecticides*, CRC Press, Boca Raton, Florida, pp. 203–207.

Fuxa, J. R., and Geaghan, J. P. 1983. Multiple-regression analysis of factors affecting prevalence of nuclear polyhedrosis virus in *Spodoptera frugiperda* (Lepidoptera: Noctuidae) populations. *Environ. Entomol.* 12:311–316.

Gard, I. E. 1975. Utilization of light traps to disseminate insect viruses for pest control. Ph.D. dissertation, University of California, Berkeley.

Gasser, C. S., and Fraley, R. T. 1989. Genetically engineered plants for crop improvement. *Science* 244:1293–1299.

Gibbs, K. E., Brautigam, F. C., Stubbs, C. S., and Zibilske, L. M. 1986. Experimental applications of *B.t.i.* for larval black fly control: persistence and downstream carry, efficacy, impact on non-target invertebrates and fish feeding, Maine Agricultural Experiment Station Technical Bulletin 123.

Gillett, J. W. (ed.). 1987. Prospects for physical and biological containment of genetically engineered organisms, Institute for Comparative and Environmental Toxicology, Cornell University, Ithaca, New York.

Gorick, B. D. 1980. Release and establishment of the baculovirus disease of *Oryctes rhinoceros* (L.) (Coleoptera: Scarabaeidae) in Papua New Guinea. *Bull. Entomol. Res.* 70:445–453.

Graham, T. L. 1988. Genetic engineering of soil microorganisms for pest control. *Agric. Ecosystems Environ.* 24:317–323.

Grison, P., Martouret, D., Servais, B., and Devriendt, M. 1976. Pesticides microbiens et environnement. *Ann. Zool.-Écol. Anim.* 8:133–160.

Hamm, J. J., Nordlund, D. A., and Marti, O. G. 1985. Effects of a nonoccluded virus of *Spodoptera frugiperda* (Lepidoptera: Noctuidae) on the development of a parasitoid, *Cotesia marginiventris* (Hymenoptera: Braconidae). *Environ. Entomol.* 14:258–261.

Hostetter, D. L. 1971. A virulent nuclear polyhedrosis virus of the cabbage looper, *Trichoplusia ni*, recovered from the abdomens of sarcophagid flies. *J. Invertebr. Pathol.* 17:130–131.

Hostetter, D. L., and Bell, M. R. 1985. Natural dispersal of baculoviruses in the environment. In K. Maramorosch and K. E. Sherman (eds.), *Viral Insecticides for Biological Control*, Academic, Orlando, Florida, pp. 249–284.

Hukuhara, T. 1973. Further studies on the distribution of a nuclear-polyhedrosis virus of the fall webworm, *Hyphantria cunea*, in soil. *J. Invertebr. Pathol.* 22:345–350.

Ignoffo, C. M., Garcia, C., Hostetter, D. L., and Pinnell, R. E. 1977. Laboratory studies of the entomopathogenic fungus *Nomuraea rileyi*: soil-borne contamination of soybean seedlings and dispersal of diseased larvae of *Trichoplusia ni*. *J. Invertebr. Pathol.* 29:147–152.

Ignoffo, C. M., Hostetter, D. L., Biever, K. D., Garcia, C., Thomas, D., Dickerson, W. A., and Pinnell, R. 1978. Evaluation of an entomopathogenic bacterium, fungus, and virus for control of *Heliothis zea* on soybeans. *J. Econ. Entomol.* 71:165–168.

Irabagon, T. A., and Brooks, W. M. 1974. Interaction of *Campoletis sonorensis* and a nuclear polyhedrosis virus in larvae of *Heliothis virescens*. *J. Econ. Entomol.* 67:229–231.

Jaques, R. P. 1969. Leaching of the nuclear-polyhedrosis virus of *Trichoplusia ni* from soil. *J. Invertebr. Pathol.* 13:256–263.

Jaronski, S., and Axtell, R. C. 1983. Persistence of the mosquito fungal pathogen *Lagenidium giganteum* (Oomycetes: Lagenidiales) after introduction into natural habitats. *Mosquito News* 43:332–337.

Jeger, M. J. 1985. Long distance transport of aerially dispersed fungal pathogens. In D. R. MacKenzie, C. S. Barfield, G. G. Kennedy, R. D. Berger, and D. J. Taranto (eds.), *The Movement and Dispersal of Agriculturally Important Biotic Agents*, Claitor's, Baton Rouge, Louisiana, pp. 107–113.

Jeger, M. J. 1985. Models of focus expansion in disease epidemics. In D. R. MacKenzie, C. S. Barfield, G. G. Kennedy, R. D. Berger, and D. J. Taranto (eds.), *The Movement and Dispersal of Agriculturally Important Biotic Agents*, Claitor's, Baton Rouge, Louisiana, pp. 279–288.

Jones, D. D. 1988. USDA safety review of biotechnology research in agriculture. American Biotechnology Laboratory, January.

Kaya, H. K. 1985. History of microbial control. In M. P. Ferguson and H. G. Alford (eds.), *Microbial/Biorational Pesticide Registration*, Univ. Calif. (Berkeley) Coop. Ext. Spec. Publ. No. 3318, pp. 6–10.

Keeler, K. H. 1988. Can we guarantee the safety of genetically engineered organisms in the environment? *CRC Crit. Rev. Biotechnol.* 8:85–97.

Khachatourians, G. G. 1986. Production and use of biological pest control agents. *Trends Biotechnol.* 4:120.

Kilbourne, E. D. 1985. Epidemiology of viruses genetically altered by man—predictive principles. In B. Fields, M. A. Martin, and D. Kamely (eds.), *Genetically Altered Viruses and the Environment*, Cold Spring Harbor Laboratory, Cold Spring Harbor, New York, pp. 103–117.

King, D. S., and Humber, R. A. 1981. Identification of the Entomophthorales. In H. D. Burges (ed.), *Microbial Control of Pests and Plant Diseases 1970–1980*, Academic, London, pp. 107–127.

King, S. R. 1985. Economic impacts of biotechnology. In A. H. Teich, M. A. Levin, and J. H. Pace (eds.), *Biotechnology and the Environment*, American Association for the Advancement of Science, Washington, D.C. pp. 29–59.

Kirschbaum, J. B. 1985. Potential implication of genetic engineering and other biotechnologies to insect control. *Annu. Rev. Entomol.* 30:51–70.

Kish, L. P., and Allen, G. E. 1978. The biology and ecology of *Nomuraea rileyi* and a program for predicting its incidence on *Anticarsia gemmatalis* in soybean, Univ. Flor. Agric. Exp. Stns. Bull. no. 795 (Tech.), Gainesville, Florida.

Klein, M. G. 1981. Advances in the use of *Bacillus popilliae* for pest control. In H. D. Burges (ed.), *Microbial Control of Pests and Plant Diseases 1970–1980*, Academic, London, pp. 183–192.

Kramm, K. R., West, D. F., and Rockenbach, P. G. 1982. Termite pathogens: transfer of the entomopathogen *Metarhizium anisopliae* between *Reticulitermes* sp. termites. *J. Invertebr. Pathol.* 40:1–6.

Krieg, A. 1987. Diseases caused by bacteria and other prokaryotes. In J. R. Fuxa and Y. Tanada (eds.), *Epizootiology of Insect Diseases*, Wiley, New York, pp. 323–355.

Kring, T. J., Young, S. Y., and Yearian, W. C. 1988. The striped lynx spider, *Oxyopes salticus* Hentz (Araneae: Oxyopidae), as a vector of a nuclear polyhedrosis virus in *Anticarsia gemmatalis* Hübner (Lepidoptera: Noctuidae). *J. Entomol. Sci.* 23:394–398.

Laigo, F. M., and Tamashiro, M. 1966. Virus and insect parasite interaction in the lawn armyworm, *Spodoptera maurita acronyctoides* (Guenée). *Proc. Hawaii Entomol. Soc.* 19:233–237.

Laird, M. 1985. Use of *Coelomomyces* in biological control: introduction of *Coelomomyces stegomyiae* into Nukunono, Tokelau Islands. In J. N. Couch and C. E. Bland (eds.), *The Genus* Coelomomyces, Academic, Orlando, Florida, pp. 369–390.

Langford, G. S., Vincent, R. H., and Cory, E. N. 1942. The adult Japanese beetle as host and disseminator of type A milky disease. *J. Econ. Entomol.* 35:165–169.

Latch, G. C. M., and Falloon, R. E. 1976. Studies on the use of *Metarhizium anisopliae* to control *Oryctes rhinoceros*. *Entomophaga* 21:39–48.

Lautenschlager, R. A., and Podgwaite, J. D. 1979. Passage of nucleopolyhedrosis virus by avian and mammalian predators of the gypsy moth, *Lymantria dispar*. *Environ. Entomol.* 8:210–214.

Lautenschlager, R. A., Podgwaite, J. D., and Watson, D. E. 1980. Natural occurrence of the nucleopolyhedrosis virus of the gypsy moth, *Lymantria dispar* [*Lep.: Lymantriidae*] in wild birds and mammals. *Entomophaga* 25:261–267.

Levin, D. B., Laing, J. E., and Jaques, R. P. 1979. Transmission of granulosis virus by *Apanteles glomeratus* to its host *Pieris rapae*. *J. Invertebr. Pathol.* 34:317–318.

Levin, M. A., Seidler, R., Borquin, A. W., Fowle, J. R., III, and Barkay, T. 1987. EPA developing methods to assess environmental release. *Bio/technology* 5:38–45.

Lindow, S. E., Panopoulos, N. J., and McFarland, B. L. 1989. Genetic engineering of bacteria from managed and natural habitats. *Science* 244:1300–1307.

Lloyd, M., White, J., and Stanton, N. 1982. Dispersal of fungus-infected periodical cicadas to new habitat. *Environ. Entomol.* 11:852–858.

Lüthy, P., Cordier, J.-L., and Fischer, H.-M. 1982. *Bacillus thuringiensis* as a bacterial insecticide: basic considerations and application. In E. Kurstak (ed.), *Microbial and Viral Pesticides*, Marcel Dekker, New York, pp. 35–74.

Maeda, S. 1988. Genetic engineering of baculoviruses. In D. W. Roberts and R. R. Granados (eds.), *Biotechnology, Biological Pesticides and Novel Plant-Pest Resistance for Insect Pest Management*, Boyce Thompson Institute, Cornell University, Ithaca, New York, pp. 17–21.

Maruniak, J. E. 1988. Genetic engineering of baculoviruses. In G. A. Herzog, S. Ramaswamy, G. Lentz, J. L. Goodenough, and J. J. Hamm, Theory and tactics of *Heliothis* population management. III. Emerging control tactics and techniques, Southern Coop. Ser. Bull. no. 337, Agricultural Pub., Oklahoma State University, Stillwater, Oklahoma, pp. 69–73.

Marx, J. L. 1987. Assessing the risks of microbial release. *Science* 237:1413–1417.

McLaughlin, R. E., and Vidrine, M. F. 1985. Factors affecting distribution of *Bacillus thuringiensis* serotype H-14 during flooding of rice fields. *J. Am. Mosq. Control Assoc.* 1:381–384.

Miller, L. K. 1987. Expression of foreign genes in insect cells. In K. Maramorosch (ed.), *Biotechnology in Invertebrate Pathology and Cell Culture*, Academic, San Diego, pp. 295–303.

Miller, L. K., Lingg, A. J., and Bulla, L. A., Jr. 1983. Bacterial, viral, and fungal insecticides. *Science* 219:715–721.

Mitchell, F. L., and Fuxa, J. R. 1990. Multiple regression analysis of factors influencing a nuclear polyhedrosis virus in populations of fall armyworm (Lepidoptera: Noctuidae) in corn. *Environ. Entomol.* 19:260–267.

Mohamed, M. A., Coppel, H. C., Hall, D. J., and Podgwaite, J. D. 1981. Field release of virus-sprayed adult parasitoids of the European pine sawfly (Hymenoptera: Diprionidae) in Wisconsin. *Great Lakes Entomol.* 14:177–178.

Mollison, D. 1986. Modelling biological invasions: chance, explanation, prediction. *Phil. Trans. R. Soc. Lond. B, Disc. Mtg., Quantitative Aspects of the Ecology of Biological Invasions*, 26–27 February 1986.

Morris, O. N. 1982. Bacteria as pesticides: forest applications. In E. Kurstak (ed.), *Microbial and Viral Pesticides*, Marcel Dekker, New York, pp. 239–287.

Office of Technology Assessment of the U.S. Congress (OTA). 1989. New developments in biotechnology: 5. Patenting life. Special Report OTA-BA-370, U.S. Government Printing Office, Washington, D.C.

Olofsson, E. 1988. Dispersal of the nuclear polyhedrosis virus of *Neodiprion sertifer* from soil to pine foliage with dust. *Entomol. Exp. Appl.* 46:181–186.

Payne, C. C. 1982. Insect viruses as control agents. *Parasitology* 84:35–77.

Perlak, F., McPherson, S. A., Fuchs, R. A., MacIntosh, S. C., Dean, D. A., and Fischhoff, D. A. 1988. Expression of *Bacillus thuringiensis* proteins in transgenic plants. In D. W. Roberts and R. R. Granados (eds.), *Biotechnology, Biological Pesticides and Novel Plant-Pest Resistance for Insect Pest Management*, Boyce Thompson Institute, Cornell University, Ithaca, New York, pp. 77–81.

Podgwaite, J. D. 1985. Strategies for field use of baculoviruses. In K. Maramorosch and K. E. Sherman (eds.), *Viral Insecticides for Biological Control*, Academic, Orlando, Florida, pp. 775–797.
Rabb, R. L., and Kennedy, G. G. (eds.). 1979. *Movement of Highly Mobile Insects: Concepts and Methodology in Research*. University Graphics, North Carolina State University, Raleigh.
Raimo, B., Reardon, R. C., and Podgwaite, J. D. 1977. Vectoring gypsy moth nuclear polyhedrosis virus by *Apanteles melanoscelus* (Hym.: Braconidae). *Entomophaga* 22: 207–215.
Rissler, J. F. 1983. Research needs for biotic environmental effects of genetically engineered microorganisms, American Association for the Advancement of Science/ USEPA, Environmental Science and Engineering Fellows Report.
Rockwood, L. P. 1950. Entomogenous fungi of the family *Entomophthoraceae* in the Pacific Northwest. *J. Econ. Entomol.* 43:704–707.
Schabel, H. G. 1982. Phoretic mites as carriers of entomopathogenic fungi. *J. Invertebr. Pathol.* 39:410–412.
Shapiro, M. 1986. In vivo production of baculoviruses. In R. R. Granados and B. A. Federici (eds.), *The Biology of Baculoviruses*, vol. 2: *Practical Application for Insect Control*, CRC Press, Boca Raton, Florida, pp. 31–61.
Sherman, K. E. 1985. Considerations in the large-scale and commercial production of viral insecticides. In K. Maramorosch and K. E. Sherman (eds.), *Viral Insecticides for Biological Control*, Academic, Orlando, Florida, pp. 757–774.
Simonsen, L., and Levin, B. R. 1988. Evaluating the risk of releasing genetically engineered organisms. In J. Hodgson and A. M. Sugden (eds.), *Planned Release of Genetically Engineered Organisms*. Trends in Biotechnology/Trends in Ecology and Evolution Spec. Publ., Elsevier, Cambridge, pp. S27–S30.
Singer, S. 1981. Potential of *Bacillus sphaericus* and related spore-forming bacteria for pest control. In H. D. Burges (ed.), *Microbial Control of Pests and Plant Diseases 1970–1980*, Academic, London, pp. 283–298.
Smirnoff, W. A., and MacLeod, C. F. 1961. Study of the survival of *Bacillus thuringiensis* var. *thuringiensis* Berliner in the digestive tracts and in feces of a small mammal and birds. *J. Insect Pathol.* 3:266–270.
Smith, C. E. G. 1971. The spread and maintenance of infections in vertebrates and arthropods. *J. Invertebr. Pathol.* 18:i–xi.
Smith, D. B., Hostetter, D. L., Law, S. E., Pinnell, R. E., and Plummer, D. D. 1984. *Heliothis* mortality from drift-deposited aerosol virus sprays. *J. Georgia Entomol. Soc.* 19:394–407.
Stairs, G. R. 1965. Artificial initiation of virus epizootics in forest tent caterpillar populations. *Can. Entomol.* 97:1059–1062.
Staples, R., St. Leger, R. J., Bhairi, S., and Roberts, D. W. 1988. Strategies for genetic engineering of fungal entomopathogens. In D. W. Roberts and R. R. Granados (eds.), *Biotechnology, Biological Pesticides and Novel Plant-Pest Resistance for Insect Pest Management*, Boyce Thompson Institute, Cornell University, Ithaca, New York, pp. 44–48.
Sterling, P. H., Kelly, P. M., Speight, M. R., and Entwistle, P. F. 1988. The generation of secondary infection cycles following the introduction of nuclear polyhedrosis virus to a population of the brown-tail moth, *Euproctis chrysorrhoea* L. (Lep., Lymantriidae). *J. Appl. Entomol.* 106:302–311.
Storey, G. K., and Gardner, W. A. 1987. Vertical movement of commercially formulated *Beauveria bassiana* conidia through four Georgia soil types. *Environ. Entomol.* 16: 178–181.
Strauss, H. S., Olson, B., Loper, J., Alexander, M., Goldhammer, A., Hirano, S., Levin, M., Rissler, J., Stern, A. M., Tiedje, J., and Wilson, C. R. 1987. Controlling the dispersal of genetically engineered bacteria and fungi during field trials. In J. W. Gillett (ed.), *Prospects for Physical and Biological Containment of Genetically Engineered Organisms*, Institute for Comparative and Environmental Toxicology, Cornell University, Ithaca, New York, pp. 22–30.
Summers, M. D., and Smith, G. E. 1985. Genetic engineering of the genome of the *Autographa californica* nuclear polyhedrosis virus. In B. Fields, M. A. Martin, and D.

Kamely (eds.), *Genetically Altered Viruses and the Environment*, Cold Spring Harbor Laboratory, Cold Spring Harbor, New York, pp. 319–329.

Suzuki, N., and Kunimi, Y. 1981. Dispersal and survival rate of adult females of the fall webworm, *Hyphantria cunea* Drury (Lepidoptera: Arctiidae), using the nuclear polyhedrosis virus as a marker. *Appl. Entomol. Zool.* 16:374–385.

Sweeney, A. W. 1981. Fungal pathogens of mosquito larvae. In E. W. Davidson (ed.), *Pathogenesis of Invertebrate Microbial Diseases*, Allanheld, Osmun, Totowa, New Jersey, pp. 403–424.

Tanada, Y., and Fuxa, J. R. 1987. The pathogen population. In J. R. Fuxa and Y. Tanada (eds.), *Epizootiology of Insect Diseases*, Wiley, New York, pp. 113–157.

Thompson, C. G. 1978. Nuclear polyhedrosis epizootiology. In M. H. Boorks, R. W. Stark, and R. W. Campbell (eds.), The Douglas-fir tussock moth: a synthesis, USDA Forest Service Science Education Agency Tech. Bull. no. 1585, Pacific Northwest Forest and Range Experiment Station, USDA Forest Service, Portland, Oregon, p. 136.

Thompson, C. G., and Scott, D. W. 1979. Production and persistence of the nuclear polyhedrosis virus of the Douglas-fir tussock moth, *Orgyia pseudotsugata* (Lepidoptera: Lymantriidae), in the forest ecosystem. *J. Invertebr. Pathol.* 33:57–65.

Thompson, C. G., and Steinhaus, E. A. 1950. Further tests using a polyhedrosis virus to control the alfalfa caterpillar. *Hilgardia* 19:411–415.

Tiedje, J. M., Colwell, R. K., Grossman, Y. L., Hodson, R. E., Lenski, R. E., Mack, R. N., and Regal, P. J. 1989. The planned introduction of genetically engineered organisms: ecological considerations and recommendations. *Ecology* 70:298–315.

Valiulis, D. 1985. Debating biotech. *Agrichem. Age.* Aug.–Sept.:20D, 28B–28C.

Van Mellaert, Van Rie, J., Hofman, C., and Reynaerts, A. 1988. Insecticidal crystal proteins from *Bacillus thuringiensis*: mode of action and expression in transgenic plants. In D. W. Roberts and R. R. Granados (eds.), *Biotechnology, Biological Pesticides and Novel Plant-Pest Resistance for Insect Pest Management*, Boyce Thompson Institute, Cornell University, Ithaca, New York, pp. 82–87.

Veber, I. 1964. Virulence of an insect virus increased by repeated passages. *Entomophaga*, Mém. hors. Sér. 2:403–405.

Weiss, S. A., and Vaughn, J. L. 1986. Cell culture methods for large-scale propagation of baculoviruses. In R. R. Granados and B. A. Federici (eds.), *The Biology of Baculoviruses*, vol. 2: *Practical Application for Insect Control*, CRC Press, Boca Raton, Florida, pp. 63–87.

Whisler, H. C. 1985. Life history of species of *Coelomomyces*. In J. N. Couch and C. E. Bland (eds.), *The Genus* Coelomomyces, Academic, Orlando, Florida, pp. 9–22.

White, R. T. 1943. Effect of milky disease on *Tiphia* parasites of Japanese beetle larvae. *J. N.Y. Entomol. Soc.* 51:213–218.

White, R. T., and Dutky, S. R. 1940. Effect of the introduction of milky diseases on populations of Japanese beetle larvae. *J. Econ. Entomol.* 33:306–309.

Wilding, N. 1970. *Entomophthora conidia* in the air-spora. *J. Gen. Microbiol.* 62:149–157.

Wilding, N. 1975. *Entomophthora* species infecting pea aphis. *Trans. R. Entomol. Soc.* 127:171–183.

Young, E. C. 1974. The epizootiology of two pathogens of the coconut palm rhinoceros beetle. *J. Invertebr. Pathol.* 24:82–92.

Young, S. Y. 1990. Influence of sprinkler irrigation on dispersal of nuclear polyhedrosis virus from host cadavers on soybean. *Environ. Entomol.* 19:717–720.

Young, S. Y., III, and Yearian, W. C. 1986. Formulation and application of baculoviruses. In R. R. Granados and B. A. Federici (eds.), *The Biology of Baculoviruses*, vol. 2: *Practical Application for Insect Control*, CRC Press, Boca Raton, Florida, pp. 157–179.

Young, S. Y., and Yearian, W. C. 1987. *Nabis roseipennis* adults (Hemiptera: Nabidae) as disseminators of nuclear polyhedrosis virus to *Anticarsia gemmatalis* (Lepidoptera: Noctuidae) larvae. *Environ. Entomol.* 16:1330–1333.

Zelazny, B. 1976. Transmission of a baculovirus in populations of *Oryctes rhinoceros*. *J. Invertebr. Pathol.* 27:221–227.

Zimmermann, G., and Bode, E. 1983. Untersuchungen zur verbreitung des insektenpathogen pilzes *Metarhizium anisopliae* (Fungi Imperfecti, Moniliales) durch bodenarthropoden. *Pedobiologia* 25:65–71.

Chapter

6

Persistence, Establishment, and Mitigation of Phytopathogenic Viruses

Sue A. Tolin

Dept. of Plant Pathology, Physiology, and Weed Science
Virginia Polytechnic Institute and State University
Blacksburg, Virginia 24061

Introduction

Plants in nature are frequently infected by viruses. In some crops, viruses may cause losses in yield so large that crops cannot be grown in a particular area. Such a situation occurs if a virus has a ready source of inoculum, an efficient means of dissemination to the susceptible crop, and is transmitted to a crop under conditions such that infection causes a severe impact on the crop.

The intent of this chapter is to review information about plant-pathogenic viruses relative to their ability to survive, persist, and become established in an ecosystem, and to spread within and between species within an ecosystem. This information has been generated by research conducted with naturally occurring plant viruses over nearly 100 years, enabling the design of management systems for disease mitigation and control and, more recently, the development of predictive models of disease development (Madden et al. 1988). It will become evident that plant virus ecology as it relates to disease control is a much reviewed topic and that nearly all of the data come from studies of crop ecosystems.

I will review how this information is used to mitigate plant virus diseases, and go on to suggest how it might be used in assessing any risks posed by genetically engineered viruses or plants engineered to be virus-resistant, and to discuss whether those risks are unique. An attempt has been made to assemble information that could be relevant to risk assessment approaches and provide an entry into the plant virus literature by reference to previous publications on mechanisms of virus survival and persistence, dispersal, and establishment in nature. The focus is nearly exclusively on plant-pathogenic viruses, and will not include viruses of fungi (Koltin and Leibowitz 1988) or invertebrates (Longworth 1978, Evans and Entwistle 1987).

It is appropriate to include a discussion of the more traditional aspects of virology in a volume on genetic engineering because some of the first and most dramatic successes of biotechnology have been the genetic engineering of plants to resist viral infection. Issues are raised relative to assessing the potential environmental impact of using genetically engineered protection and of widespread growth of virus-resistant plants developed either through genetic resistance or cross-protection, and of using genetically engineered viruses. Aspects of a more molecular nature for viruses or other related sequences are considered in Chapter 7.

The Nature of Plant Viruses

Viruses are unique microorganisms that are frequently excluded from the realm of the living because of their noncellular nature, absence of any energetic capability, and possession of little or no enzymatic activity. Yet their genetic nature and capability of directing their own replication warrants their consideration as "living" organisms having the capability of persistence and dissemination.

The simplest viruses are composed exclusively of a nucleic acid genome and a presumably protective shell constructed of protein. The particle of some of the more complex viruses contains lipid in a membranelike structure that forms an envelope. About 80% of the plant viruses have a genome that consists of single-stranded RNA containing four to eight genes (Harrison and Robinson 1988). The remaining 20% of the viruses have a DNA genome and belong to one of two groups, geminivirus and caulimovirus, or have an RNA genome and belong to one of two families, Rhabdoviridae or Reoviridae.

The single-stranded RNA viruses are classified into groups that possess a series of characteristics in common (Francki 1981, Matthews 1982). The criteria include nature of the nucleic acid, nature of the protein, size and type of symmetry of the capsid (cubic or helical), presence or absence of an envelope, site for intracellular multiplication, and biological relationships with host plants and vectors. Among the ssRNA viruses, supergroups have been proposed to include viruses that are similar in their genome organization and strategy of replication (Goldbach 1986).

Plant viruses differ from other viruses in one important way: they are incapable of infecting an intact, nonwounded plant cell. Once introduced into a cell via a wound, however, replication proceeds as it does with most viruses except that plant viruses are not released from the cell by either lysis or budding. Instead, since cells within a plant are connected via plasmodesmata, viruses pass through them to spread into other tissues. Cell-to-cell movement appears to be facilitated by modification of plasmodesmata by a virus-encoded protein (Hull 1989). Some viruses, such as members of the luteovirus group, are confined to phloem tissues, but most others spread to all tissues within the host plant except perhaps meristems and seeds. Although viruses do not lyse plant cells to accomplish their release, they may be released in vascular fluids of living plants via root exudates and water of guttation (Matthews 1981, Walkey 1985). Viruses are usually inactivated upon death of their plant host.

Experimental Methods with Viruses

One property of all plant viruses, with the possible exception of cryptic viruses (Boccardo et al. 1987), is that they can be transmitted from one plant to another. Thus Beijerinck called his unknown agent of the mosaic disease of tobacco *contagium vivum fluidum* because he had demonstrated its contagious nature. In practice today, experimental transmission is usually done by grinding a leaf from a diseased plant, in the presence of stabilizing buffer, and then using this sap to inoculate a leaf of a healthy plant by applying it to the surface of a leaf with sufficient force to assure a microwound through which the virus penetrates. If a drop of this sap were merely placed onto a healthy leaf, there would be no infection. Without some type of wound, plant viruses are not generally regarded as capable of passing through the cuticle or other structures on leaves and entering cells. Roots can also be inoculated, but this is generally not done experimentally except with soilborne viruses. Some or all members of the geminivirus, luteovirus, and closterovirus groups, and of the Reoviridae and Rhabdoviridae families, are not mechanically transmissible, and only grafting or specific vectors can effect transmission.

Most plant viruses have been purified and antibodies prepared (Gibbs and Harrison 1976, Matthews 1981). Numerous serological tests are utilized for rapid and sensitive detection of viruses from field-grown plants or from vectors. One of the early applications of enzyme-labeled immunosorbent assays (ELISA) was with plant viruses (Clark and Adams 1977). Examples of ELISA applications are numerous and include, for example, detection of tomato spotted wilt virus in individual thrip vectors (Cho et al. 1988) and of barley yellow dwarf luteovirus in aphid vectors and plants (Clement et al. 1986). Viral nucleic acid detection by either direct visualization of the double-stranded RNA which is formed during the replication of single-stranded RNA viruses (Dodds et al. 1988, Jordan and Dodds 1985), or by labeled cloned DNA probes specific for viral sequences (Palukaitis et al. 1985, Polston et al. 1989), has also been widely used for virus detection.

Ecology of Plant Viruses

Viruses are obligate parasites, a fact that greatly influences their persistence and survival in nature. Three components have traditionally been considered in discussions of ecology of plant viruses (Matthews 1981), more recently considered as mechanisms of the "pathosystem" (Robinson 1980, Tomlinson 1987). These are (1) a host in which it can multiply, (2) a means of spreading effectively to and infecting fresh plant, and (3) a supply of fresh plants to which it can spread.

Experiences gained by plant virologists over more than 60 years since the work of Doolittle and Walker (1925), have led to a rather complete understanding of the principles of ecology and epidemiology of many plant viruses. The goal desired, but not always achieved, is discovery of an action that will control the virus. Chapters on virus ecology can be found in general textbooks on plant viruses (Gibbs and Harrison 1976, Matthews 1981, Walkey 1985) as well as in any introductory plant pathology textbook.

Plant virus ecology seems to be coming of age, as it is now being included in some general publications on virus ecology (Cooper and MacCallum 1984, Mayo and Harrap 1984), although not in others (Block and Schwartzbrod 1988). Discussions on viruses are also being increasingly included in general treatises on the epidemiology of plant diseases (Fry 1982, Robinson 1987, Thresh 1983, 1987).

Many specific reviews have appeared describing ecological significances of soil and water (Harrison 1977, Koenig 1986), wild plants and weeds (Duffus 1971, Bos 1981, Hammond 1982), soilborne (Murant and Lister 1967) and aerial (Raccah 1986) vectors, and general cropping practices (Broadbent 1976, Thresh 1982) in the survival and dissemination of viruses in nature.

Viruses that infect specific crops are often covered separately: e.g., viruses of maize (Knoke and Louie 1981), ornamental trees (Cooper 1988), potato (Peters 1987), vegetable crops (Lovisolo 1980, Tomlinson 1987), and fruit crops (Bovey et al. 1980, Converse 1987, Fridlund 1989). Other publications present details on specific viruses (Irwin and Goodman 1981, Ruesink and Irwin 1986, Quiot 1980, and see also Kurstak 1981, McLean et al. 1986) or groups of viruses (Gooding 1986, Falk and Duffus 1988, Garrett et al. 1985, Kurstak 1981).

The underlying intent of most of these articles is to describe current knowledge concerning the disease cycle and how this knowledge is used to design a strategy to mitigate the disease caused by the virus and decrease plant damage and economic losses. Avoidance of infection is the primary strategy, for once a plant is infected it cannot, with few exceptions (Hansen 1988, Walkey 1985), be "cured."

The terms "ecology" and "epidemiology" are generally used in the plant virology literature. They essentially encompass all factors involved in survival and persistence of viruses, their dispersal and dissemination, and their establishment. Since viruses are not free-living and do not become established in the same way as other organisms, the terms may be somewhat overlapping.

Survival, persistence, and dispersal of viruses

Perennation in plants A virus essentially survives as long as it is associated with cells in which it can replicate. Generally, once a plant is infected it remains infected, and all progeny taken as vegetative cuttings, sprouts, etc., will maintain the virus. Vertical transmission through seeds is variable, but occurs with high frequency with some viruses (Stace-Smith and Hamilton 1988). Some very stable viruses, such as tobacco mosaic virus (TMV) will persist in dry leaf tissue for years (Gooding 1986, Matthews 1981). Early studies claimed TMV was inactivated by flue curing of tobacco leaves (Thornberry et al. 1937), yet it can be recovered from most cured tobacco products (Gooding 1969).

Association with vectors Several viruses have a persistent relationship with biological organisms that actively transmit them between plants. Viruses that belong to the Reoviridae and Rhabdoviridae are known to replicate in their insect vectors, mainly aphids and leafhoppers, and be transmitted through the eggs of the vectors (Black 1984, Sylvester 1969, Sylvester and Richardson 1970). Most other viruses do not replicate. A subgroup of the potyviruses are transmitted by mites, in which they persist for the life of the mite, but there is no evidence for viral replication (Takahashi and Orlob 1969). The nematode-borne viruses, members of nepovirus and tobravirus groups, are known to persist in the vectors in soil for long periods of time (Bitterlin and Gonsalves 1986, Cooper and Harrison 1973, Harrison 1977). Similarly, viruses transmitted by soil-inhabiting fungi persist in or on resting spores and survive long periods of drying (Jones and Harrison 1969, Rao and Brakke 1969).

Survival in an abiotic environment Once outside the safe environment of a plant cell, viruses do not generally survive for a very long time. Initial classifications of viruses had as a primary characteristic the "longevity in vitro," and inactivation times range from a few minutes for the more labile viruses to several years for stable viruses such as tobamoviruses and potexviruses (Matthews 1981).

TMV is not known to be transmitted by a biological vector, yet it survives and persists in soils (Johnson and Hoggan 1937, Gooding 1969, 1986, Gooding and Todd 1976). The medium of persistence, however, may be dead plant residues since their destruction is known to decrease soil contamination by TMV (Gooding and Lucas 1969). Johnson and Hoggan (1937) suggested that TMV was inactivated by common microorganisms, yet the mechanism has not been established. Later experiments by Cheo (1980, Cheo and Nickoloff 1980, 1981) have confirmed these early observations.

The recent review by Nienhaus and Castello (1989) presents valuable accumulated reports of viruses recovered from soil and water from forests, and discusses the epidemiology of viruses in forest ecosystems from the ecological point of view. Most of the viruses are those that do not require a mechanism for rapid dispersal for their survival, since the woody plant provides a long-lived virus reservoir. Most are WILPAD viruses (see under "Establishment in the Environment") and also "generalist" and "equilibrium" viruses. Among the viruses most frequently detected in forest soils and waters are tobamoviruses and potexviruses.

One of the most surprising sites in which plant viruses have recently been found to survive is in river waters (Koenig 1986, Tosic and Tosic 1984). And by the far the most intriguing survival environment tested is that reported in the study by Tomlinson and Faithfull (1982), who found that when purified tomato bushy stunt virus was passed through the alimentary tract of human volunteers, it remained infective.

Dispersal of viruses The way a virus is dispersed between plants and the various factors influencing this dispersal are usually considered the dominant ecological factor in virus survival in a given ecosystem. Thresh (1982) has applied the term "vertical transmission" to those viruses spread from a parent plant to its progeny, and "horizontal transmission" to those spread between plants essentially growing at the same time and in the same place.

Several volumes have been published that have dealt specifically with relations of plant viruses with their vectors (Harris and Maramorosch 1977, 1980, Maramorosch and Harris 1981) and their role in plant virus ecology and epidemiology (McLean et al. 1986, Plumb and Thresh 1983).

Establishment in the environment

Ecosystem adaptation Harrison has proposed two broad types of plant viruses based on their adaptation to become established in specific ecosystems and types of plant communities (Harrison 1981, 1983). In the first type he includes viruses in cultivated plants in which the environment favors contact-transmitted potexviruses and tobamoviruses, both of which produce high concentrations of virus particles that are relatively stable inside and outside their hosts. Another virus group that fits the pattern of the cultivated-plant-adapted (CULPAD) viruses is the ilarviruses, since they are spread in pollen to the plant pollinated and are mainly found in woody species. Nienhaus and Castello (1989) suggest these viruses are "specialists" that infect a restricted range of host-plant species that are themselves closely related.

Conversely, "generalist" viruses have a range of plants that can serve as hosts and can survive in ecosystems with a great deal of species diversity (Thresh 1983). These include viruses such as nepoviruses, tobraviruses, geminiviruses, and luteoviruses, which have a wide range of susceptible plant species and the ability to survive in any of them. These wild-plant-adapted (WILPAD) viruses often have a very specific association with their vector and long persistence times in their vectors (Harrison 1983, Nienhaus and Castello 1989). The vectors include nematodes for the first two virus groups, and either leafhoppers, whiteflies, or aphids for the latter two groups. Thus CULPAD viruses are particularly well-adapted for survival in crops, and WILPAD viruses survive in wild plants and infect crop plants only incidentally (Harrison 1983, Thresh 1983).

The survival patterns of viruses have also been distinguished recently in terms used by ecologists: "opportunist," or "r-selected," species and "equilibrium," or "K-selected," species (Thresh 1983). The "opportunist" viruses are those that multiply and spread rapidly and invade short-lived, annual crops. They rapidly and readily invade and exploit new sites, and have rapid increases and decreases in populations. This behavior would be characteristic of viruses with active, airborne arthropod vectors that search out and colonize new plantings of crops each season. The CULPAD viruses generally show these characteristics.

In contrast, the "equilibrium" viruses are those that do not invade new sites readily, but persist for long periods of time in the same site and reappear in susceptible crops whenever they are planted. Often they have no known vector or have soilborne vectors, either nematodes or fungi. These viruses survive either within or on the vector, in plants or plant parts such as seeds or roots, or perhaps even as free virus in soil and water. Most generalist viruses are

WILPAD viruses, including the following viruses or groups: nepoviruses, tombusviruses, and tobacco necrosis virus (Nienhaus and Castello 1989).

Historical aspects Smith (1934) made the observation that virus diseases of plants were of more importance than they had been even a decade before. He concluded that this was not due to the appearance of new viruses but to increasing spread of these agents with improved methods of transport and development of communications between countries, and added that they also become adapted to new host plants. He also made the statement (Smith 1934) that the first disease attack of a certain virus disease was always in association with agricultural areas, particularly disturbed virgin areas. Bawden (1943) acknowledged that many workers were convinced virus diseases were more prevalent, but countered that there was no reason to consider them to be afflictions of modern times. Bawden suggested potato viruses were actually introduced into Europe, since the degeneration was unknown before 1770.

Experimental introductions Scientists have often used the practice of field inoculations in order to evaluate germ plasm and select for resistance, and to study the inheritance of the reaction (Roane et al. 1973, 1986). Greenhouse experiments provide valuable initial screening but have not always correlated with results from field-grown plants. Differences could be due either to the susceptibility of the plant to initial infection, or to the influence of environmental factors on the development of disease symptoms. The fate of the virus from these introductions has seldom been reported, if it has been monitored.

In our work to study the genetics of resistance of maize to maize dwarf mosaic virus (MDMV), we chose a site for the tests that was free of the reservoir host Johnsongrass to avoid inadvertent infection with other MDMV strains or other viruses also known to use the same weed as a reservoir (Roane et al. 1973, 1989). Since other known hosts in this area were annual grasses, we did not expect MDMV to become established. To test this, we would have to have a means of distinguishing the introduced strain from any resident strain.

One observation virologists have often made is that the greatest diversity of viruses and the incidence of new viruses often occurs near agricultural research farms in which a very diverse array of plants has been grown. In my personal experience, a small research plot near my laboratory has a number of viruses established in wild clover plants. I have often wondered whether this high incidence was related to use of the plot for bean-breeding experiments, including growing of import beans potentially carrying viruses in the seeds, for over 30 years. This, and many other mysteries of the pattern of establishment of viruses in nature, will probably remain a matter of speculation.

Strategies for Mitigating Plant Virus Diseases

Designing the strategy

Data collection and design Control measures for plant virus diseases are developed by designing a strategy to interrupt the disease cycle and thus avoid infection of high-value plants. Such mitigation strategies are based on knowledge of and management of the survival, dissemination, and establishment of viruses. These are the only effective approaches to controlling losses caused by virus diseases of plants.

Chemical control of viruses is not practical on a field scale, even though there are several candidate compounds that successfully mitigate or eliminate a virus infection (Hansen 1988). At present, economic considerations restrict the potential use of chemotherapeuticals to particular situations, such as elimination of virus from elite mother-plant stocks and protection of extremely high value plants such as orchids. If, however, elimination of virus from plants in an experimental test area were desired, chemicals could probably be selected that would be effective.

The strategies are designed following empirical studies, often experiments in which viruses have been intentionally introduced in fields in order to monitor their spread and to test the efficacy of approaches to limit their spread. It is from such experiments that the most information is available regarding the fate of a virus when it is intentionally introduced in a research plot.

Reduction of virus inoculum

Sanitation Glasshouse crops and ornamentals that are subject to frequent handling are often infected with viruses spread by contact. Control recommendations include removing infected plants as soon as they are detected, and decontaminating surfaces on which viruses might survive (Broadbent and Fletcher 1963). Most viruses, even TMV, will be inactivated on human hands by thorough washing with soap and water. A strong solution of trisodium phosphate will usually eliminate TMV on surfaces. Cutting knives used to remove or divide plant parts should be subject to regular decontamination, since they are known to spread virus quite effectively.

Clean-stock programs Viruses in plants that are propagated from tubers or cuttings are nearly always vertically transmitted and infect all propagules. Thus for nearly all such plants that are grown commercially, there are in place programs to detect known viruses and certify planting material distributed to growers as free from certain viruses. If the virus is also horizontally transmitted, even a small percentage of infected planting stock can provide inoculum for a large acreage of the crop, and reinfection will occur over time (Bar-Joseph et al. 1989, Moran and Wilson 1985; Peters 1987). These testing and certification programs have traditionally been operated by state and/or federal agencies (Hansen 1985). Foundation plantings of virus-free mother plants are maintained under some type of isolation and used as source material for horticulturists who increase cuttings and budwood for commercial sale. Some of the large private nurseries, particularly those growing ornamental plants and potatoes, are incorporating sensitive virus detection methods into their plant production systems.

Elimination of virus from infected stocks is sometimes possible through extended heat treatment (Walkey 1985) or by treatment with chemotherapeutants (Hansen 1988). Since virus often does not invade meristems, tissue culture has also proven an effective method for freeing many plants from virus (Walkey 1985).

Clean-seed programs Programs are also in place for many crops to detect viruses in seeds in the crop, and to market only those that are virus-free or have a very low level of virus. Viruses that are commonly seedborne include those transmitted by nematodes (Murant and Lister 1967) and aphids (Demski 1975, Irwin and Goodman 1981), and also those with no known vector such as TMV (Gooding 1986) and barley stripe mosaic virus (Nutter et al. 1984). There is really no successful treatment to rid seeds of internally carried viruses. However, viruses carried on the seed coat, such as TMV on tomato seeds, can be inactivated by trisodium phosphate and sodium hypochlorite without harm to the seed (Gooding 1975).

Stace-Smith and Hamilton (1988) have summarized information on seedborne viruses. They describe how allowable threshold levels of infected seed are determined, and the pros and cons of growing-out tests versus ELISA to detect virus in seeds. A tremendous variability in allowable level of virus to be certified exists among various crops, even for those viruses with the same type of vector. For example, lettuce mosaic potyvirus must not be detectable in more than 1 in 10,000 seeds for the seeds to be certified in California.

In soybean, Ruesink and Irwin (1986) defined an acceptable level of soybean mosaic potyvirus (SMV) in soybean as that level which would cause no more than 1% yield loss and conducted experiments to determine this value. Epidemics of SMV are known to occur only when virus-infected soybeans are planted and aphid vectors effect horizontal transmission. Virus-infected seeds were planted at known incidence levels, and the spread of virus into adjacent surrounding rows seeded with virus-free seed was monitored at regular intervals by both symptom development and ELISA. The experiment was conducted in several locations in different states (Illinois, Georgia, Louisiana, Mississippi, Virginia), thus under different environmental conditions. Acceptable seed transmission rate was 1.0% under low vector intensity, but was reduced to 0.02% if vector intensity was high.

Peanut mottle potyvirus is seedborne in peanut at a very high incidence, but since it causes little yield loss, growers were not convinced certification of peanut seeds for freedom from this virus was necessary (Demski 1975).

Reduction in perennial reservoir hosts Natural reservoirs of viruses are often weeds (Bos 1981, Duffus 1971, Hammond 1982, Knoke and Louie 1981, Powell et al. 1984, Rist and Lorbeer 1989, Tomlinson and Carter 1970). A standard recommendation, therefore, for reducing virus inoculum for an annual crop is to apply weed-control measures to eliminate weeds from which vectors can acquire the virus and effect horizontal transmission (Matthews 1981, Walkey 1985). Cultivation and standard herbicide applications timed to coincide with emergence of new seedlings and peaks in resident or migrant vector populations are usually successful.

For this approach to be effective, however, the target weeds must be known and it must be possible to control them within the distance the vector can travel. If this area includes abandoned fields, roadsides, stream banks, fence rows, or residential areas, timely and effective control may be difficult.

Reduction in other perpetuating sources There are many examples in which other crops growing nearby provide the initial inoculum (Zitter and Simons 1980). Overlapping celery crops in Florida provided a perpetual source of celery mosaic potyvirus until scheduled breaks in the cropping cycle were instituted (Zitter and Simons 1980). Continuous cropping as well as volunteer plants contributed to epidemics of viruses in carrots in Washington (Howell and Mink 1977). Clovers in pastures and roadsides, particularly in the southeastern United States, have a high incidence of viruses such as peanut stunt cucumovirus (PSV) and are a source of viruses for infection of legume field crops (McLaughlin and Boykin 1988). Careful choice of field location and judicious use of volunteer plant control has provided disease mitigation, as shown by the fact that PSV is seldom found in Virginia peanuts (Tolin et al. 1970).

Crop rotation and destruction of crop residues, particularly roots, will reduce inoculum level in tobacco field soils (Gooding and Lucus 1969).

Site and practices selection For certain viruses, particularly those with soilborne vectors, there is no effective means of mitigation in sites in which the virus and its vector are well established. Nematicides or fungicides seldom provide total control since they are effective in reducing vector population but not to a level ineffective in transmitting the virus (Harrison 1977, 1981). Such sites are generally unsuitable for production of the specific susceptible crop, but would provide a good test site for breeding and selecting resistant crops or for testing the effectiveness of engineered protection. Fungusborne viruses in this category include several cereal viruses (including barley yellows, soilborne wheat mosaic, and wheat spindle streak mosaic), and rhizomania of sugar beet.

Timing of the seeding of crops to avoid peak vector populations or activity periods is also an effective practice to avoid initial infection with virus (Raccah 1986, Thresh 1985, Watson et al. 1975).

Reduction in dissemination potential

Reduction in vector populations Insecticides, nematicides, and fungicides are all utilized for vector control (Walkey 1985, Zitter and Simons 1980). The success of attempts to use vector population reduction to mitigate infection of plants by virus has been variable and quite dependent upon the relationship between the virus and its vector as well as on vector efficiency (Broadbent 1969). Vectors alighting from long-distance flights essentially cannot be prevented from inoculating initial plants unless the insecticide acts quickly enough to prevent transmission. Controlling vectors, however, has controlled secondary spread of circulative viruses (Burt et al. 1964). With nonpersistent viruses, little reduction in initial infection occurs, and in some cases spread increases because of greater vector activity (Burt et al. 1964, Raccah 1986).

Nonetheless, for crops known to be infected by nematode-transmitted viruses, control of vector nematodes is required in plantings of nurseries grown for certified virus-free stocks. A good example is grapes and fanleaf virus in California and tomato ringspot virus in the eastern United States.

Interference with vector activity Aphid vectors are known to be attracted by green and yellow but repelled by short wavelengths. Reflective mulches, particularly aluminum foil, are effective in reducing virus incidence in a number of crops (Zitter and Simons 1980). Yellow sticky board traps attract many aerial vectors and thus reduce vector number and transmission.

Planting of barrier crops to block direct vector movement has long been used for mitigation of virus disease (Matthews 1981). In Georgia (Demski 1975) and Virginia (Bays et al. 1986), PMV spreads from peanut to adjacent soybean fields where it causes a serious disease. Intercropping with corn, a nonhost for PMV, will interfere with vector movement and prevent inoculation of the soybean.

Interference with the transmission process The first report that aphid transmission of potato virus Y was prevented by oil sprayed on the surface of a leaf was by Bradley et al. (1962). Although the mechanism of interference is not understood, this method has been used to protect a number of high-value crops from viruses in sites where the quantity of inoculum source is known to be high and difficult to decrease (Zitter and Simons 1980).

The classic method during the transplanting process of controlling the spread of TMV by spraying plants or dipping workers' hands in milk is also thought to be effective by inhibiting the infection process since it does not inactivate the virus (Gooding 1986). Strangely, it has not been reported as useful with any other virus.

Reduction in crop susceptibility

Genetically mediated resistance to infection The response of a plant to a virus can be total susceptibility (showing some systemic expression of infection), hypersensitivity (necrotic local lesions), subliminal infection (confined to initially infected cells), and total immunity (nonhost, no expression of infection). Susceptible plants can be sensitive and show severe symptoms or tolerant and show very few symptoms (Walkey 1985). Any of these responses is useful in disease mitigation. Where the genetics of the resistance response have been studied, mediation by single genes has usually been found (Fraser 1987). However, the vast majority of field and, particularly, horticulture (vegetable) crops are affected severely by viruses for which there is no known or exploitable resistance (Fraser 1987, Tomlinson 1987).

To select and breed resistant plant lines, field experiments are conducted under high natural inoculum pressure or by using field inoculation. With mechanically transmissible viruses this is often done with an airbrush (Roane et al. 1973), although hand rubbing of sap onto plant leaves is also done (Kuhn et al. 1989). Manual inoculation is usually more reliable and has the advantage of assuring that all plants have been exposed to the virus. However, plants with resistance to vectors would not be selected under such experimental conditions. Natural infection would also expose plants to a broader spectrum of viruses and/or virus strains.

In some experiments, particularly with luteoviruses which are persistently transmitted, vectors raised on infected plants have been released by placing them directly on test plants in the field (Clement et al. 1986). Vectors are then killed by insecticides after an appropriate inoculation feeding time. Alternatively, cages are used to confine the aphids to the plants to be inoculated, or to protect control plants from unwanted visitation by viruliferous aphids.

Cross-protection The phenomenon of cross-protection, in which infection of a plant by one virus protects it from infection with a related virus, was first recognized more than 60 years ago (McKinney 1929). It was recognized as a potential disease-control measure soon thereafter (Deacon 1983, Fulton 1986). Since then it has been used commercially for control of tomato mosaic virus in tomato throughout the world (Brunt 1986) and of citrus tristeza virus in crops in South America (Bar-Joseph et al. 1989, Deacon 1983). Both selection of naturally occurring mild strains and deliberate mutation have been used to develop protecting strains of these two viruses. Deliberate mutation of strains or of satellites has also resulted in effective cross-protection of viruses in cucumber (Tien et al. 1987) and papaya (Yeh and Gonsalves 1984).

However, widespread use of cross-protection in commercial agriculture, particularly in the United States, has never been utilized because of uncertainties regarding the direct effect of the protecting virus, the potential of the virus to spread to other plants in which it might be more virulent, and the potential for the virus to mutate to a form that would cause a destructive disease. In addition, the mechanism of cross-protection was totally speculative (Fulton 1986, Ponz and Bruening 1986).

Resistance to dispersal The general approach to demonstrating the role of resistance to dispersal of viruses is to generate infections in the field and measure such parameters as plant response and spatial and temporal aspects of dispersal of the virus. With aphidborne, nonpersistent viruses, transmission to and among cultivars of breeding lines with resistance to virus is well documented (Raccah 1986). Numerous examples exist in which decreased incidence and severity of the virus disease is correlated with resistance of the host plant to the vector.

A common experimental design is to distribute source plants and allow natural populations of vectors to spread the virus to target plants. Lapointe et al. (1987) measured the effect of source plant on the dispersal of potato virus Y from potato to potato. Plots were established in symmetrical square arrays of susceptible target plants, arranged in five rows of five plants each, 1.8 m apart, with a mechanically inoculated source plant infested with aphids in the center. Plants were monitored for symptom development and for virus by ELISA. Dispersal of virus from the source was greatly reduced if the source plant was *Solanum berthaultii*, resistant to the aphid vector, rather than *S. tuberosum*, susceptible to the aphid vector.

Resistance to the virus as well as to the vector provides effectively resistant plants in the field and affects the ultimate dispersal of the virus in the field.

In a typical study that demonstrates this phenomenon, Gray et al. (1986) inoculated 20 randomly selected plants within each plot of 200 plants with an isolate of the potyvirus watermelon mosaic virus 2. The isolate used had been maintained by aphid transmission, since it is known that potyvirus cultures maintained by mechanical transmission often select virus with a lower probability of being aphid-transmitted. In this example, the final virus incidence was 11% lower in aphid-resistant genotypes and 33% lower in genotypes resistant to both aphids and viruses.

Potential for Genetic Engineering to Change Virus Exposures

There are several ways in which genetic engineering might be used to mitigate effects of plant viruses. The potential impacts that might be expected, and approaches to assessing their risks, are discussed in this section. The first approach to be demonstrated, and the one most widely exploited to date, is that of engineering plants for resistance to virus diseases using parts of the virus to interfere with infection (Baulcombe et al. 1987). Attempts are also being made to identify genes that mediate virus or vector resistance in plants in order to test their utility in engineering resistant plants. The possibility of engineering plant viruses to accomplish weed control, and the issues involved in addressing the inherent risks, are also discussed. Finally, there is a brief discussion of risk assessment issues that might be associated with engineering of plant viruses to be expression vectors for producing a useful product in plants (Hays et al. 1988).

Engineering virus-resistant plants

Coat-protein-induced protection Genetically engineered protection resulting from the expression of viral coat protein from viral sequences inserted into nuclear genes was first demonstrated by Powell Abel et al. (1986) with TMV in tobacco. This approach has now been extended to several other viruses and plants. The free viral protein is presumed to interfere with an early stage of infection, a mechanism that is also proposed to explain cross-protection (Register and Beachy 1988). The experiments by Beachy's group were, in fact, originally designed to test the hypothesis that viral coat protein was responsible for the cross-protection phenomenon (Beachy 1987). Field tests were first conducted with tomato engineered with TMV coat protein for TMV resistance (Nelson et al. 1988). Tests are now being conducted in many locations with several crop plants resistant to several viruses.

Protection by other viral sequences Engineered protection based on antisense RNA (Rezaian et al. 1988) and on satellite RNA (Jacquemond et al. 1988) was first demonstrated with cucumber mosaic virus in tobacco plants. These approaches, which have also been demonstrated in other plants and with other viruses, do not appear to be as effective as coat-protein-induced resistance, and are not at present being pursued as actively or by as many research groups.

Protection by introducing resistance-response genes A number of molecular responses induced in resistant plants following inoculation with a necrotizing pathogen have been identified (Collinge and Slusarenko 1987). The system most widely studied to elucidate virus resistance has been with tobacco having the *N* gene conditioning hypersensitivity to TMV (Ponz and Bruening 1986, van Loon 1987). Several new mRNAs and proteins are induced which are correlated with the appearance of necrotic lesions, but the precise role of these pathogenesis-related proteins and their relationship to the resistance genes identified by genetic crosses have yet to be demonstrated (Antoniw and White 1986, Ponz and Bruening 1986, van Loon et al. 1987). Many of the virus-induced proteins are similar to those induced by other pathogens or stresses (Legrand et al. 1987). Two proteins unique to virus infection, PR-1 and GRP, are also induced by spraying tobacco leaves with salicylic acid solutions (Linthorst et al. 1989), a treatment which renders leaves resistant to TMV (White 1979). In the most direct experiment thus far, Linthorst et al. (1989) engineered tobacco plants to express these pathogenesis-related proteins constitutively and found no effect on the susceptibility of the plants to virus. Thus it appears resistance genes activate genes for the defense-response proteins *in trans*, but may not be useful directly to engineer resistance to virus infection.

Genes for nonhost resistance As with other pathogens, resistance of plants to a particular virus is probably the rule, rather than the exception. Conversely, it is unlikely that any one plant or plant species is not infected in nature by some virus. Data on host ranges of 258 viruses, found in individual original reports of viruses and in *Descriptions of Plant Viruses*, compiled by the Commonwealth Mycological Institute/Association of Applied Biologists, indicate that about 40% infect 6 or fewer plants, and only 15% infect more than 50 plants. It is interesting to speculate on the mechanism of resistance for the virus–nonhost combination and whether a broad spectrum resistance "gene" could be identified.

Genes for vector resistance Plant molecular biologists have now begun to demonstrate that resistance to certain insects is due to production, for example, of specific proteinase inhibitors which affect the digestive systems of the insects. Whether such approaches could have, or in fact have had, a role in transmission is speculative. Issues to be addressed in assessing risk of such an approach should not, however, be different from those in which vector resistance has been utilized (Gray et al. 1986). Chances are that beetle vectors, which ingest and digest plant leaves, would be the most likely targets for this approach.

Assessing risks of using engineered resistance

The questions that might be asked are whether this type of resistance poses any unique risks to producers and consumers of the plants or to the environment in which the plants are grown. It is not within the scope of this review to consider potential risks of engineered plants to human or animal consumers. Risks to producers are those that have the potential to cause greater loss in yield and profitability than would result, for example, from the evolution of either a more invasive or a resistance-breaking virus strain which might also have greater potential for survival, dispersal, and establishment.

Evolution of resistance-breaking strains of virus The potential for the evolution of resistance-breaking strains developing from the use of genetically engineered resistance is not known, but can best be approached by analogy to experiences in using resistance for control of virus diseases (Fry 1982, Matthews 1981, Ponz and Bruening 1986, Robinson 1987, Walkey 1985). In the few cases where cross-protection has been used to induce resistance to virus, there are apparently no reports of strains overcoming the protection (Fulton 1986). However, cross-protection is often incomplete and dependent on the aggressiveness and/or the inoculum pressure of the challenging strains (Bar-Joseph et al. 1989). Since engineered protection has been shown to be effective against more than one strain, it may have distinct advantages over either classical cross-protection or use of resistant cultivars.

Resistance is widely utilized for virus control, but few systems are well studied genetically (van Loon 1987). Resistance-breaking strains are known to occur, but this phenomenon is judged to occur much less frequently than it does with fungi (Harrison 1981). Three systems—tobacco–TMV, tomato–TMV, and bean–bean common mosaic virus—are reviewed by van Loon (1987), but several other systems are also well characterized (Evered and Harnett 1987). Resistance can be expressed by hypersensitivity as for TMV, but restriction of viral replication (Kuhn et al. 1989, Ponz et al. 1987) or movement (Dawson 1967, Hull 1989, Meshi and Okada 1987, Ponz and Bruening 1986) also provide effective resistance.

Tobacco mosaic virus Tobamoviruses provide the classic example of evolution of a resistance-breaking strain, with the report of Pelham et al. (1970) that a new strain of TMV had resulted from the use of resistant varieties of tomato. Later, MacNeill and Boxall (1974) recovered a new strain of TMV following graft inoculation of resistant tomato lines, and other strains have been selected following nitrous acid mutation (Rast 1972). Strains that have been analyzed differ from wild-type virus by only one or two amino acid changes (Hull 1989, Meshi et al. 1988). After a decade or more of widespread use, strains that break resistance have not been become established and are only rarely reported, usually from plants grown at high temperatures in greenhouses.

In tobacco, the *N* gene has been used in burley tobacco for many years, but no resistance-breaking strains have been identified (Gooding 1986). Resistant cultivars of flue-cured tobacco have not been used as widely, and the common strain of TMV that prevails in tobacco fields (Ford and Tolin 1983, Gooding 1986) appears to have changed little over time.

Legume-infecting viruses Strains of viruses capable of overcoming resistance genes have been recognized in a number of important crops, and examples are given of three viruses in soybean and one in cowpea. Whether their evolution was influenced by the presence of resistance genes in soybean remains speculative.

Demonstration of strain development in cowpea chlorotic mottle virus (CCMV) on cowpeas in greenhouse experiments (Wyatt and Kuhn 1980) was followed by a report of distinguishing seven natural variants of CCMV by different disease reaction on six soybean genotypes (Paguio et al. 1988). Additionally, when one variant was passaged through a resistant soybean cultivar, two new CCMV strains were formed.

Cultivars of soybean have been used to differentiate between resistance-breaking strains of SMV (Buss et al. 1989, Cho and Goodman 1979, Kiihl and Hartwig 1979, Roane et al 1973, Ross 1969) and peanut mottle virus (PMV) (Bays et al. 1986). The SMV strains, which were originally isolated from breeders' germ plasm and plant introductions, cause systemic mosaic, severe systemic necrosis, local necrosis, or no infection, depending on the cultivar inoculated (Cho and Goodman 1979). They have been shown to overcome different alleles at the same locus and appear to follow a gene-for-gene relationship (Buss et al. 1989, Roane et al. 1986).

Resistance-breaking strains of SMV will sometimes arise out of a particular strain upon passage through resistant cultivars (Hunst and Tolin 1982). Variant strains are occasionally isolated from soybean fields or breeding plots in the United States, but these have yet to become established. Establishment may have occurred in Korea, however, since a severe necrotic disease was caused by an SMV strain that developed and became established when a resistant cultivar was planted widely (Cho et al. 1977). A variant of PMV arose following culture in an SMV-resistant cultivar (Bays et al. 1986), perhaps because genes for resistance to the two viruses are closely linked (Roane et al. 1973).

Potential for new virus strains Genetic changes in viruses occur during replication and are most likely in response to selection pressure. RNA viruses are generally thought to be less stable genetically than are DNA viruses and have a high rate of mutation (van Vloten-Doting and Bol 1988), and many plant viruses exist as numerous strains in nature. Yet resistance of plants to viruses has been much more durable than it has to fungi (Harrison 1981).

The question that should be asked to assess the potential for development of new strains is, then, whether plants engineered to be resistant would cause any change in selection pressure. It is possible, for example, that coat-protein-engineered protection may actually pose less evolutionary pressure on the virus than do other types of resistance, because no initial infection and replication appears to occur.

A second potential risk could be the creation of a new virus via interactions between the protecting elements of virus and another virus in plants that became infected with other viruses. For example, pseudorecombination, transcapsidation, or RNA recombination (Ahlquist and French 1987, Bujarski and Kaesberg 1986, van Vloten-Doting and Bol 1988) could possibly occur.

Engineering viruses for control of weeds

Biological control of weeds by viruses has been considered in general discussions of biocontrol of weeds, but it is generally concluded to be less of a potentially successful venture than is the use of other biocontrol agents (Charudatten and Walker 1982, Hatzios 1987, Wilson 1969). The basic problem encountered is that the virus would have to be specific for the target weed, yet infect and be highly virulent on all genetically diverse forms of it. It would also need to have a specific vector that would disperse the virus to all populations of weeds in the target ecosystem. Nonetheless, with increased knowledge of viral genomes and the ability to manipulate them genetically, the management of weeds by specific viral infections might one day be feasible and might first be utilized in managed ecosystems that have specific weed problems.

Viruses in weeds Weeds are often infected by many viruses in nature, and have commonly been detected in searches for reservoirs of viruses infecting many other species including crop plants and desirable species (Duffus 1971, Hammond 1982, Powell et al. 1984). The virulence of the virus is low in weed species, and is often much higher in crops. Incidence of infection is higher in crops or managed ecosystems because cropping patterns of extensive areas of monoculture of genetically uniform crop species enhance the exposure to virus and vectors (Bos 1981, Thresh 1982, Matthews 1981). When cropping systems result in uniform weed infestations and vectors are present, virus infection in the weeds may also be uniform and of high incidence (Duffus 1971, Thresh 1982).

Proposed biocontrol systems One report attempting to utilize a virus as a biocontrol agent is that of Charudattan et al. (1980), who described a potyvirus naturally infecting four species of closely related plants in Argentina. They advocated it as a safe and promising biocontrol agent of the milkweed vine, or "strangler vine," a very important problem weed in citrus in Florida. The host range of this virus was restricted to six genera within the same family to which the weed belongs, Asclepiadaceae. It did not infect 150 test cultivars representing 108 species of 75 genera in 25 plant families. The virus-infected weeds grew less rapidly than did noninfected plants, but all tests reported were done within quarantine facilities. The authors predicted that, if released, the virus would be spread by aphids to many milkweed vines and would act synergistically with endemic cucumber mosaic virus and root-infecting fungi to cause severe symptoms and weaken or kill the weed vines. The restricted host range would, however, assure the safety of its use.

Zettler and Freeman (1972) concluded that there was potential for control of vascular aquatic weeds by viruses. Once infected, such weeds in a given location should perpetuate viruses indefinitely since the weeds are largely non-seed-producing and thus would have little opportunity to produce virus-free plants. Initial infection should be possible since potential vectors, namely aquatic chytrid fungi and dorylaimid nematodes, are common in aquatic waterways. Aphids, leafhoppers, and mites have been reported on aerial portions of water hyacinths and several other aquatic plants (Zettler and Freeman 1972). However, reports of isolation of viruses from aquatic weeds are rare. This probably reflects lack of effort and support for this potentially important area rather than a lack of viruses. Viruses have, however, been isolated from algae but have not been exploited to mitigate unwanted algal blooms in aquatic environments (van Etten et al. 1987).

Assessment of risks and needs for weed biocontrol An accurate assessment of risks of controlling weeds with viruses will require a much greater knowledge of the role of specific viral sequences controlling infection, host range, virulence, and dispersal than is currently available. It is also likely that the way in which a virus is delivered to the site would affect any assessment of the exposure received by the site. For example, if the virus is delivered by release of viruliferous aerial vector, dispersal would be expected over a greater area than if a virus has a soilborne vector, is not vector-transmitted, or if its vector is not present.

The critical information needed for assessing risks posed by genetically engineered viruses should be no different from that needed to assess risks of using nonengineered virus, and might include:

- An understanding of attributes of a virus, such as virulence, host range, vector specificity, and survival and dispersal requirements, that contribute to success in biological control of weeds
- A means of detecting the desirable attributes as well as the undesirable attributes of the virus
- A means of monitoring the stability of these attributes in the virus in the environment in which it is used
- A means of monitoring the effect of the engineered virus on the genetic stability of the target weed
- An understanding of the release site and how climatic and other environmental variables and local flora and fauna affect infection, dispersal, population dynamics, and ultimately, the safety of the release

In all likelihood, greater knowledge of the above attributes would be available for an engineered virus than for a wild-type or selected strain of a virus.

Engineering viruses as vectors

Initially, many studies were conducted attempting to develop plant virus vectors to transform plant cells. Successful vectors to transform plants now include the 35S promoter of cauliflower mosaic virus (CaMV), but the whole virus is not useful because of restrictions on the capacity of DNA that can be added in a nonessential gene position (Woolston et al. 1987). However, it has been possible to infect plants with a defective but replicating geminivirus, in cytoplasm, that encodes and expresses foreign genes (Davies and Stanley 1989, Hays et al. 1988). The sequences coding for viral coat protein are deleted, since they are not required for movement throughout the plant (Gardiner et al. 1988). Thus a plant may be used for production of a useful product in plants, and potentially become a new field crop.

Exposure assessment issues for this application should be the same as for any virus infection. Provided the viral vector was engineered to preclude survival and dissemination, such an application should pose a lower risk than a normal virus disease. For example, Ward et al. (1988) utilized a geminivirus with a deletion of coat protein in order to preclude transmission by the whitefly vector.

The question regarding potential interactions of the vector DNA or RNA sequences with invading viruses would again have to be addressed. The risk would be that of the vector sequences acquiring a coat protein to enable their dissemination, or of the vector sequences modifying the invading virus to change a property affecting its ecology and, potentially, its host range (Davies and Stanley 1989). Information regarding vector specificity and host range should be addressed, since issues should be the same as those for any virus. The patterns of natural variation observed in four groups of viruses (Harrison and Robinson 1988) might be of value in predicting possible outcomes.

References

Ahlquist, P., and French, R. 1987. Molecular genetic approaches to replication and gene expression in brome mosaic and other RNA viruses. In *RNA Genetics*, vol. III: *Variability of RNA Genomes* (E. Domingo, J. J. Holland, and P. Ahlquist, eds.). CRC Press, Boca Raton, Florida, pp. 53–69.

Antoniw, J. F., and White, R. F. 1986. Changes with time in the distribution of virus and PR protein around single local lesions of TMV infected tobacco. *Plant Mol. Biol.* 6:145–149.

Bar-Joseph, M., Marcus, R., and Lee, R. F. 1989. The continuous challenge of citrus tristeza virus control. *Annu. Rev. Phytopathol.* 27:291–316.

Baulcombe, D. C., Hamilton, W. D. O., Mayo, M. A., and Harrison, B. D. 1987. Resistance to viral disease through expression of viral genetic material from the plant genome. In *Plant Resistance to Viruses* (D. E. Evered and S. Harnett, eds.). Ciba Foundation Symp. no. 133. Wiley, Chichester, U.K., pp. 170–184.

Bawden, F. C. 1943. *Plant Viruses and Virus Diseases.* Chronica Botanica. Waltham, Massachusetts, 294 pp.

Bays, D. C., Tolin, S. A., and Roane, C. W. 1986. Interactions of peanut mottle virus strains and soybean germ plasm. *Phytopathology* 76:764–768.

Beachy, R. N. 1987. Genetic engineering of plants for protection against virus diseases. In *Plant Resistance to Viruses* (D. E. Evered and S. Harnett, eds.). Ciba Foundation Symp. no. 133. Wiley, Chichester, U.K., pp. 151–169.

Bitterlin, M. W., and Gonsalves, D. 1986. Persistence of tomato ringspot virus and its vector in cold stored soil. *Acta Hort.* 193:119–124.

Black, L. M. 1984. The controversy regarding multiplication of some plant viruses in their insect vectors. In *Current Topics in Vector Research*, vol. II (K. F. Harris, ed.). Praeger, New York, pp. 1–30.

Block, J. C., and Schwartzbrod, L. 1988. *Viruses in Water Systems. Detection and Identification.* VHS Publishers, 136 pp.

Boccardo, G., Lisa, V., Luisoni, E., and Milne, R. G. 1987. Cryptic plant viruses. *Adv. Virus Res.* 32:171–214.

Bos, L. 1981. Wild plants in the ecology of virus diseases. In *Plant Diseases and Vectors: Ecology and Epidemiology* (K. Maramorosch and K. F. Harris, eds.). Academic, New York, pp. 1–33.

Bovey, R., Gartel, W., Hewitt, W. B., Martelli, G. P., and Vuittenez, A. 1980. *Virus and Virus-like Diseases of Grapevines.* Editions Payot, Lausanne, 181 pp.

Bradley, R. H. I., Wade, C. V., and Wood, F. A. 1962. Aphid transmission of potato virus Y inhibited by oils. *Virology* 18:327–328.

Broadbent, L. 1969. Disease control through vector control. In *Viruses, Vectors and Vegetation* (K. Maramorosch, ed.). Interscience, New York, pp. 593–630.

Broadbent, L. 1976. Epidemiology and control of tomato mosaic virus. *Annu. Rev. Phytopathol.* 14:75–96.

Broadbent, L., and Fletcher, J. T. 1963. Epidemiology of tomato mosaic. IV. Persistence of virus on clothing and greenhouse structures. *Ann. Appl. Biol.* 52:233–241.

Brunt, A. A. 1986. Tomato mosaic virus. In *The Viruses*, vol. 2: *The Rod-Shaped Plant Viruses* (M. H. V. van Regenmortel and H. Fraenkel-Conrat, eds.). Plenum, New York, pp. 181–204.

Bujarski, J. J., and Kaesberg, P. 1986. Genetic recombination between RNA components of a multipartite plant virus. *Nature* (London) 321:528–531.

Burt, P. E., Heathcote, G. D., and Broadbent, L. 1964. The use of insecticides to find when leaf roll and Y viruses spread within potato crops. *Ann. Appl. Biol.* 54:13–22.

Buss, G. R., Chen, P., Tolin, S. A., and Roane, C. W., 1989. Breeding soybeans for resistance to soybean mosaic virus. In *Proc. World Soybean Res. Conf. IV.*, Buenos Aires, Argentina, pp. 1144–1154.

Charudattan, R., Zettler, F. W., Cordo, H. A., and Christie, R. G. 1980. Partial characterization of a potyvirus infecting the milkweed vine, Morrenia odorata. *Phytopathology* 70:909–913.

Charudattan, R., and Walker, H. L. 1982. *Biological Control of Weeds with Plant Pathogens*. Wiley, New York, 293 pp.

Cheo, P. C. 1980. Antiviral factors in soil. *Soil Sci. Soc. Am. J.* 44:62–67.

Cheo, P. C., and Nickoloff, J. A. 1980. The priming effect on rate of tobacco mosaic virus degradation in soil columns. *Soil Sci. Soc. Am. J.* 44:883–884.

Cheo, P. C., and Nickoloff, J. A. 1981. Rate of tobacco mosaic virus degradation in a field plot after repeated application of the virus. *Soil Sci.* 131:284–289.

Cho, E. K., Chung, B. J., and Lee, S. H. 1977. Studies on identification and classification of soybean viruses in Korea. Etiology of a necrotic disease of Glycine max. *Plant Dis. Rep.* 61:313–317.

Cho, E. K., and Goodman, R. M. 1979. Strains of soybean mosaic virus: classification based on virulence in resistant cultivars. *Phytopathology* 69:467–470.

Cho, J. J., Mau, R. F. L., Hamasaki, R. T., and Gonsalves, D. 1988. Detection of tomato-spotted wilt virus in individual thrips by enzyme-linked immunosorbent assay. *Phytopathology* 78:1348–1352.

Clark, J. F., and Adams, A. N. 1977. Characteristics of the microplate method of enzyme-linked immunosorbent assay for the detection of plant viruses. *J. Gen. Virol.* 34:475–483.

Clement, D. L., Lister, R. M., and Foster, J. E. 1986. ELISA-based studies on the ecology and epidemiology of barley yellow dwarf virus in Indiana. *Phytopathology* 76:86–92.

Collinge, D. B., and Slusarenko, A. J. 1987. Plant gene expression in response to pathogens. *Plant Mol. Biol.* 9:389–410.

Converse, R. H. (ed.). 1987. *Virus Diseases of Small Fruits*. U.S. Department of Agriculture, Agric. Hndbk. USGPO no. 631, 277 pp.

Cooper, J. I. 1988. Ecology of viruses in ornamental and forest trees. *Acta Hort.* 234: 359–364.

Cooper, J. I., and Harrison, B. D. 1973. The role of weed hosts and the distribution and activity of vector nematodes in the ecology of tobacco rattle virus. *Ann. Appl. Biol.* 73:53–66.

Cooper, J. I., and MacCallum, F. O. 1984. *Viruses in the Environment*. Chapman and Hall, London, 182 pp.

Davies, J. W., and Stanley, J. 1989. Geminivirus genes and vectors. *Trends Genet.* 5: 77–81.

Dawson, J. R. 1967. The adaptation of tomato mosaic virus to resistant tomato plants. *Ann. Appl. Biol.* 60:209–214.

Deacon, J. W. 1983. *Microbial Control of Plant Pests and Diseases. Aspects of Microbiology*, 7. American Society for Microbiology, Washington, D.C., 88 pp.

Demski, J. W. 1975. Source and spread of peanut mottle virus in peanut and soybean. *Phytopathology* 65:917–920.

Dodds, J. A., Valverde, R. A., and Matthews, D. M. 1988. Detection and interpretation of dsRNA. In *Viruses of Fungi and Simple Eukaryotes* (Y. Koltin and M. J. Leibowitz, eds.). Marcel Dekker, New York, pp. 309–326.

Doolittle, S. P., and Walker, M. N. 1925. Further studies on the overwintering and dissemination of cucurbit mosaic. *J. Agric. Res.* 31:1–58.

Duffus, J. E. 1971. Role of weeds in the incidence of virus diseases. *Annu. Rev. Phytopathol.* 9:319–340.

Evans, H. F., and Entwistle, P. F. 1987. Viral diseases. In *Epizootiology of Insect Diseases* (J. Fuxa and Y. Tanada, eds.). Wiley, New York, pp. 257–322.

Evered, D. E., and Harnett, S. (eds.). 1987. *Plant Resistance to Viruses.* Ciba Foundation Symp. no. 133. Wiley, Chichester, U.K., 215 pp.

Falk, B. W., and Duffus, J. E. 1988. Ecology and control. In *The Plant Viruses*, vol. 4: *The Filamentous Plant Viruses* (R. G. Milne, ed.). Plenum, New York, pp. 275–296.

Ford, R. H., and Tolin, S. A. 1983. Genetic stability of tobacco mosaic virus in nature. *Tobacco Sci.* 27:14–17.

Francki, R. I. B. 1981. Plant virus taxonomy. In *Handbook of Plant Virus Infections* (E. Kurstak, ed.). Elsevier/North-Holland, New York, pp. 3–16.

Fraser, R. S. S. 1987. Genetics of plant resistance to viruses. In *Plant Resistance to Viruses* (D. E. Evered and S. Harnett, eds.). Ciba Foundation Symp. no. 133. Wiley, Chichester, U.K., pp. 6–22.

Fridlund, P. R. (ed.). 1989. *Virus and Viruslike Diseases of Pome Fruits and Simulating Noninfectious Disorders.* Cooperative Extension, Washington State University, Pullman, 336 pp.

Fry, W. E. 1982. *Principles of Plant Disease Management.* Academic, New York 378 pp.

Fulton, R. W. 1986. Practices and precautions in the use of cross protection for plant virus disease control. *Annu. Rev. Phytopathol.* 24:67–81.

Gardiner, W. E., Sunter, G., Brand, L., Elmer, J. S., Rogers S. G., and Bisaro, D. M. 1988. Genetic analysis of tomato golden mosaic virus: the coat protein is not required for systemic spread or symptom development. *EMBO J.* 7:899–904.

Garrett, R. G., Cooper, J. A., and Smith, P. R. 1985. Virus epidemiology and control. In *The Viruses*, vol. 1: *Polyhedral Virions with Tripartite Genomes* (R. I. B. Francki, ed.). Plenum, New York, pp. 269–297.

Gibbs, A., and Harrison, B. 1976. *Plant Virology: The Principles.* Wiley, New York, 292 pp.

Goldbach, R. 1986. Molecular evolution of plant RNA viruses. *Annu. Rev. Phytopathol.* 24:289–310.

Gooding, G. V., Jr. 1969. Epidemiology of tobacco mosaic on flue-cured tobacco in North Carolina. N. C. State Univ. Tech. Bull. 195.

Gooding, G. V. 1975. Inactivation of tobacco mosaic virus on tomato seed with trisodium orthophosphate and sodium hypochlorite. *Plant Dis. Rep.* 59:770–772.

Gooding, G. V., Jr. 1986. Tobacco mosaic virus: epidemiology and control. In *The Viruses*, vol. 2: *The Rod-Shaped Plant Viruses* (M. H. V. van Regenmortel and H. Fraenkel-Conrat, eds.). Plenum, New York, pp. 133–152.

Gooding, G. V., Jr., and Lucas, G. B. 1969. Tobacco stalk and root destruction with herbicides and their effects on tobacco mosaic virus. *Plant Dis. Rep.* 53:174–178.

Gooding, G. V., Jr., and Todd, F. A. 1976. Soil-borne tobacco mosaic virus as an inoculum source for flue-cured tobacco. *Tobacco Sci.* 20:140–142.

Govier, D. A., and Kassanis, B. 1974. A virus-induced component from plant sap needed when aphids acquire potato virus Y from purified preparations. *Virology* 57:420–426.

Gray, S. M., Moyer, J. W., Kennedy, G. G., and Campbell, C. L. 1986. Virus-suppression and aphid resistance effects on spatial and temporal spread of watermelon mosaic virus 2. *Phytopathology* 76:1254–1259.

Hall, T. J. 1980. Resistance at the Tm-2 locus in the tomato-to-tomato mosaic virus. *Euphytica* 29:1–9.

Hammond, J. 1982. Plantago as a host of economically important plant viruses. *Adv. Virus Res.* 27:103–140.

Hansen, A. J. 1985. An end to the dilemma—virus-free all the way. *HortScience* 20:852–859.

Hansen, A. J. 1988. Chemotherapy of plant virus infections. In *Applied Virology Research*, vol. 1 (E. Kurstak, R. G. Marusyk, F. A. Murphy, and M. H. V. van Regenmortel, eds.). Plenum, New York, pp. 285–299.

Harris, K. R., and Maramorosch, K. 1977. *Aphids as Virus Vectors.* Academic, New York, 559 pp.

Harris, K. R., and Maramorosch, K. 1980. *Vectors of Plant Pathogens*. Academic, New York, 467 pp.
Harrison, B. D. 1977. Ecology and control of viruses with soil-inhabiting vectors. *Annu. Rev. Phytopathol*. 15:331–360.
Harrison, B. D. 1981. Plant virus ecology: ingredients, interactions and environmental influences. *Ann. Appl. Biol*. 99:195–209.
Harrison, B. D. 1983. Epidemiology of plant virus diseases: a prologue. (R. Plumb and J. M. Thresh, eds.). Blackwell Scientific, Oxford, U.K. pp. 1–6.
Harrison, B. D., and Robinson, D. J. 1988. Molecular variation in vector-borne plant viruses: epidemiological significance. *Phil. Trans. R. Soc*. (London) 321:447–462.
Hatzios, K. K. 1987. Biotechnology applications in weed management: now and in the future. *Adv. Agron*. 41:325–375.
Hays, R. J., Petty, I. T. D., Coutts, R. H. A., and Buck, K. W. 1988. Gene amplification and expression in plants by a replicating geminivirus vector. *Nature* 334: 179–182.
Hendrie, L. K., Irwin, M. E., Liquido, N. J., Ruesink, W. G., Mueller, E. A., Voegtlin, D. J., Acktemeier, G. L., Steiner, W. M., and Scott, R. W. 1985. Conceptual approach to modeling aphid migration. In *The Movement and Dispersal of Agriculturally Important Biotic Agents* (D. R. MacKenzie, C. S. Barfield, G. G. Kennedy, R. D. Berger, and D. J. Taranto, eds.). Claitor's, Baton Rouge, pp. 541–582.
Howell, W. E., and Mink, G. I. 1977. The role of weed hosts, volunteer carrots and overlapping growing seasons in the epidemiology of carrot thin leaf and carrot motley dwarf viruses in Central Washington. *Plant Dis. Rep*. 61:217–222.
Hull, R. 1989. The movement of viruses in plants. *Annu. Rev. Phytopathol*. 27:213–240.
Hunst, P. L., and Tolin, S. 1982. Isolation and Comparison of two strains of soybean mosaic virus. *Phytopathology* 72:710–713.
Irwin, M. E., and Goodman, R. M. 1981. Ecology and control of soybean mosaic virus. In *Plant Diseases and Vectors: Ecology and Epidemiology* (K. Maramorosch and K. F. Harris, eds.). Academic, New York, pp. 181–220.
Jacquemond, M., Amselem, J., and Tepfer, M. 1988. A gene coding for a monomeric form of cucumber mosaic virus satellite RNA confers tolerance to CMV. *Mol. Plant-Microbe Interact*. 1:311–316.
Johnson, J., and Hoggan, I. A. 1937. Inactivation of the ordinary tobacco mosaic virus by microorganisms. *Phytopathology* 27:1014–1027.
Jones, A. T. 1987. Control of virus infection in crop plants through vector resistance: a review of achievements, prospects and problems. *Ann. Appl. Biol*. 111:745–772.
Jones, R. A. C., and Harrison, B. D. 1969. The behaviour of potato mop-top virus in soil, and evidence for its transmission by Spongospora subterranea (Wallr.) Lagerh. *Ann. Appl. Biol*. 63:1–17.
Jordan, R. L., and Dodds, J. A. 1985. Double-stranded RNA in detection of diseases of known and unproven viral etiology. *Acta Hort*. 164:101–108.
Kiihl, R. S. A., and Hartwig, E. E. 1979. Inheritance of reaction to soybean mosaic virus in soybeans. *Crop Sci*. 19:372–375.
Knoke, J. K., and Louie, R. 1981. Epiphytology of maize virus diseases. In *Virus and Viruslike Diseases of Maize in the United States* (D. T. Gordon, J. K. Knoke, and G. Scott, eds.). So. Coop. Ser. Bull. no. 247. June 1981, Wooster, OH, pp. 92–102.
Koenig, R. 1986. Plant viruses in rivers and lakes. *Adv. Virus Res*. 31:321–333.
Koltin, Y., and Leibowitz, M. J. (eds.). 1988. *Viruses of Fungi and Simple Eukaryotes*. Marcel Dekker, New York, 434 pp.
Kuhn, C. W., Nutter, F. W., Jr., and Padgett, G. B. 1989. Multiple levels of resistance to tobacco etch virus in pepper. *Phytopathology* 79:814–818.
Kurstak, E. (ed.). 1981. *Handbook of Plant Virus Infections and Comparative Diagnosis*. Elsevier/North-Holland Biomedical, Amsterdam, 943 pp.
Lapointe, S. L., Tingey, W. M., and Zitter, T. A. 1987. Potato virus Y transmission reduced in an aphid-resistant potato species. *Phytopathology* 77:819–822.
Legrand, M., Kauffmann, S., Geoffroy, P., and Fritig, B. 1987. Biological function of pathogenesis-related proteins: four tobacco pathogenesis-related proteins are chitinases. *Proc. Natl. Acad. Sci. USA* 84:6750–6754.

Lehman, S. G. 1934. Contaminated soil and cultural practices as related to occurrence and spread of tobacco mosaic. *N.C. Agric. Exp. Stn. Bull.* 46:43.
Linthorst, H. J. M., Meuwissen, R. L. J., Kauffmann, S., and Bol, J. F. 1989. Constitutive expression of pathogenesis-related proteins PR-1, GRP and PR-S in tobacco has no effect on virus infection. *Plant Cell* 1:285–291.
Longworth, J. F. 1978. Small isometric viruses of invertebrates. *Adv. Virus Res.* 23: 140–147.
Lovisolo, O. 1980. Virus and viroid diseases of cucurbits. *Acta Hort.* 88:33–82.
McKinney, H. H. 1929. Mosaic diseases in the Canary Islands, W. Africa, and Gibraltar. *J. Agric. Res.* 39:557–578.
McLaughlin, M. R., and Boykin, D. L. 1988. Virus diseases of seven species of forage legumes in the southeastern United States. *Plant Dis.* 72:539–542.
McLean, G. D., Garrett, R. G., and Ruesink, W. G. 1986. *Plant Virus Epidemics: Monitoring, Modelling and Predicting Outbreaks*. Academic, Sydney, Australia 550 pp.
MacNeill, G. H., and Boxall, M. 1974. The evolution of a pathogenic strain of tobacco mosaic virus in tomato: a host passage phenomenon. *Can. J. Bot.* 52:1305–1307.
Madden, L. V., Reynolds, K. M., Pirone, T. P., and Raccah, B. 1988. Modeling of tobacco virus epidemics as spatio-temporal autoregressive integrated moving-average processes. *Phytopathology* 78:1361–1366.
Maramorosch, K., and Harris, K. F. (eds.). 1981. *Plant Diseases and Vectors: Ecology and Epidemiology*. Academic, Orlando, Florida, 368 pp.
Matthews, R. E. F. 1981. *Plant Virology*, 2d ed. Academic, New York, London, 987 pp.
Matthews, R. E. F. 1982. Classification and nomenclature of viruses. *Intervirology* 17: 1–199.
Mayo, M. A., and Harrap, K. A. (eds.). 1984. *Vectors in Virus Biology*. Society for General Microbiology/Academic Press, London 188 pp.
Meshi, T., and Okada, Y. 1987. Systemic movement of viruses. In *Plant-Microbe Interactions. Molecular and Genetic Perspectives*, vol. 2 (T. Kosuge and E. W. Nester, eds.). Macmillan, New York, pp. 285–304.
Meshi, T., Motoyoshi, F., Adachi, A., Watanebe, Y., Takanatsu, N., and Okada, Y. 1988. Two concomitant base substitutions in the putative replicase genes of tobacco mosaic virus confer the ability to overcome the effects of a tomato resistance gene, Tm-1. *EMBO J* 7:1575–1581.
Moran, J. R., and Wilson, J. M. 1985. Rates of reinfection with virus in commercial carnation, chrysanthemum and gladiolus crops. *Acta Hort.* 164:325–332.
Murant, A. F., and Lister, R. M. 1967. Seed-transmission in the ecology of nematode-borne viruses. *Ann. Appl. Biol.* 59:63–76.
Nelson, R. S., McCormick, S. M., Delannay, X., Dube, P., Layton, J., Anderson, E. J., Kaniewska, M., Proksch, R. K., Horsch, R. B., Rogers, S. G., Fraley, R. T., and Beachy, R. N. 1988. Virus tolerance, plant growth, and field performance of transgenic tomato plants expressing coat protein from tobacco mosaic virus. *Bio/Technology* 6:403–409.
Nienhaus, F., and Castello, J. D. 1989. Viruses in forest trees. *Annu. Rev. Phytopathol.* 27:165–186.
Nutter, F. W., Jr., Pederson, V. D., and Timian, R. G. 1984. Relationship between seed infection by barley stripe mosaic virus and yield loss. *Phytopathology* 74:363–366.
Odell, J. T., Nagy, F., and Chua, N.-H. 1985. Identification of DNA sequences required for activity of the cauliflower mosaic virus 35S promoter. *Nature* 313:810–812.
Paguio, O. R., Kuhn, C. W., and Boerma, H. R. 1988. Resistance-breaking strains of cowpea chlorotic mottle virus in soybean. *Plant Dis.* 72:768–770.
Palukaitis, P., Cotts, S., and Zaitlin, M. 1985. Detection and identification of viroids and viral nucleic acids by "dot-blot" hybridization. *Acta Hort.* 164:109–117.
Pelham, J., Fletcher, J. T., and Hawkins, J. H. 1970. The establishment of a new strain of tobacco mosaic virus resulting from the use of resistant varieties of tomato. *Ann. Appl. Biol.* 65:293–297.
Peters, D. 1987. Spread of viruses in potato crops. In *Viruses of Potatoes and Seed-Potato Production* (J. A. de Bokx and J. P. H. van der Want, eds.). Pudoc, Wageningen, The Netherlands, pp. 126–145.

Plumb, R. T., and Thresh, J. M. 1983. *Plant Virus Epidemiology: The Spread and Control of Insect-Borne Viruses*. Blackwell Scientific, Oxford, U.K., 377 pp.

Polston, J. E., Dodds, J. A., and Perring, T. M. 1989. Nucleic acid probes for detection and strain discrimination of cucurbit geminiviruses. *Phytopathology* 79:1123–1127.

Ponz, F., and Bruening, G. 1986. Mechanisms of resistance to plant viruses. *Annu. Rev. Phytopathol.* 24:355–381.

Ponz, F., Glascock, C. B., and Bruening, G. 1987. An inhibitor of polyprotein processing with the characteristics of a natural virus resistance factor. *Mol. Plant-Microbe Interact.* 1:25–31.

Powell Abel, P., Nelson, R. S., De, B., Hoffmann, N., Rogers, S. G., Fraley, R. T., and Beachy, R. N. 1986. Delay of disease development in transgenic plants that express the tobacco mosaic virus coat protein gene. *Science* 232:738–743.

Powell, C. A., Forer, L. B., Stouffer, R. F., Cummins, J. N., Gonsalves, D., Rosenberger, D. A., Hoffman, D. A., and Lister, R. M. 1984. Orchard weeds as hosts of tomato and tobacco ringspot viruses. *Plant Dis.* 68:242–244.

Quiot, J. B. 1980. Ecology of cucumber mosaic virus in the Rhône Valley of France. *Acta Hort.* 88:9–21.

Raccah, B. 1986. Nonpersistent viruses: epidemiology and control. *Adv. Virus Res.* 31: 387–429.

Rao, A. S., and Brakke, M. K. 1969. Relation of soil-borne wheat mosaic virus and its fungal vector, *Polymyxa graminis*. *Phytopathology* 59:581–587.

Rast, A. T. B. 1972. MII-16, An artificial symptomless mutant of tobacco mosaic virus. *Neth. J. Plant Pathol.* 73:147–156.

Register, J. C., III, and Beachy, R. N. 1988. Resistance to TMV in transgenic plants results from interference with an early event in infection. *Virology* 166:524–532.

Rezaian, M. A., Skene, K. G. M., and Ellis, J. G. 1988. Anti-sense RNAs of cucumber mosaic virus in transgenic plants assessed for control of the virus. *Plant Mol. Biol.* 11:463–471.

Rist, D. L., and Lorbeer, J. W. 1989. Occurrence and overwintering of cucumber mosaic virus and broad bean wilt virus in weeds growing near commercial lettuce fields in New York. *Phytopathology* 79:65–69.

Roane, C. W., Tolin, S. A., and Buss, G. R. 1973. Inheritance of reaction to two viruses in the soybean cross York X Lee 68. *J. Hered.* 74:289–291.

Roane, C. W., Tolin, S. A., and Buss, G. R. 1986. Application of the gene-for-gene hypothesis to soybean-soybean mosaic virus interactions. *Soybean Gen. Newsl.* 13:136–139.

Roane, C. W., Tolin, S. A., Aycock, H. 1989. Genetics of reaction to maize dwarf mosaic virus strain A in several maize inbred lines. *Phytopathology* 79:1364–1368.

Robinson, R. A. 1980. New concepts in breeding for disease resistance. *Annu. Rev. Phytopathol.* 18:189–210.

Robinson, R. A. 1987. *Host Management in Crop Pathosystems*. MacMillan, New York, 263 pp.

Robinson, D. J., Hamilton, W. D. O., Harrison, B. D., and Baulcombe, D. C. 1987. Two anomalous tobravirus isolates: evidence for RNA recombination in nature. *J. Gen. Virol.* 68:2551–2561.

Ross, J. P. 1969. Pathogenic variation among isolates of soybean mosaic virus. *Phytopathology* 59:829–832.

Ruesink, W. G., and Irwin, M. E. 1986. Soybean mosaic virus epidemiology: a model and some implications. In *Plant Virus Epidemics. Monitoring, Modelling and Predicting Outbreaks* (G. D. McClean, R. G. Garrett, and W. G. Ruesink, eds.). Academic, Sydney, Australia, pp. 295–313.

Sherwood, J. L. 1987. Mechanisms of cross-protection between plant virus strains. In *Plant Resistance to Viruses* (D. E. Evered and S. Harnett, eds.). Ciba Foundation Symp. No. 133. Wiley, Chichester, U.K., pp. 136–150.

Smith, K. M. 1934. *Recent Advances in the Study of Plant Viruses*. Blakiston's, Philadelphia, 423 pp.

Stace-Smith, R., and Hamilton, R. I. 1988. Inoculum thresholds of seedborne pathogens: viruses. *Phytopathology* 78:875–880.

Sylvester, E. S. 1969. Evidence of transovarial passage of sowthistle yellow vein virus in the aphid *Hyperomyzus lactucae*. *Virology* 38:440–448.

Sylvester, E. S., and Richardson, J. 1970. Infection of *Hyperomyzus lactucae* by sowthistle yellow vein virus. *Virology* 42:1023–1042.

Takahashi, Y., and Orlob, G. 1969. Distribution of wheat streak mosaic virus-like particles in *Aceria tulipae. Virology* 38:230–240.

Thornberry, H. H., Valleau, W. D., and Johnson, E. M. 1937. Inactivation of tobacco-mosaic virus in cured tobacco leaves by dry heat. *Phytopathology* 27:129–134.

Thresh, J. M. 1982. Cropping practices and virus spread. *Annu. Rev. Phytopathol.* 20: 193–218.

Thresh, J. M. 1983. Plant virus epidemiology and control: current trends and future prospects. In *Plant Virus Epidemiology: The Spread and Control of Insect-Borne Viruses* (R. T. Plumb and J. M. Thresh, eds.). Blackwell Scientific, Oxford, U.K., pp. 349–360.

Thresh, J. M. 1985. Plant virus dispersal. In *The Movement and Dispersal of Agriculturally Important Biotic Agents* (D. R. MacKenzie, C. S. Barfield, G. G. Kennedy, R. D. Berger, and D. J. Taranto, eds.). Claitor's, Baton Rouge, Louisiana, pp. 51–106.

Thresh, J. M. 1987. The population dynamics of plant virus diseases. In *Populations of Plant Pathogens: Their Dynamics and Genetics* (M. S. Wolfe and C. E. Caten, eds.). Blackwell Scientific, Oxford, U.K., pp. 135–148.

Tien, P., Zhang, X., Qiu, B., Qin, B., and Wu, G. 1987. Satellite RNA for the control of plant diseases caused by cucumber mosaic virus. *Ann. Appl. Biol.* 111:143–152.

Tolin, S. A., Isakson, O. W., and Troutman, J. L. 1970. Association of white clover and aphids with peanut stunt virus in Virginia. *Plant Dis. Rep.* 54:935–938.

Tomlinson, J. A. 1987. Epidemiology and control of virus diseases of vegetables. *Ann. Appl. Biol.* 110:661–681.

Tomlinson, J. A., and Carter, A. L. 1970. Studies in the seed transmission of cucumber mosaic virus in chickweed (*Stellaria media*) in relation to the ecology of the virus. *Ann. Appl. Biol.* 66:381–386.

Tomlinson, J. A., and Faithfull, E. 1982. Isolation of infective tomato bushy stunt virus after passage through the human alimentary tract. *Nature* 300:637–638.

Tosic, M., and Tosic, D. 1984. Occurrence of tobacco mosaic virus in water of the Danube and Sava Rivers. *Phytopathol. Zeit.* 110:200–202.

van Etten, J. L., Xia, Y., and Meints, R. H. 1987. Viruses of a Chlorella-like green alga. In *Plant-Microbe Interactions. Molecular and Genetic Perspectives*, vol. 2 (T. Kosuge and E. W. Nester, eds.). Macmillan, New York, pp. 307–325.

van Loon, L. C. 1987. Disease induction by plant viruses. *Adv. Virus Res.* 33:205–255.

van Loon, L. C., Gerritsen, Y. A. M., and Ritter, C. E. 1987. Identification, purification, and characterization of pathogenesis-related proteins from virus-infected Samsun NN tobacco leaves. *Plant Mol. Biol.* 9:593–609.

van Vloten-Doting, L., and Bol, J. F. 1988. Variability, mutant selection, and mutant stability in plant RNA viruses. In *RNA Genetics*, vol. III: *Variability of RNA Genomes* (E. Domingo, J. J. Holland, and P. Ahlquist, eds.). CRC Press, Boca Raton, Florida, pp. 37–51.

Walkey, D. G. A. 1985. *Applied Plant Virology*. Wiley, New York, 329 pp.

Ward, A., Etessami, P., and Stanley, J. 1988. Expression of a bacterial gene in plants mediated by infectious geminivirus DNA. *EMBO J.* 7:1583–1587.

Watson, M. A., Heathcote, G. D., Lauckner, F. B., and Sowray, P. 1975. The use of weather data and counts of aphids in the field to predict the incidence of yellowing viruses of sugarbeet crops in England in relation to the use of insecticides. *Ann. Appl. Biol.* 81:181–198.

White, R. F. 1979. Acetylsalicylic acid (aspirin) induces resistance to tobacco mosaic virus in tobacco. *Virology* 99:410–412.

Wilson, C. L. 1969. Use of plant pathogens in weed control. *Annu. Rev. Phytopathol.* 7:411–434.

Woolston, C. J., Czaplewski, L. G., Markham, P. G., Goad, A. S., Hull, R., and Davies, J. W. 1987. Location and sequence of a region of cauliflower mosaic virus gene 2 responsible for aphid transmissibility. *Virology* 160:246–251.

Wyatt, S. D., and Kuhn, C. W. 1980. Derivation of a new strain of cowpea chlorotic mottle virus from resistant cowpeas. *J. Gen. Virol.* 49:289–296.

Yeh, S.-D., and Gonsalves, D. 1984. Evaluation of induced mutants of papaya ringspot virus for control by cross protection. *Phytopathology* 74:1086–1091.
Zettler, F. W., and Freeman, T. E. 1972. Plant pathogens and biocontrols of aquatic weeds. *Annu. Rev. Phytopathol.* 10:455–470.
Zitter, T. A., and Simons, J. N. 1980. Management of viruses by alteration of vector efficiency and by cultural practices. *Annu. Rev. Phytopathol.* 18:289–310.

Chapter

7

Virus-Mediated Genetic Transfer in Plants

Peter Palukaitis

Department of Plant Pathology
Cornell University
Ithaca, New York 14853

Introduction

Although plant viruses interact with their hosts in a number of ways that alter the biology and ecological relationships of their hosts, they do not transfer genetic materials from host to host in any stable fashion. Thus, the use of recombinant DNA vis-à-vis viruses is not viewed as something that is likely to increase or alter such pathogen-host interactions. There are, however, a number of potential issues that need to be addressed when either using plant viruses as vectors for gene expression in higher plants or inserting viral sequences into the genomes of higher plants (by transformation procedures) as a method of protecting such plants from infection by related viruses. These issues will be addressed in this chapter.

Plant Viruses As Vectors

There has been relatively little attention given to plant viruses as vectors for several reasons: (1) The infectious nature of the viral pathogen requires either the identification and removal of pathogenic sequences without affecting the ability of the vector to multiply and spread within the inoculated host plant or the removal of sequences involved in the transmission of the virus vector from plant to plant. (2) The application of the virus vector to each plant in a field is very labor-intensive, and since most plant viruses are not seed-transmissible, the virus would have to be applied after each planting. (3) Most plant viruses infect a rather narrow spectrum of plant species. Thus, multiple vectors would have to be created for use in a broad spectrum of plants. Conversely, if viruses with broad host ranges are used as vectors, then all potential hosts would have to be tested for pathogenic responses. (4) The conservative utilization of nucleic acid, restrictive mechanisms of regulation of gene expression, and/or packaging limitations of nucleic acid into virus particles place severe constraints on where other genes can be inserted and the size of the introduced gene. And finally, (5) there are questions about the stability of the inserted sequences; i.e., the maintenance of genetic integrity and function in the absence of selection pressures (Brisson et al. 1984, Siegel 1985, van Vloten-Doting et al. 1985, Hayes et al. 1989).

Each of the above points raises further issues when it is explored. For example, sequences involved in pathogenicity in one host may not be involved in pathogenicity in a second host (Visvader and Symons 1986, Palukaitis 1988). Therefore, for a virus with a broad host range, it would be virtually impossible to localize all potentially pathogenic sequences in a virus. Moreover, many sequences involved in pathogenicity are localized within regions of the molecule that are indispensable for particular viral functions (Daubert 1988, Stratford and Covey 1989, and M. Shintaka, D. E. Sleat, and P. Palukaitis, unpublished data). In short, there seem to be very few advantages to using plant viruses as vectors and many disadvantages. Nevertheless, some research has gone into the construction of plant viruses as vectors.

For technical reasons, the virus groups containing DNA as their genetic material were considered to offer the best potential as vectors. Thus, considerable work went into constructing a vector from cauliflower mosaic virus (CaMV) (reviewed by Howell 1982, 1985). Two small genes were found to be dispensable, including one involved in the transmission of the virus from plant to plant, but only a limited amount of genetic material could be inserted into this region for the DNA to be packaged into virus particles (Gronenborn et al. 1981, Howell et al. 1981, Daubert et al. 1983, Dixon et al. 1983). Thus, while it was shown that this virus could be used as a vector, and one which would remain localized in the inoculated plants, the genetically modified virus was still a pathogen and was capable of carrying only limited genetic information (Brisson et al. 1984, Lefebvre et al. 1987, Fütterer et al. 1988, De Zoeten et al. 1989).

There has been some work on other DNA-containing viruses (African cassava mosaic virus and tomato golden mosaic virus) belonging to a different taxonomic group, the geminiviruses, which shows that these viruses too could be potential vectors (Hayes et al. 1988, 1989, Ward et al. 1988). In these cases, the coat protein gene of the virus has been replaced with genes encoding resistances to antibiotics. These viruses are normally transmitted by whiteflies. Thus, by removing the coat protein, the viruses would be restricted to the inoculated plant. However, questions related to the pathogenicity of such vectors have not been addressed.

In the case of viruses with RNA genomes, recombinant DNA technology has been used to construct viral genomes in which the coat protein gene has been replaced with marker genes (French et al. 1986, Takamatsu et al. 1987). While such constructs are capable of replication and cell-to-cell movement in plants, they are not capable of long-distance movement to other leaves. They are certainly not capable of movement from plant to plant, but at least one of the recombinant viruses was shown to still be pathogenic (Takamatsu et al. 1987).

In all of the above cases where the virus has been disabled, preventing its movement to other plants, plant-to-plant spread could still occur if the plants became infected with the wild-type form of the same virus, since the missing gene functions can be complemented.

After several weeks of replication, a plant may contain up to 10^{18} virus particles (Matthews 1981), whereas the initial inoculum may have contained as few as 5 to 10 particles (T. P. Pirone, personal communication). Thus, there has been an enormous amount of virus replication, with considerable opportunity for mutation, selection, and recombination with other viruses containing similar sequences. Although there are selection pressures associated with maintaining viral functions, one may question what selection pressures will maintain the integrity of the introduced gene. Moreover, will the introduced gene itself, or one of its mutated derivatives, induce a pathogenic response in the host? All of these questions (the imponderables involved in developing safe, nonpathogenic, and efficacious vectors), and the question as to the actual need for such vectors, have led most researchers involved in plant virology to eschew the development of functional plant virus vectors for introducing other genes into plants. Most of those still actively pursuing this goal appear to be more interested in showing what can be done, rather than developing an environmentally safe plant virus–vector system capable of competing with the Ti-plasmid-based transformation systems. In the end, the marketplace will determine whether there will be any place for a nonintegrative vector system with questionable stability.

Viral Regulatory Sequences in Transgenic Plants

The presence of highly active promoters of RNA synthesis in the DNA virus CaMV has led to the use of such regulatory sequences to enhance gene expression in transformed plants (Paszkowski et al. 1984, Bevan et al. 1985, Nagy et al. 1985). Few natural plant promoters are as active as those from CaMV, and these natural promoters are usually functional only after induction by some other signal, and are often tissue- or species-specific for expression (Nagy et al. 1985, McElroy et al. 1990). The two CaMV promoters are constitutive (i.e., they are functional all the time and do not require special inducers) and show little tissue or species specificity (Benfey et al. 1989). The two CaMV promoters, the "26S" and the "35S" promoters, are so-called because the numbers represent the sedimentation coefficient of the RNAs synthesized from the two different promoters (Guilley et al. 1982). Both of these promoters have been used in a large number of studies involving transgenic plants expressing genes of various origins (Harpster et al. 1988). While some of these plants may have shown either unusual or abnormal responses, it has in every case been possible to delimit these host abnormalities to the expression of the gene and not to the presence of a promoter of viral origin. There is no evidence that the sequences of the CaMV promoters are in themselves inducers of pathogenicity. In fact, the data available suggest that the major inclusion body protein synthesized from the 26S RNA of CaMV is involved in eliciting a pathogenic response in host as well as nonhost plants (Baughman et al. 1988, Stratford and Covey 1989, Takahashi et al. 1989). Thus, the major gene product rather than the well-characterized regulatory signals on the CaMV DNA are involved in the induction of pathogenicity in plants.

Recent work with the geminiviruses has also indicated their potential use as donators of regulatory sequences (Fenoll et al. 1988, Hanley-Bowdoin et al. 1988, Hayes et al. 1988, 1989, Ward et al. 1988). Some of these may be more useful than the CaMV promoter sequences for gene expression in monocots (e.g., corn, wheat), which are the natural hosts of some geminiviruses (Stanley 1985). More work on their full potential will be forthcoming. Additional work will also need to be done on delimiting the pathogenicity domains of these viruses, to determine whether the regulatory sequences in themselves are capable of inducing pathogenic responses in a variety of plants.

The sequences preceding the first gene of tobacco mosaic virus (TMV) have been cloned adjacent to several genes such that the expressed RNA would contain the same TMV sequences preceding the gene of interest (Gallie et al. 1987). These TMV sequences act as "translational enhancers"; i.e., they result in a higher level of production of the protein encoded by the gene. Translational-enhancer sequences from other viral genes expressed at high levels in plants have been and will continue to be tested for their utility in the expression to high(er) levels of various genes in plants (Jobling and Gehrke 1987). None of the data accumulated thus far indicates any direct role in pathogenicity for these translational-enhancer sequences. In fact, with TMV the domains for various pathogenic responses have all mapped to specific genes (Saito et al. 1987, Takamatsu et al. 1987, Dawson et al. 1988, Knorr and Dawson 1988). Nevertheless, if more regulatory sequences from other plant viruses containing RNA genomes are used to enhance gene expression, those sequences will each have to be scrutinized for their pathogenicity potential. Such analyses are part of the screening process for efficacy and/or effects on crop yields (Nelson et al. 1988).

In contrast to the situation with viruses used as vectors, the use of viral regulatory sequences in transgenic plants does not involve infectious molecules capable of escape and inducing disease responses in other plants. Thus, only the "target species" needs to be examined for its potential pathogenic response to the regulatory sequences. Some of the potential responses that need to be examined are discussed below.

Viral Sequences Involved in Pathogenicity

Over the last 5 years, we have learned much about the kinds of sequence changes that lead to altered pathogenic responses. With subviral pathogens such as viroids and satellite RNAs, a few specific nucleotide changes in a defined region of the molecule can have quite dramatic effects on pathology (Schnölzer et al. 1985, Visvader and Symons 1986, Sleat and Palukaitis 1990).

For example: (1) A few nucleotide changes in one region of the potato spindle tuber viroid molecule can alter its pathogenicity on tomato from mild to severe or even lethal (Schnölzer et al. 1985). Sequence changes in the same region of the citrus exocortis viroid molecule also affect its pathogenicity to tomato, but do not appear to affect its pathogenicity on citrus (Visvader and Symons 1986). (2) A few nucleotide changes in one region of a satellite RNA of cucumber mosaic virus can alter its pathological properties in one host without affecting those properties in other hosts (Sleat and Palukaitis 1990). Furthermore, a single nucleotide change in a different region of the satellite RNA molecule can alter the host specificity of the pathogenic response; i.e., with one sequence the satellite RNA may be pathogenic in tomato and nonpathogenic in tobacco, while a very specific single nucleotide change makes the satellite RNA nonpathogenic in tomato and pathogenic in tobacco (D. E. Sleat and P. Palukaitis, unpublished data). On all other plant species tested, these nucleotide sequence changes do not have any effect on the pathogenicity of the satellite RNA (D. E. Sleat and P. Palukaitis, unpublished results).

In the cases of plant viruses, pathogenicity domains have also been delimited using recombinant viruses. These domains have either been delimited to certain gene products encoded by the plant virus (Daubert 1988, Schoelz and Shepherd 1988, Stratford and Covey 1989, and M. Shintaku and P. Palukaitis, unpublished data), or to specific amino acid alterations within a given gene product (Knorr and Dawson 1988). Once again, the alterations in pathogenicity brought about by specific amino acid changes are usually host-specific; i.e., the appearance or development of disease is altered in one host but not in others. How any of these changes in either nucleotide sequence or amino acid sequence elicit a disease response on the part of a plant remains completely unknown. Thus, we cannot predict in what host a given viral sequence is likely to be pathogenic, or even what viral sequences are likely to be pathogenic in *any* host. In the absence of such predictive capabilities, we have to rely on experience as a guide.

Various viral sequences have been expressed in transgenic plants (Baulcombe et al. 1987, Beachy et al. 1987). Most of these sequences encode viral coat proteins, while others either encode various nonstructural viral proteins or contain only viral DNA and no viral-encoded proteins were expressed. There have been no instances of pathogenic plants produced as the result of expression of viral coat proteins or viral RNA segments in transgenic plants; however, specific pathogenic effects have been observed during the expression of viral inclusion-body proteins in transgenic plants (Baughman et al. 1988, Takahashi et al. 1989, and J. G. Shaw, personal communication). This was true whether or not the particular plant species analyzed was a host for the virus which forms the inclusion bodies. Thus, plants appear to be capable of responding to the presence of some viral-encoded proteins expressed in transgenic plants, while not to others—even when the other proteins (e.g., coat proteins) are in many cases associated with a particular disease development. Whether this is due to particular disease elicitation, either by different thresholds of expression by various viral genes or by interactions among several viral components (nucleic acids and/or proteins), still needs to be determined.

Cross-Protection and Viruses Modified by Recombinant DNA Techniques

The use of cross-protection as a mode of controlling virus disease is well documented (see Chap. 6). It needs to be reconsidered vis-à-vis viruses modified by recombinant DNA techniques only to the extent that there are some differences between such viruses and chemically induced mutants; viz., the recombinant virus is probably safer to use!

In cross-protection, plants of economic importance with little or no natural resistance to a particular virus are deliberately infected with a "mild strain" of the same virus. The infected plant shows little or no symptomatology, and there is usually no effect on the crop yield; however, such plants are "protected" against infection by severe strains of the same virus which do produce symptoms and affect crop yields (reviewed by Hamilton 1980, Palukaitis and Zaitlin 1984, Fulton 1986). The mechanism or mechanisms of cross-protection are not understood, although it is not an "immunization," since plants do not have an immune system. Moreover, cross-protection has been known to "break down," i.e., offer only a temporary inhibition (Fletcher and Rowe 1975, Fulton 1986). However, this delay is often sufficient to prevent the crop from being affected. Thus, where it has been applied, cross-protection has generally been successful.

In some cases, strains of a virus which produce "mild" or no symptoms on a particular plant species are not known. In these instances, mild strains were created or selected after treatment of a severe strain of a virus with nitrous acid (Rast 1972, Yeh and Gonsalves 1984). These chemically mutagenized mild strains are then inoculated onto plants to induce cross-protection. There has been some concern expressed about the possibility of such mild strains reverting to severe strains (Palukaitis and Zaitlin 1984, Fulton 1986). Although, as far as is known, this has not been observed, it is still a potential hazard. However, unless the reversion to a severe form took place soon after application of the initial virus, it probably would not have been noticed, due to cross-protection.

Use of a virus that has been modified by recombinant DNA methodology could obviate the above problem if the alteration in sequence were due to a deletion, since deletions cannot revert. However, for technical reasons it may be more difficult to obtain a viable virus carrying a deletion. Nevertheless, if cross-protection were to see expanded use in fields and greenhouses, recombinant-DNA-modified viruses would be safer to use—both for the reasons given above and because the organism modified by recombinant DNA technology would be better characterized and consist of a single pure species. This is not true of (chemically modified) natural viruses.

In all likelihood, cross-protection using recombinant-DNA-modified viruses will not see widespread use, mostly for the same reasons that cross-protection using either natural or chemically modified virus has not seen widespread use: (1) all of the (crop) plants are infected with a virus and thus act as a reservoir for the transmission of this virus to other potential hosts, in which the same strain may be more pathogenic; (2) in a field situation, every plant would have to be inoculated, which could become quite labor-intensive and costly; (3) there are often cheaper and less hazardous alternative means of (usually genetic) resistance available; and (4) cross-protection is often incomplete and has been known to break down (Fletcher and Rowe 1975). In addition, the idea of spreading recombinant-DNA-modified viruses in the field is just too radical a suggestion for many to go along with, even if it would be safer than the current practices.

It is also interesting to note that the major criticism of applying cross-protection to the field—the generation of plants harboring viruses—is not voiced by plant breeders, who often generate plants resistant to the disease induced by a virus but not resistant to the multiplication of the pathogen itself—a state known as "tolerance," in which virus reservoirs are also created (Matthews 1981, Fraser 1986). [Apparently genes for tolerance are more common in plants, and thus easier to find in wild species related to important crop species, than genes for true resistance or immunity to virus infection (R. Provvidenti, personal communication).]

Protection Using Viral Genome Segments

Resistance mediated via the expression of viral coat protein

A few years ago, Roger Beachy and his colleagues were attempting to test various hypotheses concerning the mechanism of cross-protection (Beachy et al. 1985). They transformed tobacco plants with DNA containing the coat protein gene of TMV adjacent to a CaMV 35S RNA promoter. The resultant transgenic plants expressed the coat protein mRNA and produced coat protein. The level of viral coat protein expressed in such plants was orders of magnitude lower than the level of viral coat protein expressed in a naturally infected plant. These transgenic plants expressed some degree of resistance to infection by TMV at low inoculum doses (Powell Abel et al. 1986). Since this initial discovery of this phenomenon, which some refer to as "coat-protein-mediated resistance" and others erroneously refer to as "cross-protection" (Tumer et al. 1987), Beachy and his colleagues, as well as many other workers at other universities and/or biotechnology-oriented companies, have shown that the expression of the coat proteins of a number of different plant viruses results in a protective effect against closely related strains of the same virus (Loesch-Fries et al. 1987, Tumer et al. 1987, Van Dun et al. 1987, Cuozzo et al. 1988, Hemenway et al. 1988, Stark and Beachy 1989, Hoekema et al. 1989, Angenent et al. 1990, Lawson et al. 1990). Protection against either more distantly related viruses or unrelated viruses is either less, marginal, or not observed. Because only a small portion of the viral genome is expressed in the transgenic plants, a productive infection does not occur as in cross-protection. Finally, because the viral coat protein gene is incorporated into the genetic material of the plant, it can be passed through the seed to successive generations just as a true genetic resistance. Thus, in virtually all respects, coat-protein-mediated resistance is superior to cross-protection.

Heterologous encapsidation and pathogen dissemination

Horizontal transmission When two viruses infect the same plant, a number of interactions can occur between them (reviewed by Bennett 1953): (1) virus

A may suppress the replication and/or pathogenicity of virus B [e.g., in both cross-protection and systemic acquired resistance (Gianninazzi 1984, Palukaitis and Zaitlin 1984)]; (2) virus A may enhance the replication of virus B (Goodman and Ross 1974*a*,*b*); (3) viruses A and B may act synergistically to induce a worse disease than either virus alone (Garces-Orejuela and Pound 1957); or (4) some of the genome of virus A may be encapsidated into the coat protein of virus B (see Fig. 7.1). This latter phenomenon is called "genomic masking," "transcapsidation," or "heterologous encapsidation."

Heterologous encapsidation is a well-recognized phenomenon that has been extensively reviewed (Rochow 1972, 1977, Dodds and Hamilton 1976, Falk and Duffus 1981). It occurs with viruses infecting prokaryotes, animals, and plants. It occurs more often between "related viruses" than between "unrelated viruses" (Dodds and Hamilton 1976), and when it occurs between viruses belonging to different taxa, heterologous encapsidation is not always reciprocal: e.g., TMV RNA could be encapsidated in barley stripe mosaic virus (BSMV) coat protein, but BSMV RNA was not detected encapsidated in TMV coat protein (Dodds and Hamilton 1974). In many cases, heterologous encapsidation has been demonstrated in vitro as well as in vivo (reviewed by Dodds and Hamilton 1976); in other situations, only one of the two has been demonstrated.

Other instances of heterologous encapsidation or other virus-assisted transmission that have been described include those of the tobacco rosette virus complex (Smith 1945, 1946), parsnip yellow fleck virus and anthriscus yellows

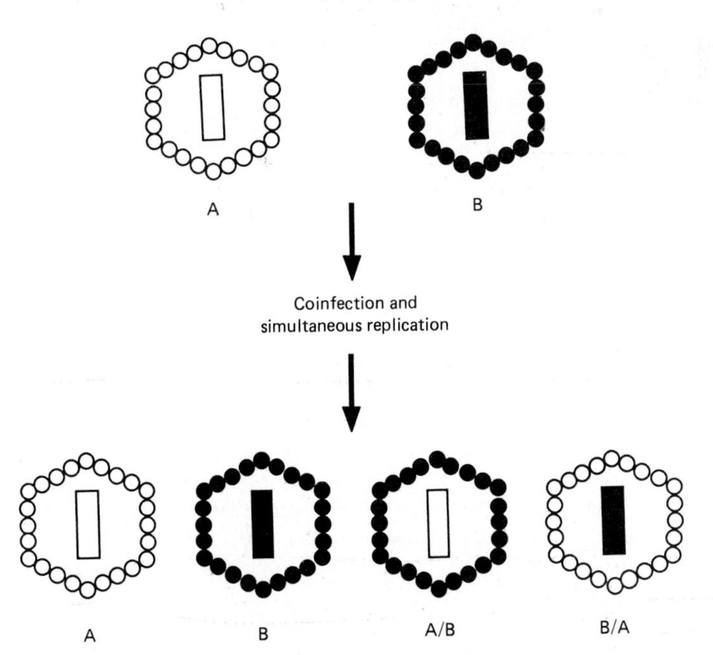

Figure 7.1 Homologous and heterologous encapsidation. Two viruses (A and B), replicating in the same cell, can produce either their "normal progeny" (A and B), or transcapsidated particles containing either the genome of virus A and the protein coat of virus B (A/B), or the genome of virus B and the protein coat of virus A (B/A).

virus (Murant and Goold 1968), groundnut rosette and its assistor virus (Hull and Adams 1968), tobacco yellow vein and its assistor virus (Adams and Hull 1972), carrot red leaf virus and carrot mottle virus (Elnagar and Murant 1978), bean yellow vein-banding virus and pea enation mosaic virus (Cockbain 1978), beet western yellows virus and lettuce speckles mottle virus (Falk et al. 1979), and also the complex of a DNA-containing virus dependent on the assistance of a symptomless, picornalike virus for the establishment and transmission of the tungro disease of rice (R. Hull, personal communication). Except for rice tungro disease, which is quite severe and endemic to southeast Asia, the Philippines, and Indonesia, most of the diseases produced by such complexes of virus vary considerably in their frequency of occurrence and usually are limited to a small, defined geographical area (e.g., Salinas Valley or Scotland). Presumably, this is due to the requirement for the presence of two specific viruses, certain host species, and particular vector species.

The efficiency of heterologous encapsidation is quite variable, depending upon the virus combinations being examined, their relative ratios in vivo, and whether other interactions occur between such viruses (see Dodds and Hamilton 1976). Nevertheless, in most cases where the efficiency of heterologous encapsidation has been measured, from less than 1% up to 10% of the encapsidated RNA is derived from the heterologous genome. In vitro, the level of heterologous encapsidation can be higher, especially when only coat protein of virus B and the genetic material of virus A are present (Hiebert et al. 1968, Fritsch et al. 1973, Chen and Francki 1990). This is due to the absence of the preferred interacting species; viz., the homologous nucleic acid and coat protein.

The relevancy of heterologous encapsidation to transgenic plants expressing viral coat protein genes is as follows: A plant transgenic for the coat protein gene of virus A should contain the coat protein of virus A, but not particles of virus A. In most cases, either that part of the virus genome expressing the viral coat protein doesn't contain the nucleotide sequences required to initiate encapsidation, or the coat protein mRNA contains insufficient mass to be encapsidated (and is normally coencapsidated with larger segments of the viral genome). Thus, when a virus (B) of a different taxonomic group enters a plant transgenic for virus A coat protein, the plant is not protected against infection by virus B (Loesch-Fries et al. 1987). Furthermore, at an early stage of the infection process, prior to the production of the coat protein of virus B (a "late" gene), the infected cells will contain the coat protein of virus A and the genetic material of virus B. Thus, under these conditions—i.e., the absence of the homologous interacting species—heterologous encapsidation is likely to occur at a higher frequency than in a dual infection (see Dodds and Hamilton 1976). Moreover, for the reasons given above, plants transgenic for viral coat protein production are more likely to demonstrate heterologous encapsidation with virus combinations previously not demonstrated during dual infections.

In addition, most heterologous encapsidation in dual infections is swamped out by the presence of large amounts of both parental (homologous) viruses. Furthermore, in many of the instances cited above, one of the viruses in each combination could only be transmitted in the presence of the helper or assistor virus. Thus, virus B was never found alone, but only in combination with virus A (which could be found alone).

Over the years, there have been anecdotal descriptions of field infections by TMV (not insect-transmissible) and the potato spindle tuber viroid (PSTV, which has no coat protein and is not insect-transmissible per se) coinciding with infestation by insects. However, in carefully controlled experiments, no evidence for insect transmission of these pathogens could be obtained (reviewed by Harris and Maramorosch 1977, Schumann et al. 1980, Dickson 1979, Diener 1979, Langrish 1981, Gooding 1986). It is possible that some of these incidents might have been due to heterologous encapsidation during mixed infections.

It has been shown that *Nicotiana clevelandii* plants infected with velvet tobacco mottle virus (VTMoV) and PSTV could produce particles of VTMoV which contained PSTV RNA (Francki et al. 1986). Inoculation of such VTMoV preparations to tomato produced an infection by PSTV, but no infection by VTMoV (tomato is not a host for VTMoV). Thus, in a field situation, no trace of the "helper virus" (VTMoV) would remain in the uninfected crop, but it would probably be present in some reservoir plants.

Although heterologous encapsidation may seem an esoteric process, of interest only to those specializing in the specificity of nucleic acid–protein interactions, in the context of the widespread use of plants expressing one or more virus coat protein genes, there exists an identifiable risk of increase in the dissemination of other viral pathogens. Specifically, if plants transgenic for virus A are able to encapsidate the genetic material of virus B, then the invertebrate vector of virus A should be able to transmit virus B (masked in virus A coat protein; see Fig. 7.1) horizontally—i.e., from plant to plant. When one considers that most plant viruses are transmitted by insect vectors, principally aphids (Harris 1981, D'Arcy and Nault 1982), then considering the millions of aphids that are likely to move across a field, the probability of dissemination of virus B is very high, even if the level of heterologous encapsidation is only a few percent: i.e., this becomes a "numbers game."

Recently, Chen and Francki (1990) showed that TMV RNA could be encapsidated in vitro by the coat protein of cucumber mosaic virus (CMV). In addition, such particles could be transmitted by aphids and initiate an infection with TMV. Thus, all the pieces of the puzzle exist. However, whether they fit together to pose a biohazard remains to be determined.

For some viruses the potential for enhanced dissemination as the result of heterologous encapsidation is either minimal or nonexistent. For example:

1. TMV has no natural vector. Thus, encapsidation of heterologous RNA into TMV particles will not enhance the dissemination of some other virus.

2. Potyviruses and caulimoviruses require an additional viral-encoded gene product (helper component) beside the viral coat protein for their transmission by their aphid vectors (Pirone and Thornbury 1984). Thus, encapsidation into potyvirus or caulimovirus particles will in itself not enhance the transmission of heterologous viral genomes.

The above issues clearly should be investigated: i.e., does heterologous encapsidation occur in transgenic plants, and does this lead to an increase in the dissemination of heterologous virus? However, what is the "acceptable threshold" of such dissemination? How many plants and aphids should be used to determine the frequency of aphid transmission of heterologously encapsidated virus?

The difficulty in addressing these questions lies in making probability assessments with low input numbers, and in getting sufficient data to make a risk assessment that could cover any and all contingencies. For example, if one were to set up a test involving 20 transgenic plants expressing the coat protein of virus A and inoculated with virus B, how many aphids should be used per transgenic plant or per recipient plant to measure the rate of transmission of heterologous virus? If 100 aphids are fed onto each transgenic plant and transferred to a similar number of recipient plants and no transmission is detected, one cannot assume that there is no risk of dissemination of heterologous virus A by this mechanism, since the threshold may be 1 in 1000. A field of plants may see hundreds of thousands or even millions of aphids during one growing season. Thus, how useful are negative results obtained using limited numbers of aphids in rather labor-intensive, controlled experiments?

A two-stage test may provide sufficient data to permit a risk assessment to be made: First, a controlled experiment done in a greenhouse to assess whether the probability of transmission of heterologously encapsidated virus is less than 1%; second, a field test done over one growing season in at least two climatic zones. The greenhouse test will provide an assessment of the risk associated with doing the field test. The field test will provide an assessment of the probability of occurrence of any transmission of heterologous virus.

How many different viruses should one have to test to demonstrate an absence of transmission of heterologous virus? For the greenhouse experiments one should consider a subclass of those viruses capable of infecting other crop and native plants in the same ecosystem as the transgenic plants, where these plants are also hosts for the insect vector being analyzed. For the field experiments, either those viruses that cause the most severe disease or those viruses with the greatest host range need to be tested.

Vertical transmission Although most plant viruses are not transmitted vertically, i.e., through the seed, there are many plant viruses that are vertically transmitted, albeit each in only a few species and often to only a low percentage (Phatak 1974, Neergaard 1977, Stace-Smith and Hamilton 1988). [There are exceptions, especially nepoviruses, which often show a high rate of seed transmissibility, and legumes, which tend to transmit viruses through the seed more efficiently than many other crops (Matthews 1981).] Nevertheless, because plants produced from infected seed can act as reservoirs for horizontal transmission to other field plants, there are quarantine regulations on the percentage of infected seed that may be used as seed stock for planting. Thus, it is important to determine whether transgenic plants expressing coat protein show any increase in the level of seed transmission of heterologous viruses (which themselves may not be seed-transmissible), either by heterologous encapsidation or via some other mechanism.

An example of the type of potential effect can be seen in the case of CMV. CMV itself is seed-transmissible in at least 19 species of plants, including tomato (0.2%), cowpeas (4 to 28%), lupines (0 to 34%), chickweeds (1 to 40%), beans (0 to 54%), and wild cucumbers (9.1 to 55%) (Neergaard 1977, Phatak 1974). The ranges from different studies either by the same or different authors indicate the effects of both cultivars and incidence of initial infection. Thus, while the expression of CMV coat protein in one cultivar of a given plant species may not increase the level of seed transmission of various heterologous viruses, it is not possible to extrapolate this observation to include other cultivars. However, if this phenomenon is tested in either several cultivars of an important crop, or in several species of plants, and there is no increase in the level of seed transmissibility of heterologous viruses, then the potential risk can be assessed. At that point, determinations can be made about the necessity for carrying out such tests with other specific cultivars or species expressing various coat protein genes.

Synergistic infection

Another aspect of virus interaction that has not been addressed vis-à-vis plants transgenic for virus coat proteins is the issue of virus synergism (see Bennett 1953, and Hamilton 1980). For example, although it may be known that virus A and virus B interact to cause a worse disease than either virus alone [e.g., CMV and TMV (Garces-Orejuela and Pound 1957)], it is not known what part (or genome segment) of virus A interacts with virus B to effect the synergistic response. If the expression of the coat protein gene of virus A alone is sufficient for the synergistic reaction with virus B, then a field of plants transgenic for the coat protein of virus A could be destroyed if infected by virus B. Many of the virus combinations that elicit synergistic responses in important crop plants have been described in the literature. However, to my knowledge, the determination of the ability of the coat protein produced by transgenic plants to interact with some other virus, either to elicit a synergistic response or to encapsidate the genetic material of that virus, has not been determined.

Resistance mediated via the expression of other viral sequences

To date, the only viral sequences expressed in transgenic plants, beside the coat protein genes, that have produced any significant resistance to virus infection are the 54K gene of TMV (M. Zaitlin, personal communication) and various satellite RNAs (Gerlach et al. 1987, Harrison et al. 1987, Jacquemond et al. 1988). Antisense RNAs to some plant viruses have also been tested, but these have shown at best only marginal protective effects (Cuozzo et al. 1988, Hemenway et al. 1988, Rezaian et al. 1988, Powell et al. 1989). The idea of using either viral nonstructural genes or satellite RNAs has some advantages over the expression of coat protein: viz., the mitigation or obviation of the various potential hazards described above. However, resistance engendered by the 54K gene of TMV [a gene that is presumed to encode a 54,000-Da protein thought to be part of the viral replication machinery (Palukaitis and Zaitlin 1986)] has only been observed with one virus and shows a strong strain specificity for protection (M. Zaitlin, personal communication). Thus it is not yet clear how widespread this type of resistance may be, with respect to other viruses. Nevertheless, further work along these lines with other viruses will determine the prospects for the widespread use of this form of transgenic plant–mediated resistance.

Some satellite RNAs of plant viruses contain sequences capable of self-cleavage (Buzayan et al. 1986, Forster and Symons 1987, Feldstein et al. 1989). One part of this self-cleavage domain contains the ability to cause a strand scission in another part of the molecule (Uhlenbeck 1987). Such sequences are referred to as "RNA enzymes," or "ribozymes" (Cech and Bass 1986). It has been suggested that ribozymes capable of binding to a viral RNA, by incorporating complementary flanking sequences, could cleave the viral RNA and prevent its replication (Haseloff and Gerlach 1988). After considering various options to improve the efficiency of the system, Wayne Gerlach and his colleagues have combined the ribozyme approach with an antisense RNA approach. Thus, they have produced transgenic plants expressing antisense RNA of part of a plant virus genome, containing three ribozymes spanning the antisense RNA. This type of resistance appears to be more efficient than the use of antisense RNA alone (Gerlach et al. 1990).

It should be pointed out that the possibility of synergistic interactions (described above) may not be limited to the interaction of the coat protein of virus A with some part of the genome of another virus (B). Thus, transgenic plants expressing antisense RNA, nonstructural genes, and satellite RNA sequences should also be assessed for their potential for synergistic interactions.

Resistance mediated via the expression of satellite RNAs

Satellite RNAs of plant viruses generally reduce the virus titer of their associated helper virus, and in many cases they attenuate the symptoms normally induced by their helper virus (reviewed by Francki 1985). Thus, several groups have introduced cDNA clones of satellite RNAs of plant viruses into the tobacco genome (Gerlach et al. 1987, Harrison et al. 1987, Jacquemond et al. 1988). These transgenic plants express the satellite RNAs, which are then replicated by the inoculated helper virus. The plants show little if any symptoms of the virus infection; i.e., the transgenically expressed satellite RNA shows the same protective effect as the application of natural satellite RNA. Satellite RNA of CMV, mixed in with helper virus, has also been applied to field-grown plants in China (Tien et al. 1987). These plants showed significant protection against the ravages of natural CMV infection, which is quite severe in China. Moreover, the protective effect was due primarily to the satellite RNA and not just to cross-protection induced by the associated helper virus (Wu et al. 1989).

There are, however, two primary concerns among a number of researchers vis-à-vis the use of satellite RNAs in transgenic plant–mediated resistance: (1) The transgenically expressed satellite RNA is replicated and encapsidated during an infection by the helper virus, and the encapsidated form of the satellite RNA can be transmitted to other hosts by the vector of the helper virus

(Baulcombe et al. 1986, Harrison et al. 1987). Thus, the transgenically expressed satellite RNA cannot be contained within the original target plant. (2) Some satellite RNAs, such as the satellites of CMV, are pathogenic on some hosts yet attenuate disease symptoms on other hosts tested (Waterworth et al. 1979, Palukaitis 1988). The differences between pathogenic and nonpathogenic variants may be as little as a single nucleotide change (Sleat and Palukaitis 1990). The mutation rate of satellite RNAs can be very rapid, and is determined by the sequence of the satellite RNA itself, the strain of helper virus, and the host species (Kurath and Palukaitis 1990, and M. Roossinck and P. Palukaitis, unpublished data). Once transgenic plants expressing satellite RNA sequences are placed in a field, there is no control over what strains of helper virus will infect those plants, or to what plant species the satellite RNA will be mobilized by the aphid vector of CMV. CMV is transmitted by some 60 species of aphid (Francki et al. 1979), and the host range of CMV encompasses almost 800 species of plants in 85 families (Douine et al. 1979). Given these facts, a number of researchers are concerned about the use of satellite RNAs in transgenic plants as a form of resistance (Colwell and Kuc 1987).

In southern Italy and eastern Spain, there has been a severe disease of tomato during the 1988 to 1990 growing seasons that has destroyed thousands of hectares of tomato crops (Gallitelli et al. 1988*a,b*, F. García-Arenal, personal communication). The epidemic appears to be caused by CMV and one of its satellite RNAs. Given the devastation, the lack of any genetic resistance to CMV in tomato, and the insistence of farmers in the area on growing tomatoes, various biocontrol measures are being considered, including (in Italy) the application of CMV containing an ameliorative satellite RNA, and the use of transgenic tomato plants expressing satellite RNA sequences. The argument is made that the potential environmental risk is negligible, since without some form of protection, they will lose their entire tomato crop anyway (D. Gallitelli and G. Martelli, personal communication).

The satellite RNA of tobacco ringspot virus (TobRV) is being introduced into transgenic walnut plants as a method of protecting such plants against the severe disease caused by a related virus, cherry leafroll virus (Ponz et al. 1987, G. Bruening, personal communication). While no pathogenic effects have been observed due to the presence of the satellite RNAs of TobRV, this virus has a wide host range (Stace-Smith 1985) and experiments involving TobRV and its satellite RNAs have only been done in a few species. Nevertheless, the slow mode of transmission of TobRV [via a nematode vector (McGuire and Schneider 1973)] significantly reduces any potential risk to other, nontarget plants, although some additional studies on the potential for pathogenicity by satellite RNAs of TobRV on a variety of other hosts of TobRV should be done to further reduce any potential environmental risks.

Effects on Plant and Animal Metabolism

There appears to be some concern about the effects of the expression of viral sequences, and in particular the expression of the coat protein, in the edible parts of crop plants. The concern is expressed relative to the presence of plant viruses in the food supply. This concern is unwarranted and based on a rather inaccurate assumption: that our food supply currently is free of "contamination" by plant viruses.

While some viruses may be limited to certain tissues (e.g., the vascular system) or organs (e.g., the roots), most plant viruses are found throughout the various organs and tissues of plants (Matthews 1981). Viruses are usually not found in the cells of the tips of the growing points of plants (i.e., the apical meristem), and, as mentioned before, most viruses are usually not found in the seed (Matthews 1981). However, viruses are found in the fruit, leaves, and stems of most plants. In some cases viruses can cause distortions or marring of the fruit, making it unsalable; however, in most cases viruses cause stunting of the plant, and a reduction in the yield of the crop. There is usually less fruit set, and the size of the fruit (or tuber) is smaller. Viruses may also affect the storage life of crops (Matthews 1981).

At the beginning of this century, virtually every commercial cultivar of potato grown in North America and Europe was infected with either one or some complex of potato viruses (Beemster and de Bokx 1987). Since some of these viruses produce an essentially symptomless infection (unless the plants are co-infected with other potato viruses producing a synergistic effect), their effect on the reduction of potato tuber yield was thought to be a genetic problem within the plant. Potato varieties grown today have various "resistance" genes incorporated into their genome to reduce either the incidence of infection or the disease produced by various viruses (Matthews 1981, Fraser 1986).

Breeding programs over most of this century (including the present) have had as their objective to incorporate "resistance" to the effects of pathogenic agents into the production of the crop. This has led to the widespread use of genes for "tolerance" rather than true resistance to the pathogen (Matthews 1981). Thus, many commercially available seeds grow into plants as susceptible to infection as any other variety, but which produce a normal (or enhanced) yield of unmarred crop. These plants often contain as much virus in the crop as the susceptible plant (Matthews 1981). Thus, plant viruses have been a part of our food supply probably as long as we have been eating plant matter. Homegrown fruit and vegetables probably have more virus in them than commercially grown produce, since there is usually a reservoir of virus in nearby weeds and ornamental plants, and home gardeners are not as conscious of methods to control either vectors or the spread of disease by mechanical methods (Bawden 1964). In some cases, plant pathologists have assayed their home gardens and found viruses in most of their plants (R. I. B. Francki and T. J. Morris, personal communication).

Although I cannot cite a specific study in which the percentage of fruit or vegetables obtained from various supermarkets containing one or more viruses was determined, it is generally known that plants infected with viruses produce fruit which is sold. There are many reports from various states on the presence of particular viruses in the fields during various growing seasons. This has been known by plant pathologists and has formed much of the literature of this field during most of this century. (Such reports have been published in the journal *Plant Disease*, formerly *Plant Disease Reporter*, for the last 50 years). This situation has not raised any question concerning the safety of the food supply, since it has always been there.

Transgenic plants express coat proteins of plant viruses at two to three orders of magnitude lower concentrations than plants naturally infected by viruses (Matthews 1981, Powell Abel et al. 1986, Loesch-Fries et al. 1987, Tumer et al. 1987, Cuozzo et al. 1988). If the plants display abnormal effects which alter their metabolism, this becomes obvious during the regeneration of the plants and such plants are discarded. (These effects are probably not due to plant–viral gene interactions, but rather to some effect on the site of insertion of the viral gene within the chromosomes of the plant.)

Could the expression of viral genes in transgenic plants alter the production of some secondary metabolite within the plant? This is unlikely for the following reasons: (1) While infection by viruses does alter the plant's metabolism and respiration (as is evident by the production of symptoms), studies analyzing a variety of gene products and/or metabolic intermediates are inconsistent with such a concept (Goodman et al. 1986, Collinge and Slusarenko 1987). (2) For any given virus, the extent to which the host responds is dependent upon a number of factors (e.g., host genotype, strains of the virus), including the level of replication of the virus—i.e., if there is more of a particular virus present, there is usually more disease (Matthews 1981, Fraser 1986). Thus, transgenic plants, producing very low levels of viral gene products compared with infected plants, are much less likely to have altered metabolism.

Plants expressing viral RNAs (i.e., satellites, ribozymes, antisense RNA) are also unlikely to present any problem in the food supply, since nucleic acids that are consumed are not usually metabolized. Whether such expressed nucleic acids have the capability of altering a plant's metabolism without the production of discernable symptoms has not been established. However, these systems are still in their early development phase. When (and if) any reach the deployment stage, these questions will become relevant.

Conclusion

The application of biotechnology to plant viruses has occurred along several lines. While the use of plant viruses as vectors has seen little application and in itself may have insurmountable problems that will prevent extensive application in the future, the use of viral sequences in transgenic plants has seen considerable application. Viral regulatory sequences are being used in transgenic plants to enhance the expression of other genes. This is an application with negligible associated risk. Sequences of viral genes are also being expressed in transgenic plants as a method of inducing resistance to infection by related viruses. This is an application with some potential environmental risk, the level of which can be assessed by testing. Satellite RNAs of plant viruses are also being expressed in transgenic plants as a form of biocontrol. In some cases, the potential environmental risk is low, while in other cases, the potential for damage to the environment is high. Nevertheless, the use of viral genome segments or satellite RNAs expressed in transgenic plants is likely to be safer than current practices involving either the application of whole viruses to plants to induce a form of resistance called "cross-protection," or the use of plants bred for tolerance to virus disease, which produces reservoirs of infected plants. None of the above applications, all of which result in viral nucleic acid or protein "contaminating" such plants, is likely to have any effect on either the plants themselves or animals that consume such plants or their parts, as virus-infected plants have always been part of our food supply.

Acknowledgments

The author would like to thank the following for providing unpublished data, mostly during symposium presentations: Drs. G. Bruening, R. I. B. Francki, D. Gallitelli, F. García-Arenal, W. L. Gerlach, R. Hull, G. Martelli, T. J. Morris, T. P. Pirone, R. Provvidenti, J. G. Shaw, and M. Zaitlin. The work described in this chapter from this laboratory, was supported by grant no. DE-FG02-86ER13505 from the Department of Energy, and a grant from the Cornell Biotechnology Program, which is sponsored by the New York State Science and Technology Foundation, a consortium of industries, the U.S. Army Research Office, and the National Science Foundation.

References

Adams, A., and Hull, R. 1972. Tobacco yellow vein, a virus dependent on assistor viruses for its transmission by aphids, *Ann. Appl. Biol.* 71:135–140.

Angenent, G. C., van den Ouweland, J. M. W., and Bol, J. F. 1990. Susceptibility to virus infection of transgenic tobacco plants expressing structural and non-structural genes of tobacco rattle virus, *Virology* 175:191–198.

Baughman, G. A., Jacobs, J. D., and Howell, S. H. 1988. Cauliflower mosaic virus gene VI produces a symptomatic phenotype in transgenic tobacco plants, *Proc. Natl. Acad. Sci. USA* 85:733–737.

Baulcombe, D. C., Saunders, G. R., Bevan, M. W., Mayo, M. A., and Harrison, B. D. 1986. Expression of biologically active viral satellite RNA from the nuclear genome of transformed plants, *Nature* (London) 321:446–449.

Baulcombe, D. C., Hamilton, W. D. O., Mayo, M. A., and Harrison, B. D. 1987. Resistance to viral disease through expression of viral genetic material from the plant genome, in *Plant Resistance to Viruses* (D. Evered and S. Harnett, eds.), Wiley, Chichester, U.K., pp. 170–184.

Bawden, F. C. 1964. *Plant Viruses and Virus Disease*, 4th ed., Ronald, New York, 361 pp.

Beachy, R. N., Abel, P., Oliver, M. J., De, B., Fraley, R. T., Rogers, S. G., and Horsch, R. B. 1985. Potential for applying genetic transformation to studies of viral pathogenesis and cross-protection, in *Biotechnology in Plant Science. Relevance to Agriculture in the Eighties* (M. Zaitlin, P. Day, and A. Hollaender, eds.), Academic, New York, pp. 265–275.

Beachy, R. N., Powell Abel, P., Nelson, R. S., Register, J., Tumer, N., and Fraley, R. T. 1987. Genetic engineering of plants for protection against virus diseases, in *Plant Resistance to Viruses* (D. Evered and S. Harnett, eds.), Wiley, Chichester, U.K., pp. 151–169.

Beemster, A. B. R., and de Bokx, J. A. 1987. Survey of properties and symptoms of potato viruses, in *Viruses of Potatoes and Seed Potato Production* (J. A. de Bokx and J. P. H. vanderWant, eds.), PuDOC, Wageningen, The Netherlands, pp. 84–93.

Benfey, P. N., Ren, L., and Chua, N.-H. 1989. The CaMV 35S enhancer contains at least two domains which can confer different developmental and tissue-specific expression patterns, *EMBO J.* 8(8):2195–2202.

Bennett, C. W. 1953. Interactions between viruses and virus strains, *Adv. Virus Res.* 1:39–67.

Bevan, M. W., Mason, S. E., and Goelet, P. 1985. Expression of tobacco mosaic virus coat protein by a cauliflower mosaic virus promoter in plants transformed by *Agrobacterium*, *EMBO J.* 4(8):1921–1926.
Brisson, N., Paszkowski, J., Penswick, J. R., Gronenborn, B., Potrykus, I., and Hohn, T. 1984. Expression of a bacterial gene in plants by using a viral vector, *Nature* (London) 310:511–514.
Buzayan, J. M., Gerlach, W. L., and Bruening, G. 1986. Satellite tobacco ringspot virus RNA: a subset of the RNA sequence is sufficient for autolytic processing, *Proc. Natl. Acad. Sci. USA* 83:8859–8862.
Cech, T. R., and Bass, B. L. 1986. Biological catalysis by RNA, *Annu. Rev. Biochem.* 55:599–629.
Chen, B., and Francki, R. I. B. 1990. Cucumovirus transmission by the aphid *Myzus persicae* is determined solely by the properties of its coat protein, *J. Gen. Virol.* 71: 939–944.
Cockbain, A. J. 1978. Bean yellow vein-banding virus, in Rothamsted Experimental Station Report for 1977, Harpenden, Herts U.K., pp. 221–222.
Collinge, D. B., and Slusarenko, A. J. 1987. Plant gene expression in response to pathogens, *Plant Mol. Biol.* 9:389–410.
Colwell, R. K., and Kuc, J. 1987. Regulatory considerations: genetically-engineered plants, summary of a workshop at the Boyce Thompson Institute for Plant Research. Center for Science Information, San Francisco, pp. 53–56.
Cuozzo, M., O'Connell, K. M., Kaniewski, W., Fang, R-X., Chua, N-H., and Tumer, N. E. 1988. Viral protection in transgenic tobacco plants expressing the cucumber mosaic virus coat protein or its antisense RNA, *Bio/Technology* 6:549–557.
D'Arcy, C. J., and Nault, L. R. 1982. Insect transmission of plant viruses and mycoplasmalike and rickettsialike organisms, *Plant Dis.* 66:99–104.
Daubert, S. 1988. Sequence determinants of symptoms in the genomes of plant viruses, viroids, and satellites, *Mol. Plant-Microbe Interact.* 1(8):317–325.
Daubert, S., Shepherd, R. J., and Gardner, R. C. 1983. Insertional mutagenesis of the cauliflower mosaic virus genome, *Gene* 25:201–208.
Dawson, W. O., Bubrick, P., and Grantham, G. L. 1988. Modifications of the tobacco mosaic virus coat protein gene affecting replication, movement, and symptomatology, *Phytopathology* 78:783–789.
De Zoeten, G. A., Penswick, J. R., Horisberger, M. A., Ahl, P., Schultze, M., and Hohn, T. 1989. The expression, localization, and effect of a human interferon in plants, *Virology* 172:213–222.
Dickson, E. 1979. Viroids: infectious RNA in plants, in *Nucleic Acids in Plants* (T. C. Hall and J. W. Davies, eds.), CRC Press, Boca Raton, Florida, pp. 153–193.
Diener, T. O. 1979. *Viroids and Viroid Diseases*, Wiley, New York, 252 pp.
Dixon, L. K., Koenig, I., and Hohn, T. 1983. Mutagenesis of cauliflower mosaic virus, *Gene* 25:189–199.
Dodds, J. A., and Hamilton, R. I. 1974. Masking the RNA genome of tobacco mosaic virus by the protein of barley stripe mosaic virus in doubly infected barley, *Virology* 59:418–427.
Dodds, J. A., and Hamilton, R. I. 1976. Structural interactions between viruses as a consequence of mixed infections, *Adv. Virus Res.* 20:33–86.
Douine, L., Quiot, J. B., Marchoux, G., and Archange, P. 1979. Recensement des espèces végétales sensibles au virus de la mosaïque du concombre (CMV) etude bibliographique, *Ann. Phytopathol.* 11:439–475.
Elnagar, S., and Murant, A. F. 1978. Aphid-injection experiments with carrot mottle virus and its helper virus, carrot red leaf, *Ann. Appl. Biol.* 89:245–250.
Falk, B. W., and Duffus, J. E. 1981. Epidemiology of helper-dependent persistent aphid transmitted virus complexes, in *Plant Diseases and Vectors: Ecology and Epidemiology* (K. Maramorosch and K. F. Harris, eds.), Academic, New York, pp. 161–179.
Falk, B. W., Duffus, J. E., and Morris, T. J. 1979. Transmission, host range, and serological properties of the viruses that cause lettuce speckles disease, *Phytopathology* 69:612–617.
Feldstein, P. A., Buzayan, J. M., and Bruening, G. 1989. Two sequences participating in

the autolytic processing of satellite tobacco ringspot virus complementary RNA, *Gene* 82:53–61.
Fenoll, C., Black, D. M., and Howell, S. H. 1988. The intergenic region of maize streak virus contains promoter elements involved in rightward transcription of the viral genome, *EMBO J.* 7(6):1589–1596.
Fletcher, J. T., and Rowe, J. M. 1975. Observations and experiments on the use of an avirulent mutant strain of tobacco mosaic virus as a means of controlling tomato mosaic, *Ann. Appl. Biol.* 81:171–179.
Forster, A. C., and Symons, R. H. 1987. Self-cleavage of plus and minus RNAs of a virusoid and a structural model for the active sites, *Cell* 49:211–220.
Francki, R. I. B. 1985. Plant virus satellites, *Annu. Rev. Microbiol.* 39:151–174.
Francki, R. I. B., Mossop, D. W., and Hatta, T. 1979. Cucumber mosaic virus, CMI/AAB Descriptions of Plant Viruses. no. 213, Cambrian News, Wales, U.K.
Francki, R. I. B., Zaitlin, M., and Palukaitis, P. 1986. *In vivo* encapsidation of potato spindle tuber viroid by velvet tobacco mottle virus particles, *Virology* 155:469–473.
Fraser, R. S. S. 1986. Genes for resistance to plant viruses, *CRC Crit. Rev. Plant Sci.* 3(3):257–294.
French, R., Janda, M., and Ahlquist, P. 1986. Bacterial gene inserted in an engineered RNA virus: efficient expression in monocotyledonous plant cells, *Science* 231:1294–1297.
Fritsch, C., Stussi, C., Witz, J., and Hirth, L. 1973. Specificity of TMV RNA encapsidation: *in vitro* coating of heterologous RNA by TMV protein, *Virology* 56:33–45.
Fulton, R. W. 1986. Practices and precautions in the use of cross protection for plant virus disease control, *Annu. Rev. Phytopathol.* 24:67–81.
Fütterer, J., Bonneville, J. M., Sanfaçon, H., Torruella, M., Pisan, B., Gordon, K., Penswick, J. R., Grimsley, N., Hohn, B., and Hohn, T. 1988. Use of plant viruses and plant viral expression signals for gene expression in plants and plant protoplasts, in *Viral Vectors* (Y. Gluzman, ed.), Cold Spring Harbor Laboratory, Cold Spring Harbor, New York, pp. 178–182.
Gallie, D. R., Sleat, D. E., Watts, J. W., Turner, P. C., and Wilson, T. M. A. 1987. The 5′-leader sequence of tobacco mosaic virus RNA enhances the expression of foreign gene transcripts *in vitro* and in *vivo*, *Nucleic Acids Res.* 15(8):3257–3273.
Gallitelli, D., Di Franco, A., Vovlas, C., Crescenzi, A., and Ragozzino, A. 1988. Una grave virosi del pomodoro in Italia meridionale, *L'Informatore Agrario* XLIV(3):67–70.
Gallitelli, D., Di Franco, A., Vovlas, C., and Kaper, J. M. 1988. Infezioni miste del virus del mosaico del cetriolo (CMV) e di potyvirus in colture ortive di Puglia e Basilicata, *Informatore Fitopatologico* XXXVIII(12)57–64.
Garces-Orejuela, C., and Pound, G. S. 1957. The multiplication of tobacco mosaic virus in the presence of cucumber mosaic virus or tobacco ringspot virus in tobacco, *Phytopathology* 57:232–239.
Gerlach, W. L., Llewellyn, D., and Haseloff, J. 1987. Construction of a plant disease resistance gene from the satellite RNA of tobacco ringspot virus, *Nature* (London) 328: 802–805.
Gerlach, W. L., Haseloff, J. P., Young, N. J., and Bruening, G. 1990. Use of plant virus satellite RNA sequences to control gene expression, in *Viral Genes and Plant Pathogenesis* (T. P. Pirone and J. G. Shaw, eds.), Springer-Verlag, New York, pp. 177–186.
Gianninazzi, S. 1984. Genetic and molecular aspects of resistance induced by infections or chemicals, in *Plant Microbe Interactions: Molecular and Genetic Perspectives* (T. Kosuge and E. W. Nester, eds.), vol. I, Macmillan, New York, pp. 321–342.
Gooding, G. V. 1986. Tobacco mosaic virus: epidemiology and control, in *The Plant Viruses*, vol 2: *The Rod-Shaped Plant Viruses* (H. M. V. Van Regenmortel and H. Fraenkel-Conrat, eds.), Plenum, New York, pp. 133–152.
Goodman, R. M., and Ross, A. F. 1974*a*. Enhancement of potato virus X synthesis in doubly infected tobacco occurs in doubly infected cells, *Virology* 58:16–24.
Goodman, R. M., and Ross, A. F. 1974*b*. Enhancement by potato virus Y of potato virus X synthesis in doubly infected tobacco depends on the timing of invasion by the viruses, *Virology* 58:263–271.

Goodman, R. N., Király, Z., and Wood, K. R. 1986. *The Biochemistry and Physiology of Plant Disease*, University of Missouri Press, Columbia, 433 pp.

Gronenborn, B., Gardner, R. C., Schaefer, S., and Shepherd, R. J. 1981. Propagation of foreign DNA in plants using CaMV as a vector, *Nature* (London) 294:773–776.

Guilley, H., Dudley, R. K., Jonard, G., Balàzs, E., and Richards, K. 1982. Transcription of cauliflower mosaic virus DNA: detection of promoter sequences, and characterization of transcripts, *Cell* 30:763–773.

Hamilton, R. I. 1980. Defenses triggered by previous invaders: viruses, in *Plant Disease: An Advanced Treatise* (J. G. Horsfall and E. B. Cowlings, eds.), vol. V, Academic, New York, pp. 279–303.

Hanley-Bowdoin, L., Elmer, J. S., and Rogers, S. G. 1988. Transient expression of heterologous RNAs using tomato golden mosaic virus, *Nucleic Acids Res.* 16(22): 10511–10528.

Harpster, M. H., Townsend, J. A., Jones, J. D. G., Bedbrook, J., and Dunsmiur, P. 1988. Relative strengths of the 35S cauliflower mosaic virus, 1′,2′, and nopaline synthase promoters in transformed tobacco, sugarbeet and oilseed rape callus tissue, *Mol. Gen. Genet.* 212:182–190.

Harris, K. F. 1981. Arthropod and nematode vectors of plant viruses, *Annu. Rev. Phytopathol.* 19:391–426.

Harris, K. F., and Maramorosch, K. 1977. *Aphids as Virus Vectors*, Academic, New York.

Harrison, B. D., Mayo, M. A., and Baulcombe, D. C. 1987. Virus resistance in transgenic plants that express cucumber mosaic virus satellite RNA, *Nature* (London) 328:799–802.

Haseloff, J., and Gerlach, W. L. 1988. Simple RNA enzymes with new and highly specific endoribonuclease activities, *Nature* (London) 334:585–591.

Hayes, R. J., Petty, I. T. D., Coutts, R. H. A., and Buck, K. W. 1988. Gene amplification and expression in plants by a replicating geminivirus vector, *Nature* (London) 334: 179–182.

Hayes, R. J., Coutts, R. H. A., and Buck, K. W. 1989. Stability and expression of bacterial genes in replicating geminivirus vectors in plants, *Nucleic Acids Res.* 17(7): 2391–2403.

Hemenway, C., Fang, R-X., Kaniewski, W. K., Chua, N-M., and Tumer, N. E. 1988. Analysis of the mechanism of protection in transgenic plants expressing the potato virus X coat protein or its antisense RNA, *EMBO J.* 7(5):1273–1280.

Hiebert, E., Bancroft, J. B., and Bracker, C. E. 1968. The assembly *in vitro* of some small spherical viruses, hybrid viruses, and other nucleoproteins, *Virology* 34:492–508.

Hoekema, A., Huisman, M. J., Molendijk, L., van den Elzen, P. J. M. and Cornelissen, B. J. C. 1989. The genetic engineering of two commercial potato cultivars for resistance to potato virus X, *Bio/Technology* 7:273–278.

Howell, S. H. 1982. Plant molecular vehicles: potential vectors for introducing foreign DNA in plants, *Annu. Rev. Plant Physiol.* 33:609–650.

Howell, S. H. 1985. The molecular biology of plant DNA viruses, *CRC Crit. Rev. Plant Sci.* 2:287–316.

Howell, S. H., Walker, L. L., and Walden, R. M. 1981. Rescue of *in vitro* generated mutants of cloned CaMV genome on infected plants, *Nature* (London) 293:483–486.

Hull, R., and Adams, A. N. 1968. Groundnut rosette and its assistor virus, *Ann. Appl. Biol.* 62:139–145.

Jacquemond, M., Amselem, J., and Tepfer, M. 1988. A gene coding for a monomeric form of cucumber mosaic virus satellite RNA confers tolerance to CMV, *Mol. Plant-Microbe Interact.* 1:311–316.

Jobling, S. A., and Gehrke, L. 1987. Enhanced translation of chimaeric messenger RNAs containing a plant viral untranslated leader sequence, *Nature* (London) 325: 622–625.

Knorr, D. A., and Dawson, W. O. 1988. A point mutation in the tobacco mosaic virus capsid protein gene induces hypersensitivity in *Nicotiana sylvestris, Proc. Natl. Acad. Sci. USA* 85:170–174.

Kurath, G., and Palukaitis, P. 1990. Serial passage of infectious transcripts of a cucumber mosaic virus satellite RNA clone results in sequence heterogeneity, *Virology* 175: (in press).

Langrish, S. M. 1981. Transmission of potato spindle tuber viroid in potatoes, M.S. thesis, Cornell University, Ithaca, New York, 1981, 70 pp.

Lawson, C., Kaniewski, W., Haley, L., Rozman, R., Newell, C., Sanders, P., and Tumer, N. E. 1990. Engineering resistance to mixed virus infection in a commercial potato cultivar: resistance to potato virus X and potato virus Y in transgenic Russet Burbank, *Bio/Technology* 8:127–134.

Lefebvre, D. D., Miki, B. L., and Laliberté, J. F. 1987. Mammalian metallothionein functions in plants, *Bio/Technology* 5:1053–1056.

Loesch-Fries, L. S., Merlo, D., Zinnen, T., Burhop, L., Hill, K., Krahn, K., Jarvis, N., Nelson, S., and Halk, E. 1987. Expression of alfalfa mosaic virus RNA 4 in transgenic plants confers virus resistance, *EMBO J.* 6(7):1845–1851.

Matthews, R. E. F. 1981. *Plant Virology*, 2d ed., Academic, New York, 897 pp.

McElroy, D., Zhang, W., Cao, J., and Wu, R. 1990. Isolation of an efficient action promoter for use in rice transformation, *Plant Cell* 2:163–171.

McGuire, J. M., and Schneider, I. R. 1973. Transmission of satellite of tobacco ringspot virus by *Xiphinema americanum, Phytopathology* 63:1429–1430.

Murant, A. F., and Goold, R. A. 1968. Purification, properties and transmission of parsnip yellow fleck, a semi-persistent, aphid-borne virus, *Ann. Appl. Biol.* 62: 123–137.

Nagy, F., Odell, J. T., Morelli, G., and Chua, N-H. 1985. Properties of expression of the 35S promoter from CaMV in transgenic tobacco plants, in *Biotechnology in Plant Science. Relevance to Agriculture in the Eighties* (M. Zaitlin, P. Day, and A. Hollaender, eds.), Academic, New York, pp. 227–235.

Neergaard, P. 1977. *Seed Pathology*, Wiley, New York, 1187 pp.

Nelson, R. S., McCormick, S. M., Delannay, X., Dubé, P., Layton, J., Anderson, E. J., Kaniewska, M., Proksch, R. K., Horsch, R. B., Rogers, S. G., Fraley, R. T., and Beachy, R. N. 1988. Virus tolerance, plant growth, and field performance of transgenic tomato plants expressing coat protein from tobacco mosaic virus, *Bio/Technology* 6:403–409.

Palukaitis, P. 1988. Pathogenicity regulation by satellite RNAs of cucumber mosaic virus: minor nucleotide sequence changes alter host responses, *Mol. Plant-Microbe Interact.* 1:175–181 (1988).

Palukaitis, P., and Zaitlin, M. 1984. A model to explain the "cross-protection" phenomenon shown by plant viruses and viroids, in *Plant Microbe Interactions: Molecular and Genetic Perspectives* (T. Kosuge and E. W. Nester, eds.), vol. I, Macmillan, New York, pp. 420–429.

Palukaitis, P., and Zaitlin, M. 1986. Tobacco mosaic virus infectivity and replication, in *The Plant Viruses*, vol 2: *The Rod-Shaped Plant Viruses* (H. M. V. Van Regenmortel and H. Fraenkel-Conrat, eds.), Plenum New York, pp. 105–131.

Paszkowski, J., Shillito, R. D., Saul, M., Mandák, V., Hohn, T., Hohn, B., and Potrykus, I. 1984. Direct gene transfer to plants, *EMBO J.* 3(12):2717–2722.

Phatak, H. G. 1974. Seed-borne plant viruses—identification and diagnosis in seed health testing. *Seed Sci. Technol.* 2:3–155.

Pirone, T. P., and Thornbury, D. W. 1984. The involvement of a helper component in nonpersistent transmission of plant viruses by aphids, *Microbiol. Sci.* 1:191–193.

Ponz, F., Rowhani, A., Mircetich, S. M., and Bruening, G. 1987. Cherry leafroll virus infections are affected by a satellite RNA that the virus does not support, *Virology* 160:183–190.

Powell Abel, P., Nelson, R. S., De, B., Hoffmann, N., Rogers, S. G., Fraley, R. T., and Beachy, R. N. 1986. Delay of disease development in transgenic plants that express the tobacco mosaic virus coat protein gene, *Science* 232:738–743.

Powell, P. A., Stark, D. M., Sanders, P. R., and Beachy, R. N. 1989. Protection against tobacco mosaic virus in transgenic plants that express tobacco mosaic virus antisense RNA, *Proc. Natl. Acad. Sci. USA* 86:6949–6952.

Rast, A. Th. B. 1972. MII-16, an artificial symptomless mutant of tobacco mosaic virus for seedling inoculation of tomato crops, *Neth. J. Plant Pathol.* 78:110–112.

Rezaian, M. A., Skene, K. G. M., and Ellis, J. G. 1988. Anti-sense RNAs of cucumber

mosaic virus in transgenic plants assessed for control of the virus, *Plant Mol. Biol.* 11:463–471.
Rochow, W. F. 1972. The role of mixed infections in the transmission of plant viruses by aphids, *Annu. Rev. Phytopathol.* 10:101–124.
Rochow, W. F. 1977. Dependent virus transmissions from mixed infections, in *Aphids as Virus Vectors* (K. F. Harris and K. Maramorosch, eds.), Academic, New York, pp. 253–273.
Saito, T., Meshi, T., Takamatsu, N., and Okada, Y. 1987. Coat protein gene sequence of tobacco mosaic virus encodes a host response determinant, *Proc. Natl. Acad. Sci. USA* 84:6074–6077.
Schnölzer, M., Haas, B., Ramm, K., Hofmann, H., and Sänger, H. L. 1985. Correlation between structure and pathogenicity of potato spindle tuber viroid (PSTV), *EMBO J.* 4(9):2181–2190.
Schoelz, J. E., and Shepherd, R. J. 1988. Host range control of cauliflower mosaic virus, *Virology* 162:30–37.
Schumann, G. L., Tingey, W. M., and Thurston, H. D. 1980. Evaluation of six insect pests for transmission of potato spindle tuber viroid, *Am. Potato J.* 57:205–211.
Siegel, A. 1985. Plant-virus-based vectors for gene transfer may be of considerable use despite a presumed high error frequency during RNA synthesis, *Plant Mol. Biol.* 4(5): 327–329.
Sleat, D. E., and Palukaitis, P. 1990. Site-directed mutagenesis of a plant viral satellite RNA changes its phenotype from ameliorative to necrogenic, *Proc. Natl. Acad. Sci. USA* 87:2946–2950.
Smith, K. M. 1945. Transmission by insects of a plant virus complex, *Nature* (London) 155:174.
Smith, K. M. 1946. The transmission of a plant virus complex by aphides, *Parasitology* 37:131–134.
Stace-Smith, R. 1985. Tobacco ringspot virus, CMI/AAB Description of Plant Viruses. no. 309, Spottiswoode Ballantyne, Warwick, U.K.
Stace-Smith, R., and Hamilton, R. I. 1988. Inoculum thresholds of seedborne pathogens: viruses, *Phytopathology* 78:875–880.
Stanley, J. 1985. The molecular biology of geminiviruses, *Adv. Virus Res.* 30:139–177.
Stark, D. M., and Beachy, R. N. 1989. Protection against potyvirus infection in transgenic plants: evidence for broad spectrum resistance, *Bio/Technology* 7:1257–1262.
Stratford, R., and Covey, S. N. 1989. Segregation of cauliflower mosaic virus symptom genetic determinants, *Virology* 172:451–459.
Takahashi, H., Shimamoto, K., and Ehara, Y. 1989. Cauliflower mosaic virus gene VI causes growth suppression, development of necrotic spots and expression of defence-related genes in transgenic tobacco plants, *Mol. Gen. Genet.* 216:188–194.
Takamatsu, N., Ishikawa, M., Meshi, T., and Okada, Y. 1987. Expression of bacterial chloramphenicol acetyltransferase gene in tobacco plants mediated by TMV-RNA, *EMBO J.* 6(2):307–311.
Tien, P., Zhang, X., Qiu, B., Qin, B., and Wu, G. 1987. Satellite RNA for the control of plant diseases caused by cucumber mosaic virus, *Ann. Appl. Biol.* 111:143–152.
Tumer, N. E., O'Connell, K. M., Nelson, R. S., Sanders, P. R., Beachy, R. N., Fraley, R. T., and Shah, D. M. 1987. Expression of alfalfa mosaic virus coat protein gene confers cross-protection in transgenic tobacco and tomato plants, *EMBO J.* 6(5):1181–1188.
Uhlenbeck, O. C. 1987. A small catalytic oligoribonucleotide, *Nature* (London) 328:596–600.
Van Dun, C. M. P., Bol, J. F., and Van Vloten-Doting, L. 1987. Expression of alfalfa mosaic virus and tobacco rattle virus coat protein genes in transgenic tobacco plants, *Virology* 159:299–305.
van Vloten-Doting, L., Bol, J. F., and Cornelissen, B. 1985. Plant-virus-based vectors for gene transfer will be of limited use because of the high error frequency during viral RNA synthesis, *Plant Mol. Biol.* 4(5):323–326.
Visvader, J., and Symons, R. H. 1986. Replication of *in vitro* constructed viroid mutants: location of the pathogenicity-modulating domain of citrus exocortis viroid, *EMBO J.* 5(9):2051–2055.

Ward, A., Etessami, P., and Stanley, J. 1988. Expression of a bacterial gene in plants mediated by infectious geminivirus DNA, *EMBO J.* 7(6):1583–1587.

Waterworth, H. E., Kaper, J. M., and Tousignant, M. E. 1979. CARNA 5, the small cucumber mosaic virus-dependent replicating RNA, regulates disease expression, *Science* 204:845–847.

Wu, G., Kang, L., and Tien, P. 1989. The effect of satellite RNA on cross-protection among cucumber mosaic virus strains, *Ann. Appl. Biol.* 114:489–496.

Yeh, S.-D., and Gonsalves, D. 1984. Evaluation of induced mutants of papaya ringspot virus for control by cross protection, *Phytopathology* 74:1086–1091.

Chapter

8

The Implications of Horizontal Gene Transfer for the Environmental Impact of Genetically Engineered Microorganisms

Betty H. Olson, O. A. Ogunseitan, P. A. Rochelle, C. C. Tebbe, and Y. L. Tsai

Program in Social Ecology
University of California, Irvine
Irvine, California 92717

Introduction

Concern with genetically engineered bacteria began in the 1970s with the initial studies on how to engineer bacteria safely to protect against accidental release and proliferation outside the laboratory. At the Asilomar Conference in 1975 these concerns formed the basis of the first guidelines for the manufacture and use of genetically engineered bacteria and the prevention of their accidental release into the environment (Berg et al. 1975).

As technology advanced in molecular biology, it became possible to envision a myriad of beneficial uses of engineered microbes in a variety of environments. However, the notion of the planned release of engineered microorganisms into the environment, whether they be viruses, bacteria, or fungi, produced societal fears regarding the unprecedented breeding of chimeric and potentially dangerous organisms due to exchange of genetic information between engineered microorganisms and natural bacteria. Because of the potential ramifications of horizontal gene transfer and the fact that its occurrence was well documented in the literature, special interest was paid to the potential for the horizontal transfer of exogenous genetic determinants from engineered bacteria into natural populations. The fears of negative effects from low-probability events resulting in high-cost outcomes appeared somewhat justified, because of historical data on the introduction of exotic species dating back to the turn of the twentieth century (Elton 1958). However, given the complexities of the system, probability determinations for exchange and invasiveness of genetic information for bacteria were virtually unknown at that time. The court injunction obtained by Jeremy Rifkin in 1985 against the release of an "Ice$^-$" *Pseudomonas syringae* brought a potentially booming industry quickly into the world of intense public scrutiny and regulation. Because of the public concern regarding the release of genetically engineered bacteria, their regulation has come under the authority of the U.S. Environmental Protection Agency (USEPA), the U.S. Department of Agriculture (USDA), and the Food and Drug Administration (FDA). A number of federal legislative acts regulate release. The USEPA acts under the Toxic Substances Control Act (TSCA) and the Federal Insecticide, Fungicide, and Rodenticide Act (FIFRA).

The purpose of this chapter is (1) to review the design characteristics of genetically engineered bacteria used to minimize the potential for horizontal genetic exchange and (2) to discuss the importance of the various genetic exchange mechanisms in bacteria in relation to the invasive capacity of engineered genes in microbial communities.

Design of Genetically Engineered Microorganisms

The rapid advance of recombinant DNA technology made it possible to incorporate desirable hereditary traits into genomes of a variety of organisms, and the beneficial uses of such organisms indicated their imminent release into many natural habitats. The design of genetically engineered microorganisms (GEMs) to limit the potential for gene transfer in the environment has been a primary focus in their development. This section discusses why certain systems used to introduce heterologous DNA into GEMs have been emphasized to restrict the possibility of subsequent DNA transfer into or from GEMs that are constructed for direct release into the environment.

Plasmids

A variety of plasmid types have been used in gene cloning. Most plasmids used for these purposes are small, covalently closed, circular double-stranded DNA containing the necessary sequences for DNA replication and a region of DNA into which foreign DNA may be inserted without damage to plasmid functions for replication. A selective marker, usually for antibiotic resistance, is inserted into these plasmid vectors at specific sites. Heterologous DNA coding for the genes of interest is inserted into the plasmid vectors by cutting the DNA with restriction enzymes and enzymatically resealing with ligase. Only some of the resulting plasmid vectors will contain inserted DNA. Selective markers are utilized to identify vectors containing foreign DNA, and selective pressure is applied to these characteristics to enable only plasmids with the desired DNA fragment to grow.

The most useful plasmid vectors in genetic engineering were themselves constructed from a variety of plasmids. The development of plasmid vectors is summarized in the following three phases: (1) introduction of selective markers into plasmids (Cohen et al. 1973, Hershfield et al. 1974, Covey et al. 1976, Bolivar et al. 1977*a,b*); (2) generation of smaller plasmid vectors for the accommodation of larger segments of foreign DNA (Twigg and Sherratt 1980, Soberon et al. 1980, Sambrook et al. 1989, Hanahan 1983); (3) incorporation of ancillary sequences used for a variety of purposes, such as generation of single-stranded DNA templates for DNA sequencing (Sambrook et al. 1989), transcription of foreign DNA sequences in vitro (Melton et al. 1984), and direct selection of recombinant clones (Maloy and Nunn 1981, Craine 1982, Balbas et al. 1986).

Plasmids not suitable for genetically engineered microorganisms

Shuttle and conjugative plasmids. Many plasmids are inappropriate as vectors for foreign DNA because of potentially high transfer rates. These unsuitable plasmid types include "shuttle" and conjugative plasmids. "Shuttle vectors" are those which are expressed in two different bacterial genera. The very nature of a shuttle vector, such as two replication origins (one for each organism), increases the potential frequencies of horizontal gene transfer between species. Shuttle plasmid vectors, although useful in the study of gene regulation and enzyme production in fastidious organisms, are not recommended in the construction of GEMs because of their ability to replicate and express in a variety of microorganisms (Squires et al. 1984, Roberts et al. 1988).

Conjugative plasmids, with their ability to transfer genetic elements and to mobilize other plasmids, represent the greatest risk for gene transfer within the indigenous populations of microorganisms following the release of GEMs into environments. Again, the use of these types of plasmids for the introduction of genetic information into engineered bacteria is considered unacceptable because of the potential for horizontal transfer. In the laboratory, the production of transconjugants is directly proportional to the product of donor and recipient concentrations in both log-phase and stationary-phase cultures of *Escherichia coli* (Levin et al. 1979), and conjugative plasmids have been introduced into and maintained in diverse bacterial genera and species. It has been shown that some broad-host-range plasmids are capable of intergeneric transfer within Gram-negative bacteria (Tardiff and Grant 1983, Kumari and Vidaver 1988, Richaume et al. 1989) and of mobilizing genetic information between bacteria and yeast (Heinemann and Sprague 1989). The narrow-host-range F plasmids of *E. coli* can transfer to many of the *Enterobacteriaceae* and some species of *Pseudomonas* (Mergeay and Gerits 1978, Leary et al. 1984). This observation convincingly advocates strongly restricting the usage of conjugative plasmids in construction of GEMs in order to reduce gene transfer in situ.

Nonconjugative plasmids. Nonconjugative plasmids, which themselves cannot transfer their DNA, are more commonly found than conjugative plasmids in certain environments. For example, approximately 24% of enterobacteria isolated between 1917 and 1954 contained nonconjugative plasmids (Hughes and Datta 1983). However, nonconjugative plasmids are also inappropriate vectors for foreign DNA in engineered bacteria because they can still be mobilized to a recipient bacterium if they are coresident with a conjugative plasmid (Ditta et al. 1985, Priefer et al. 1985, Buchanan-Wollaston et al. 1987). Both conjugative and nonconjugative plasmids can and do occur in the same bacterial cell (Roberts and Falkow 1979), and conjugative plasmids can mobilize a nonconjugative plasmid if it contains *mob* genes (Buchanan-Wollaston et al. 1987). Nonconjugative plasmids may be mobilizable despite Hamer's (1977) demonstration that the insertion of foreign DNA into a nonconjugative plasmid appeared to have no significant effect on the mobilizability of the plasmid.

Suicide plasmids. "Suicide plasmids" self-destruct based on a trigger chemical or response. The release of GEMs into the environment has repeatedly raised concern among most ecologists and environmentalists that horizontal transfer of recombinant DNA within indigenous organisms will take place; the construction of an appropriate suicide vector could act to prevent transfer and to contain the GEMs within specified bounds. Suicide plasmids contain inducible "suicide," or "conditional lethal," genes which will, once triggered, kill the microorganism carrying these plasmids and prevent its spread in the open environment. Thus, suicidal GEMs could be designed to be viable in certain specific environments, such as chemically polluted areas, with their suicidal component then triggered if the environmental conditions are changed (pollutant concentration reduced). Alternatively, a GEM could be designed to survive until a special inducer is introduced into the environment to trigger its death.

Bej et al. (1988) developed a model suicide vector for the potential containment of GEMs by constructing a plasmid with the *hok* (host-killing) gene under the control of the *lac* promoter. Induction of *hok* gene overexpression causes rapid death of the host cells (Gerdes et al. 1986). Soil microcosm experiments showed the ability of a suicide vector to restrict the growth of a GEM (Bej et al. 1988). However, to provide an absolute margin of safety, more restrictive suicide systems are necessary.

Chromosome insertion via transposons

Transposable elements are special genetic entities which are capable of translocating themselves from one site to another within or between replicons without extended DNA homology at the insertion sites; they are discussed in more detail later in this chapter.

The ability of some transposable elements to insert into DNA molecules in an almost random manner suggested that they could be powerful tools for introducing foreign DNA into the bacterial chromosome, and simultaneously minimize the potential for horizontal transfer. For specifics on the numerous techniques and concepts concerning transposons in manipulation of genetic elements in bacteria, the reader is referred to Berg and Howe (1989). The process referred to as "transposon mutagenesis" relies on the insertion of transposons containing selectable drug-resistance markers and the trait of interest (Simon et al. 1983, Vanstockem et al. 1987). Thus, a cell carrying this

TABLE 8.1 Construction of Genetically Engineered Microorganisms for Their Containment in Natural Environments

Construction mode	Effect on potential horizontal spread	Methods used to constrain	Rating [a]
Conjugative plasmid	High	Introduction of conditional lethal genes	3
Nonconjugative plasmid	Moderate	Introduction of conditional lethal genes	2
Chromosome	Moderate–low	Destroying transpositional elements after insertion	1

[a] 1 = best, 2 = better, 3 = good

type of mutation can be identified by traditional selective techniques. The reduction of risk of horizontal gene transfer is achieved by eliminating the transposable functions which were inserted with the trait either by mutagenesis or some other means. The question that is unanswered about this type of construction is how the introduced trait may be influenced by other transpositional units within the bacterium or those transposable elements that may be introduced through horizontal genetic exchange once in the environment.

The mechanisms that can be used to introduce foreign DNA into bacterial cells and the various advantages and drawbacks of each in terms of construction and minimizing the potential for horizontal transfer are summarized in Table 8.1. Currently, the use of transpositional elements to introduce material followed by the destruction of the unit appears the most promising. However, this may stem from very limited knowledge of transposition, especially in natural microbial communities.

Mechanisms of Genetic Exchange

There are three well-described mechanisms of intercellular genetic exchange: transduction, transformation, and conjugation. An intracellular mechanism, transposition, allows genetic material to move amongst genetic elements in the cell and can lead to enhanced intercellular transfer. Each process is described, and the effect of each on horizontal genetic exchange is discussed. Extrapolation from known data in order to predict the potential for horizontal transfer is difficult, because the amount of environmental information on many of these processes in nature or microcosm systems is limited.

Transduction

The mechanism allowing the transfer of genes from one bacterial strain to another by means of virus particles was first observed in *Salmonella* by Zinder and Lederberg (1952), who subsequently used the term "transduction" to describe the phenomenon. Through transduction, recombinant bacteria can arise naturally that have derived a limited part of their genome from a donor cell. A transducing virus can carry the prophage of an entirely different virus particle from a lysogenic donor to a nonlysogenic recipient cell, usually in association with a contiguous region of the donor chromosome, demonstrating the simultaneous exchange of both foreign and indigenous genetic material (Jacobs 1955).

Limited data are available concerning the ecological impacts of this genetic transfer process in nature. However, certain observations, including the rapid proliferation of traits such as multiple antibiotic resistance and mineralization of recently introduced exotic chemicals, constitute a priori evidence for the efficiency of genetic exchange mechanisms in nature. It is axiomatic among bacteriophage microbiologists that there exists somewhere in nature an infecting phage for each species of bacterium. The ecological and evolutionary implications of such phage distribution and contribution to microbial community gene flow remain, to a large extent, uninvestigated (Ogunseitan et al. 1990). The current and future applications of GEMs in waste management and agriculture will probably require release of the altered bacteria in large quantities to the environment to be effective (Hartl 1985).

The occurrence of bacteriophages in environments where deliberate releases of GEMs are targeted may influence the tracking of GEMs after introduction in a number of ways. First, obligately virulent infection may predominate due to an initial low phage-to-bacteria ratio (PBR), resulting in a drastic reduction of the population of introduced organisms. Second, nonvirulent infections may occur which could lead to phenotype conversions in the lysogenized cells, thereby confounding detection procedures (Jain et al. 1988). Finally, lysogenic infections may enhance transduction of genes from introduced GEMs to indigenous populations of related strains, causing widespread problems with unwanted dissemination of the GEMs.

Transduced genes may alter the recipient's intracellular functions by homologous recombinational replacement of segments of the host chromosome. Transduction has been demonstrated between various groups of bacteria and appears to be the main process for the transmission of resistance to various antimicrobial drugs in certain groups of bacteria (Sonea 1988).

Transduction of genes located on chromosomes and plasmids has been observed at frequencies ranging between 10^{-8} to 10^{-6} per *Pseudomonas* recipient in microcosms incubated directly in natural fresh waters (Morrison et al. 1978, Saye et al. 1987, 1990). The transduction of genes from introduced donors to resident, untagged bacteria has been difficult to demonstrate, because of the inability to identify markers in natural populations and ensure that markers among recipients and donors are unique. However, transductants are readily recovered in moderate numbers when introduced recipients are used to circumvent the above problem. These findings indicate the stability of occurrence of the event in natural populations. Transductants are detected by DNA probes composed of genetically engineered genes or specific segments of the bacteriophage genome in cases where full lysogeny is involved in the gene-transfer process (Ogunseitan 1988).

The major restriction on transduction in natural communities of bacteria appears to be the limited host range of bacteriophages. Typically, a given virus will infect only a highly specific group of hosts, sometimes only a particular strain of a given species. However, genetically engineered phages have been constructed that have the ability to infect and transduce genes across species boundaries (Chakrabarty and Gunsalus 1969). From these laboratory experiments with trans-species phages it is possible to quantitatively characterize the invasiveness of a particular gene carried on an introduced GEM based on transduction events alone. Theoretical models have been constructed to predict the possibility or frequency of transduction from introduced GEM donors to native species if invasion of GEMs and their genes is the cause for concern. Models and empirical data already exist on the "predator-prey" relationship between bacteriophage and bacteria in controlled and natural environments (Lenski and Levin 1985, Wiggins and Alexander 1985, Ogunseitan et al. 1990) to characterize particular aspects of the survival of GEMs when released.

Although the results of tests carried out in one habitat may not be used as strict guidelines for assessing the response of other systems, (1) fresh water and (2) soils have dominated research conducted to assess the role of transduction in risk analyses of GEM release (Saye et al. 1987, 1990, Ogunseitan 1988, Stotzky and Babich 1984) and are being used to describe the general importance of transduction in release.

Transformation

"Transformation" is the process of uptake of naked DNA into a competent cell. "Competence" is the physiological state of a cell which enables it to take up DNA. The factors that induce competence in the laboratory are well understood, at least for some genera, but little has been reported in the literature regarding this subject in natural settings. Competence has been shown to be constitutive in at least one genus, *Neisseria* (Sparling 1966). Successful transformation requires uptake and subsequent integration and expression of the transferred DNA by the recipient cell. Transformation was first discovered by Griffith (1928), whose experiment became a milestone, because it lead to the discovery of DNA as the carrier of genetic information (Avery et al. 1944), and was also the first report of bacterial gene transfer in a natural environment.

The process of natural transformation needs to be clearly separated from artificial transformation, which is a widely used technique in genetic engineering. Whereas natural transformation requires the active involvement of the transforming organism in the process (e.g., synthesis of specific proteins), artificial transformation is achieved by manipulating the cells chemically or physically in vitro, making them permeable to extraneous DNA. Drastic methods of manipulation are used to make cells competent in artificial transformation (e.g., calcium chloride treatment, heat shock, protoplast formation, electroporation). The possible occurrence of such competence-inducing conditions in the environment warrants detailed investigation.

Transformation in the laboratory appears to be dependent on the sophistication of the method applied, while in nature transformation is restricted to those bacteria carrying genes specifically encoding the processes involved. Approximately 20 different genera are known to perform natural transformation (Stewart and Carlson 1986), and many of these are widely abundant in the environment—e.g., *Bacillus, Acinetobacter, Azotobacter, Micrococcus, Moraxella, Streptomyces*, and *Pseudomonas*. Therefore transformation may be a major process of gene transfer for these bacteria under natural conditions. Minor differences generally exist in the process of transformation between Gram-positive and Gram-negative bacteria, such as the uptake of DNA as single-stranded or double-stranded molecules, respectively (Ingraham et al. 1983, Reanney et al. 1983, Stewart and Carlson 1986).

Two main factors are likely to control the importance of transformation in the horizontal transfer of genetic information from GEMs in the environment: competence and availability of naked DNA for uptake. Competence may be under external control, mediated by the critical concentration of a soluble, proteinous competence factor which is produced and excreted by the organism itself (Ayad and Shimmin 1974). This competence factor can induce the generation of special active sites ("transformasomes") along the cell membrane for DNA uptake, as described for *Bacillus subtilis* (Istock 1989). Competence may also be under internal control, for instance in *Pseudomonas stutzeri* (Carlson et al. 1983) and *Azotobacter vinelandii* (Page 1982). Here competence occurred as growth conditions changed and organisms underwent the transition between logarithmic and stationary growth, probably induced by a growth-limiting factor. It has been shown in *E. coli* that a dramatic change in the nature of the proteins synthesized occurs as this organism enters stationary phase. Such changes may also occur in other bacteria and signal the production of proteins that induce cell competence (Schultz et al. 1988). In aquatic environments external induction of competence may be unlikely because of rapid dilution of possible competence factors as a result of diffusion by currents.

In soils, the occurrence of protected "microniches" may enhance the accumulation of competence factor to optimum concentrations (Stotzky 1989). The inactivation capacity of proteases capable of degrading competence factors could be mitigated through an increased production of these protective compounds. Induction of competence through growth-limiting conditions (internal control) is favored in natural environments. The state of starvation due to limiting concentrations of specific nutrients or growth factors would be prevalent in the environment, whereas balanced growth conditions resulting in exponential growth would be infrequent.

Since the development of competence in the natural environment is highly plausible, the occurrence of the second controlling factor, naked DNA, in the same environment needs to be characterized. The sources of extracellular DNA in natural environments include release of DNA from cells of dead organisms following lytic phage infection during certain stages of the bacterial growth cycle (Redfield 1988). The abundance of DNases in the environment would argue that extracellular DNA would be rapidly degraded and the resulting nucleotides utilized by the microorganism as carbon, nitrogen, or phosphorous sources for growth. Though secreted DNases have been shown to be stable in aquatic and terrestrial environments, extracellular DNA has been detected in these environments (DeFlaun et al. 1986). DNA seems to be protected from degradation by adsorption to sand particles (Lorenz and Wackernagel 1987). Thus particle-sorbed DNA seems to provide a source of DNA available for transformation of bacteria located on soil particles. The exchange of DNA would depend on its binding affinity to the particle and on physical or biological factors causing the release of DNA so that uptake could occur. Higher transformation frequencies have actually been observed for cells attached to sand grain particles than for those existing in liquid media (Graham and Istock 1978, Lorenz et al. 1988).

The presence of naked DNA and competent organisms in the environment makes it probable that transformation of bacteria occurs as a natural mechanism of gene transfer. Horizontal transfer of introduced genes by means of transformation will largely depend on the specificity with which competent cells take up, recombine, and express these genes. Certain unresolved questions are extremely important in determining the likelihood and frequency of such events. For example, it is not known whether transformation is restricted to DNA from very closely related organisms or whether any foreign DNA may be incorporated. In order for transformation to be of concern in horizontal transfer, there must be the recombination of an incoming gene into a homologous site of the host DNA. To date this is only known to occur for genes from closely related species, usually in a process similar to the SOS-type DNA repair system (Stewart and Carlson 1986, Duncan et al. 1978, Love and Yasbin 1984). Transformation of *B. subtilis* has been demonstrated in model soil ecosystems (Graham and Istock 1978). The transformation process found in soil differed from the observed process in controlled cultures in that transformation in soil is resistant to DNase activity and excess of heterologous DNA (Lorenz et al. 1988, Istock 1989).

Transformation is an important means of gene transfer from genetically engineered bacteria released into natural environments because many of the constraints necessitated by other mechanisms of gene exchange are reduced (i.e., species differentiation, host recognition, cell–cell contact, etc). The universal chemical composition of DNA makes it theoretically possible to transform an organism with DNA from any species. Experimental evidence exists supporting the notion that species differentiation may not necessarily prevent gene flow by transformation in *Bacillus*. Although cross-species transformation with chromosomal as well as plasmid DNA between *B. thuringiensis* and *B. cereus* has been demonstrated (Aronson and Beckman 1987), it is more likely that DNA from a GEM would transform only a species related to the original host (Stewart 1989).

Very few studies have investigated the occurrence of transformation in nonsterile soil microcosms. The use of sterile soil samples provides only estimates of the frequency of occurrence of bacterial transformation. It is not fully known what effect microbial competition, extracellular enzymes, and prevailing physicochemical conditions will have on the frequency or integrity of transformation and stability of transforming DNA. It is also possible that the process of soil sterilization destroys some labile compounds that may enhance the development of bacterial competency in nature or reduce substances destructive to DNA. Thus, the extrapolation of data gathered by studying transformation in microcosms or model systems utilizing amended soils must be limited.

Experimental data collected to date suggest that only a certain spectrum of microflora is able to undergo natural transformation. Within this group transformation with genes from closely related species is possible and may be restricted only by the proximity of transforming DNA and cell competency. A recent study suggests that transformation between distantly related genera may occur when the foreign DNA is transposon-associated (Trieu-Cuot and Courvalin 1986).

Conjugation

"Conjugation" is the process of genetic exchange between bacteria that requires cell-to-cell contact; it was the first mechanism of horizontal transfer to be studied for GEMs. Conjugation has been demonstrated in a diverse range of both Gram-negative and Gram-positive organisms and in most of these cases involves a plasmid encoding the transfer functions, although both plasmid and chromosomal genes can be transferred. Chromosomally mediated conjugation systems have been identified in *Streptococcus, Staphylococcus, Clostridium*, and *Bacteroides* (Shoemaker et al. 1980, El Solh et al. 1986, Odelson et al. 1987, Schaberg and Zervos 1986).

The mechanism of plasmid-encoded conjugation involves single-stranded transfer of DNA from the donor cell to the recipient across a mating bridge which requires the interaction of a number of cell surface components such as conjugative pili and cell membranes (Willets and Wilkins 1984). It is generally considered that for a plasmid to be conjugative it must be at least 25 kb in length. However, a 4.3-kb plasmid encoding tetracycline resistance has been shown to transfer by a conjugation mechanism in *Bacillus* spp. (Van Elsas et al. 1987). Conjugative plasmids can also mobilize smaller nonconjugative plasmids if they are present in the same cell (Guiney and Lanka 1989). Thus, introduced genes could be transferred from a GEM if they were placed on a conjugative or nonconjugative plasmid. Also this mechanism could result in transfer even if the foreign genes are located on the chromosome in cells that contain no plasmids; such transfer could occur through the introduction of plasmids into GEMs released into the environment.

In addition to the transfer functions, a wide array of phenotypes are encoded by nonconjugative and conjugative plasmids; these include antibiotic and heavy metal resistances, catabolic functions, symbiosis, and plant pathogenicity (Jacoby 1986, Rochelle et al. 1989*a*, Harayama and Timmis 1989, Schofield et al. 1987, Beck von Bodman et al. 1989). Conjugative plasmids have been detected in bacteria isolated from a number of environments, and Table 8.2 demonstrates common traits that are transferred via conjugation.

Far less work has been done to determine the ability of isolates from natural environments to receive plasmid DNA through conjugation. Schilf and Klingmüller (1983) reported that 17% of soil and water isolates could receive the broad-host-range plasmid RP4 and 38% of aquatic isolates had the capacity to act as recipients for the plasmid R68 (Genthner et al. 1988). Work in our

TABLE 8.2 Occurrence of Conjugative Plasmids in Environmental Isolates

Isolated	Proportion of isolates transferring plasmids (%)	Reference
Antibiotic-resistant coliforms	30	Grabow et al. 1975
Hg-resistant soil isolates	5	Kelly and Reanney 1984
Antibiotic-resistant fish isolates	9	Toranzo et al. 1984
Hg-resistant pseudomonads	26	Gauthier et al. 1985
Hg-resistant estuarine bacteria	25	Jobling et al. 1988
Hg-resistant epilithic bacteria	24	Rochelle et al. 1989*a*

laboratory has shown that up to 100% of mercury-sensitive isolates from a contaminated site could act as recipients for a plasmid encoding mercury resistance isolated from the same site (Acacio and Rochelle 1989). This indicates the importance of examining GEMs as recipients as well as donors, a possibility which to date has received relatively little attention.

These demonstrations of transferable plasmids and bacteria with recipient ability clearly indicate that bacteria in the environment have the potential to exchange genetic material through conjugation. This clearly indicates the need to restrict the placement of foreign genes on plasmids, especially conjugative ones.

In an attempt to gain further understanding of plasmids and their transfer behavior under more realistic conditions, many workers have studied conjugation in microcosms or model systems. Much of this work was undertaken to provide additional information on the impact of conjugation on introduced GEMs.

One of the simplest model systems used for such studies is sterile soil. Conjugation in sterile soil was first reported almost 20 years ago (Weinberg and Stotzky 1972). Conjugation was between auxotrophic strains of *E. coli*, an organism not indigenous to the soil environment. Further work with *E. coli* demonstrated the transfer of resistance plasmids in sterile soil only when the soil was amended with nutrients (Trevors and Oddie 1986). Actinomycetes are more representative of bacteria found in soil, and the conjugal transfer and mobilization of plasmids encoding thiostrepton resistance between *Streptomyces* spp. in sterile soil has recently been reported (Rafii and Crawford 1988). Transfer frequencies were as high as 7.0×10^{-1} transconjugants per recipient. Further plasmid-containing recombinant *Streptomyces* are likely to persist in soil, even in the presence of indigenous soil microorganisms, indicating that there is high potential for gene transfer after the release of genetically engineered *Streptomyces* spp. into soil (Rafii and Crawford 1989).

Rhizobium spp. are important soil bacteria from an agricultural standpoint, and their genetics have recently become the subject of intense investigation (Broughton et al. 1987, Schofield et al. 1987, Hirsch and Spokes 1988, Kaijalainen and Lindstrom 1989). Richaume et al. (1989) investigated the influence of soil variables on the transfer of an antibiotic-resistance plasmid between *E. coli* and *Rhizobium fredii* in sterile soil. Maximum transfer frequencies (1.8×10^{-4} per recipient) were observed after 5 days in soil with a clay content of 15%, 5% organic matter, pH of 7.25, moisture content of 8%, and at a temperature of 28°C. Evidence is also available, derived from plasmid-specific DNA probes, which indicates that genetic exchange of Sym plasmids occurred between *Rhizobium* spp. in a natural soil population (Schofield et al. 1987). In one instance, the hybridization pattern indicated that in situ recombination of two different Sym plasmids had also occurred.

Although experiments in sterile soil yield valuable results on plasmid transfer, they are not ideal; soil lacking an indigenous competing microflora does not accurately simulate the soil ecosystem. Therefore, experiments have been performed in nonsterile systems. Van Elsas et al. (1988) reported that the conjugal transfer of the resistance plasmid RP4 between *Pseudomonas* spp. in nonsterile soil occurred only in the presence of exogenous nutrients at temperatures above 10°C. Transfer frequencies were higher in the wheat rhizosphere than in bulk soil. The same authors found that the plasmid pFT30 (4.3 kb) only transferred between *Bacillus* spp. in nonsterile soil when bentonite clay was added, but were unable to demonstrate transfer of this plasmid in the wheat rhizosphere. Rhizosphere and phyllosphere microcosms have been used to test the predictive capacity of computer simulations of conjugative transfer of the plasmid R388::*Tn1721* between strains of *P. cepacia* (Knudsen et al. 1988), and it is worth noting that the computer simulation did remarkably well in predicting the genetic interactions in microcosms. In all experiments, transconjugants were isolated from microcosms containing indigenous microflora after 1 day, and the highest numbers of transconjugants were generally recovered 1 to 2 days after the microcosms were inoculated with donor and recipient strains. A rhizosphere model was also used to study transfer of the symbiotic plasmids pJB5J1 and pSymR1897 between *Rhizobium* spp. (Broughton et al. 1987), but in this case, no indigenous microflora were present. Interspecific transfer of pJB5J1 occurred, generating symbiotically proficient transconjugants.

Aquatic microcosms have also been used to study plasmid transfer. In nonsterile, static microcosms designed to simulate the epilithon of freshwater rivers, transfer frequencies of 4.9×10^{-6} (per recipient) were obtained for the transfer of a mercury-resistance plasmid, pQM1, between strains of *P. aeruginosa* (Bale et al. 1987). The same plasmid was used in a rotating-disc microcosm to study genetic interactions between *Pseudomonas* spp. within epilithic films (Rochelle et al. 1989*b*). The highest number of transconjugants were generally observed 2 to 4 days after inoculation, and the maximum transfer frequency was 3.6×10^{-3}, about 100-fold lower than under ideal conditions in filter matings. Much higher numbers of transconjugants were recovered from the surface of the rotating discs than from the surrounding water, indicating the preference of surfaces for bacterial growth and genetic exchange. Results reported by Fulthorpe and Wyndham (1989), who studied the survival and activity of a 3-chlorobenzoate (3CB) catabolic genotype in flow-through lake microcosms, suggest that transfer of the catabolic genes may have occurred from the introduced strain, an *Alcaligenes* sp., to indigenous bacteria within the microcosms. In that study the 3CB-degrading genes were carried on a recently isolated, unstable plasmid, shown to be conjugative in laboratory matings.

Because GEMs, whether released accidentally or deliberately, are likely to reach wastewater treatment plants, a number of studies have been performed to assess the likelihood of plasmid transfer within such systems. The mobilization of a recombinant plasmid, pHSV106, from a laboratory strain of *E. coli* was demonstrated in a laboratory-scale waste treatment facility (Mancini et al. 1987, Gealt 1988). The mobilizing organisms were laboratory and indigenous wastewater strains, and neither nutrient composition, nutrient concentration, influent flow rate, or the mode of addition or origin of the parental strains had a significant effect on the numbers of transconjugants recovered. Most of the transconjugants were recovered from the bottom of the primary clarifier and the sludge lining the bottom of the clarifier (Gealt 1988), suggesting that higher levels of particulate surfaces are necessary for gene transfer to occur. A model activated-sludge unit (ASU) was used to investigate the survival of a 3CB-degrading strain of *P. putida* harboring a recombinant plasmid, pD10 (McClure et al. 1989). The plasmid pD10 was small (20 kb) and nonconjugative, but colonies of the introduced *P. putida* strain recovered from the ASU were able to transfer pD10 to a second strain of *P. putida*. These results demonstrate that the introduced genetically engineered strain had acquired mobilization functions and/or plasmids through genetic exchange with sludge bacteria within the sludge unit. At least two of the transconjugants tested were shown to contain large plasmids in addition to pD10, and in one instance pD10 could be transferred to strains of *E. coli* and *Alcaligenes eutrophus*, indicating that the mobilizing plasmid acquired in the ASU encoded broad-host-range transfer functions.

The human gut has also been used as a site for recombinant DNA risk assessment. An in vivo study of plasmid mobilization in the human intestine demonstrated that a nonconjugative plasmid (pBJK5) could be mobilized by indigenous plasmids in normal colon flora (Levine et al. 1983). However, the indigenous colon flora were unable to mobilize a nonconjugative plasmid with a defective *mob* gene (pBR325) despite the selective pressure of tetracycline.

Thus, even though recombinant DNA may be introduced into an environment in the form of a nonconjugative plasmid (i.e., biological containment), it could easily be mobilized throughout the population by indigenous plasmids, although the use of nonconjugative plasmids with defective *mob* genes would seem to minimize the risk of transfer.

Natural systems are complex, and it is difficult to design laboratory experiments that accurately model the conditions in even the simplest environment. Therefore, the ultimate step in understanding genetic behavior in nature is to carry out experiments in natural environments. Transfer of plasmidborne antibiotic resistance has been demonstrated between pure cultures of enterobacteria in membrane diffusion chambers suspended in sewage treatment facilities (Altherr and Kasweck 1982, Mach and Grimes 1982) and dialysis sacs suspended in fresh water (Grabow et al. 1975, Gowland and Slater 1984). Transfer frequencies in the sewage treatment plants were 10-fold to 100-fold lower than in laboratory matings (Altherr and Kasweck 1982,

Mach and Grimes 1982), and in one instance transfer frequencies were higher when the water temperature was 22°C than when it was 29.5°C (Altherr and Kasweck 1982). Gowland and Slater (1984) demonstrated the transfer of two plasmids, TP120 and R1, in sterile pond water in dialysis sacs suspended in fresh water. No transconjugants were detected before 8 days for TP120 or before 15 days for R1, and transfer frequencies were 10^4-fold lower than in laboratory matings, which was probably a reflection of low temperatures and poor nutritional status of the water.

Pseudomonas spp. are commonly occurring aquatic bacteria (Holder-Franklin et al. 1978, Jones et al. 1986), and recent experiments have used representatives of this genus to demonstrate in situ conjugal transfer of plasmids. O'Morchoe et al. (1988) observed the transfer of plasmids R68.45 and FP5 between strains of *P. aeruginosa* in Teflon bags suspended in lake water. Transfer occurred in the absence and presence of indigenous lakewater bacteria, although the transfer frequencies were 10-fold to 100-fold lower when a natural microbial community was present.

To date, only one group has investigated in situ plasmid transfer in unenclosed natural environments (Bale et al. 1987, 1988). These researchers demonstrated the conjugal transfer of the mercury-resistance plasmid, pQM1, between strains of *P. aeruginosa* on the surface of stones carrying an intact epilithon in a river. The maximum transfer frequency observed in the presence of an intact epilithon was 4.9×10^{-6} transconjugants per recipient (Bale et al. 1987), which was about 10^5-fold lower than in laboratory matings. The maximum in situ transfer frequency observed when sterile stones were used was 10^{-2} (per recipient), and there was a linear relationship between river temperature and transfer frequency, with a 2.6°C increase in temperature giving rise to a 10-fold increase in transfer frequency, over the range 6 to 21°C (Bale et al. 1988).

These experimental results clearly indicate that conjugation occurs in natural environments, and that if GEMs are to be released they should be designed in such a way as to minimize their transfer via conjugation.

Transposition

To fully appreciate the significance of transposition in horizontal transfer, a brief review is valuable. "Transposition" is the mechanism by which specific mobile DNA sequences change their positions within a bacterial genome. Almost all bacterial genera have been shown to possess such mobile genetic elements. Transposition is significant because replication of transposable elements is independent of chromosomally encoded recombination systems and homologous DNA segments and because increased copy number of specific genes usually results. Further, transposable elements are probably the most important in horizontal gene transfer between distantly related microorganisms (e.g., Gram-positive and Gram-negative bacteria) because many of these elements are expressed in a wide variety of genera (see Table 8.3).

The different types of transposable elements in bacteria include "insertion sequences" ("IS elements"), which carry only genes encoding for the enzymes

TABLE 8.3 Some Transposons and Their Characteristics

Name of transposon	Length (kbp)	Associated markers[a]	Host organisms described	Transposition frequency (per cell generation)	Site specificity
			Class I: Composite Transposons		
Tn5	5.8	*kan, ble, str*	Diverse enteric and nonenteric Gram-negative organisms (*E. coli, Klebsiella*, etc.)	10^{-3}–10^{-2}	"Hot spots"[b]
Tn9	2.6	*cam*	Enteric bacteria		"Hot spots"[b]
Tn10	9.3	*tet*	*E. coli, Klebsiella, Salmonella typhimurium, Proteus, Pseudomonas, Shigella, Vibrio, Haemophilus*	10^{-7}	"Hot spots"[b]
			Class II: Noncomposite Transposons		
Tn7	14	*tmp, str*	*E. coli, Vibrio, Proteus, Rhizobium Pseudomonas, Xanthomonas, Klebsiella, Salmonella, Agrobacterium, Caulobacter*	Almost 100%	Single specific site (*attTn7*) in *E. coli* and other species
Tn3 family: *Tn3, Tn501, Tn21, Tn1000, Tn2603, Tn551 Tn917, Tn443, Tn1721*, and others	5–23.5	*amp, tet, ox, str, su, ery, Hg*	*E. coli, Shigella, Pseudomonas, Staphylococcus, Streptococcus, Bacillus*	10^{-7}–10^{-5}; in *tnpR*[c]-inactivated mutants 10 to 100 times higher	AT-rich sequences and regions with similarities to *Tn* ends are preferred

[a]Resistance against *kan*, kanamycin; *ble*, bleomycin; *str*, streptomycin; *cam*, chloramphenicol; *tet*, tetracycline; *tmp*, trimethoprim; *amp*, ampicillin; *ox*, oxacillin; *su*, sulfonamide; *ery*, erythromycin; *Hg*, mercury ion.
[b]Large number of sites on host DNA, but not random.
[c]*tnpR*, transposition repressor.
SOURCE: Compiled from data obtained by Berg and Howe (1989) and Berg (1989).

performing transposition (resolvase and transposase), and "transposons" (*Tn*), which include both these genes and additional genes encoding for specific phenotypic properties, such as antibiotic resistances.

Even though the term "transposition" describes an intracellular process, its effect on the occurrence and increased frequency of certain genes and their expression in microbial communities can be substantial. The dramatic global spread of antibiotic resistances via transpositional and genetic-exchange events indicates how these two genetic processes may work together to increase gene occurrence. (Levy and Marshall 1988).

There are several ways in which transposition could impact chromosomally inserted DNA. Mobile DNA elements can induce mutations such as gene deletions, inversions, amplifications, and replicon fusions. The insertion of mobile DNA elements can be disruptive to cell functions by inactivation of operons. Mobile DNA elements can also activate operons when inserted next to them, by inducing or enhancing gene expression. Transposable elements can increase the occurrence of certain genes by moving them from a site of low transfer frequency to a site of higher transfer frequency, e.g., from the chromosome to a conjugative plasmid.

A special group of transposable elements, so far only found in Gram-positive strains, are self-transmissible (Franke and Clewell 1981, Clewell and Gawron-Burke 1986, Courvalin and Carlier 1987). These elements carry transposition genes, resistance genes, and genes for transfer by conjugation, thus having a high potential for horizontal transfer.

Different structural properties have resulted in two classes of transpositional units (Class I and II). Transposons of Class I are flanked by IS elements, while those of Class II are not. It is not known whether either of these classes has an adaptation advantage, nor the impact of these on the likelihood of transfer of genetic information into or from a GEM.

Transposition is either "replicative" or "conservative." In replicative transposition the transposon stays at its original site, and the replicated unit is inserted into a new, specifically designated site. The conservative mechanism, also called "cut-and-paste," describes a process where the transposable element itself leaves its original site and inserts into a new site. The donor DNA molecule may be damaged or lost after excision of the mobile DNA molecule, or repair of the donor molecule or substitution of the lost molecule by replication of another copy of the same molecule in the cell may result in increased copy numbers by this mode also (Kleckner 1989, Berg 1989).

Another important factor which could influence horizontal transfer by transposition is the specificity with which transposable DNA integrates into specific sites of the host DNA. Most transposable elements have affinities for specific regions of a host DNA. The transposons *Tn7* and *Tn554* insert preferentially into a single site of the chromosome, while IS1 prefers certain regions of the host DNA that are enriched in adenine and thymine. In addition topological and functional factors of the host DNA are involved. Regions of the host DNA which are favored for insertion are called "hot spots" and tend to be specific to a transposon or its family (Table 8.3). Some transposons such as *Tn5* or *Tn10* may integrate into a hundred different sites of any gene, but no transposon has been described to insert randomly (Berg et al. 1989). For most transposable elements it is not clear what determines this site specificity. To accurately predict their potential to enhance horizontal spread of a trait in a microbial community, it is crucial to know more about the mechanisms that determine insertion.

Beside specificity of insertion varying dramatically, so does the frequency of transposition, from almost 100% for *Tn7* and *Tn554*, to 10^{-7} to 10^{-3} per cell generation (Table 8.3). Many environmental factors may also influence the frequency of transposition; for example, temperature influences the frequency of translocation of *Tn3* (Kretschmer and Cohen 1979). The presence of a plasmid may alter transpositional properties, as shown for *Tn7* (Hauer and Shapiro 1984). Therefore, the role of transposition in mobilizing DNA from GEMs by horizontal transfer is still unknown.

Most commonly described transposon-associated genes encode for antibiotic resistances. However, other genes associated with transposons have been identified, such as catabolic genes encoding for degradation of halogenated alkanoic acids (Slater et al. 1985) and of toluene (Tsuda and Iino 1987, 1988). In fact, any gene could become part of a transposon and thereby increase its mobility and occurrence by means of transposition. However, the abundance of transpositional elements in an undisturbed microbial community may be so low that they cannot be detected until selective pressure favors the growth of bacteria containing such a transposon.

There are studies suggesting that the presence of a transposon provides an advantage for cells even in the absence of selective pressure. This has been demonstrated in chemostat experiments with *E. coli* cells containing the composite transposons *Tn5* or *Tn10* (Biel and Hartl 1983, Chao et al. 1983). Energetically the possession of a transposon is a disadvantage because additional DNA must be replicated. However, it may have other benefits such as increased adaptive potential through increased mutational ability (Chao et al. 1983). Regardless, the energetic burden is often very low, as shown for *E. coli* harboring *Tn10* (Nguyen et al. 1989). Transposon-carrying strains exhibited only a 0.3% decreased fitness, and one can expect that under in situ conditions, where generation times are longer, a complete elimination of the transposon might take years. Thus to date the question of whether transposons increase or decrease fitness is unanswered and points to our lack of understanding of the role of transposition in fitness and horizontal transfer. Further, even if a GEM could be developed without any transpositional elements, the evidence suggests that it would not remain so in the environment.

The introduction of transposon-associated genes in the construction of GEMs and their subsequent destruction does not guarantee containment because other transposons may be introduced at the same site due to insertion specificity, or repair may occur as in conservative replication. So much is unknown regarding transposition that the placement of a gene or operon on the chromosome of a GEM as a method of biological containment may not be successful. Not enough information is available on factors that influence the translocation of genes from chromosome to plasmid and vice versa to predict the fate of an engineered gene located on a chromosome.

Laboratory studies may also lead to false security regarding the behavior of GEMs in the environment. The plasticity of a genome may be much higher under in situ conditions than as determined in laboratory experiments, as demonstrated by the fact that gene rearrangements on plasmids of transconjugants of *P. aeruginosa* were 10% under laboratory conditions but 50% for in situ studies (O'Morchoe et al. 1988).

Transpositional events contribute greatly to the plasticity of genomes of microorganisms. Therefore the mechanism of transpositional events may quickly change the genome of a GEM once released into the environment. The changes that could occur are (1) higher transfer frequencies of introduced genes to other species, (2) amplifications, (3) gene rearrangements leading to higher gene expression, or (4) deletions leading to gene inactivation. All these events may contribute to a low survival rate of a sophisticated laboratory-constructed organism, but could also result in changes to a different genotype optimized to a specific set of environmental conditions, but different from its ancestor. To predict these changes from pure-culture investigations is impossible because of complexities and interactions of the biotic and abiotic environment. However, microcosm and mesocosm studies provide a valid and sound approach to aid us in determining the risk of horizontal transfer associated with transpositional events.

Effect of Horizontal Transfer Potential on GEMs and Microbial Communities

There have been a number of concerns raised relative to the feasibility and the potential perturbations of the ecosystem from the release of GEMs (Curtiss 1976, Liang et al. 1982, Rissler 1984, Brill 1985). Assessment of the risks involved in the release of GEMs requires monitoring the environmental fate of the organisms to determine their survival and growth and the dissemination and persistence of the recombinant genes. It is generally recognized that the various microbial habitats, including soils, sediments, water, sewage, plants, and animal surfaces, vary widely in their species composition and in the physical-chemical parameters that define their components. All these ecosystems are possible targets for GEM applications, and each is controlled by different environmental factors. Factors that affect the frequency of horizontal gene transfer include moisture, nutrient, and mineral composition (Stotzky 1989, Saye and Miller 1989).

In this section we try to direct the discussion to factors which enhance genetic stability and, therefore, may influence our ability to accurately predict genetic exchange through horizontal transfer. An introduced trait is usually viewed as nonadvantageous unless external selective pressure favors its existence (Lenski and Nguyen 1988), and the introduced traits in GEMs are viewed in the same manner. However, our understanding of how genetic systems function in the environment to date precludes such firm assumptions for several reasons. It has been shown that a genetic trait can become advantageous for survival after a number of generations, even though it was deleterious initially (Bouma and Lenski 1988). What controls this change and whether it relates to a given environment or to some type of genetic rearrangement is presently unknown. Further, similar findings exist for transpositional elements (Nguyen et al. 1989). Without a solid understanding of such mechanisms, it is difficult to accurately predict outcomes. Further complicating our ability to understand the functioning of GEMs in the environment is the increasing awareness of viable but noncultureable states of bacteria (Roszak and Colwell 1987). Thus, the mere absence of the organism using culture methods may not be indicative of the absence of the genetic material. This places more importance on the assessment of genetic elements of interest and not just the organism. It further emphasizes the need for direct-extraction procedures from environmental samples and improved application of the polymerase chain reaction.

Multiplicative genotype-environment interactions may not be predictable based on traditional quantitative genetics theory (Falconer 1981). For example, expected outcomes in regard to adaptation in the presence or absence of a selective pressure may result in reversed-response outcomes (Gimelfarb 1986). Under these circumstances, the predicted outcome is not followed genetically. The usual explanations are genetic drift or cataclysmic environmental events. However, the frequency with which reversed-response outcomes can occur suggests that other factors control such processes. The occurrence of these events is theoretically based, and our understanding of their function, if any, in real environments is extremely limited. However, factors such as transposition might provide a genetic basis of explanation for some of these events (i.e., unexplained mutations). Findings such as those stated above suggest that introduced genetic material has a likelihood of being able to survive in the environment. The question for which little knowledge exists is the importance of survival in the absence of growth, and the relationship between that surviving DNA and horizontal transfer. Thus, there is a need to increase our understanding of unique host conditions such as viable but noncultureable states (Roszak and Colwell 1987) and changing status of introduced genetic information from disadvantageous to beneficial to organism survival (Bouma and Lenski 1988).

The relationship between the external environment into which these organisms enter and alteration of physiological state due to growth-limiting conditions is an almost unexplored field of research. Systems involved in storage polymers, metabolic enzymes, protomotive force factors, nutrient transport, and binding proteins, as well as tactic responses, become important in influencing or perhaps controlling survival, and thus in protracting the occurrence of foreign DNA in the ecosystem, where it may be able to transfer at some future time. Our understanding of these systems is very limited and argues that not only should there be an understanding of horizontal genetic exchange, but also of the physiological conditions that allow survival so that transfer could occur subsequently. The argument employed to minimize the importance of the above factors in allowing horizontal transfer in the environment is the low frequency at which transfer occurs, especially interspecies transfer (Trevors et al. 1987). Yet, even here, our knowledge is severely limited (Tiedje et al. 1989). For example, the importance of temperature, community effect, or geographical origin of the GEM in horizontal transfer is almost unknown. Data have been reported that indicate a higher ability to transfer genetic information via conjugation amongst bacteria of the same geographical origin (Acacio and Rochelle 1989). Thus, selecting a bacterium for genetic modification based on its suitability to a specific geographical location may enhance the potential for genetic exchange between the GEM and the natural community. The GEM may receive genetic determinants (i.e., transposons) from natural populations that could promote genetic exchange. This brings the discussion to the totally unstudied area of "multiple events," defined here to mean a combination of genetic events that promote horizontal transfer. An example would be the insertion of a transpositional element next to recombinant DNA and the subsequent movement of the foreign DNA onto a plasmid which also had been introduced into the GEM from the natural bacterial population. One documented multiple event reported in the literature probably occurred due to a transformation event which was followed by a transpositional event (Trieu-Cuot and Courvalin 1986). Because the probability of each event is small, concern over horizontal transfer via multiple events would be very limited.

Currently our knowledge, like the technology, is relatively limited. The success of GEMs and the best way to construct them for survival and to confine possibilities of horizontal transfer are likely to be further defined in the future. In order to ensure the success of this biotechnology, it is necessary to continue to study the modes of horizontal transfer available to bacteria at ever expanding levels of physiological and environmental sophistication.

Acknowledgment

The authors wish to acknowledge funding from cooperative agreement CR813411-01-2 with the USEPA, through which several of the ideas expressed here were developed.

References

Acacio, B. and Rochelle, P. A. 1989. Conjugation in bacterial from mercury contaminated sites. Abstract Q79. 89th annual meeting, American Society for Microbiology. New Orleans.

Altherr, M. R., and Kasweck, K. L. 1982. In situ studies with membrane diffusion chambers of antibiotic resistance transfer in *Escherichia coli*. *Appl. Environ. Microbiol.* 44: 838–843.

Aronson, A. I., and Beckman, W. 1987. Transfer of chromosomal genes and plasmids in *Bacillus thuringiensis*. *Appl. Environ. Microbiol.* 53:1525–1530.

Avery, O. T., McLeod, C. M., and McCarty, M. 1944. Studies on the chemical nature of the substance inducing transformation of pneumococcal types. I. Induction of transformation by a DNA fraction isolated from pneumococcus type III. *J. Exp. Med.* 79: 137–159.

Ayad, S. R., and Shimmin, E. R. A. 1974. Properties of the competence inducing factor of *Bacillus subtilis*. *Biochem. Genet.* 11:455–474.

Balbas, P., Soberon, X., Merino, E., Zurita, M., Lomeli, H., Valle, F., Flores, N., and Bolivar, F. 1986. Plasmid vector pBR322 and its special-purpose derivatives—a review. *Gene* 50:3–40.

Bale, M. J., Fry, J. C., and Day, M. J. 1987. Plasmid transfer between strains of *Pseudomonas aeruginosa* on membrane filters attached to river stones. *J. Gen. Microbiol.* 133:3099–3107.

Bale, M. J., Fry, J. C., and Day, M. J. 1988. Transfer and occurrence of large mercury resistance plasmids in river epilithon. *Appl. Environ. Microbiol.* 54:972–978.

Beck von Bodman, S., McCutchan, J. E., and Farrand, S. K. 1989. Characterization of conjugal transfer functions of *Agrobacterium tumefaciens* Ti plasmid pTiC58. *J. Bacteriol.* 171:5281–5289.

Bej, A. K., Perlin, M. H., and Atlas, R. M. 1988. Model suicide vector for containment of genetically engineered microorganisms. *Appl. Environ. Microbiol.* 54:2472–2477.

Berg, C. M., Berg, D. E., and Groisman, E. A. 1989. Transposable elements and the genetic engineering of bacteria. In D. E. Berg and M. M. Howe (eds.), *Mobile DNA*. American Society for Microbiology, Washington D.C., pp. 879–925.

Berg, D. E. 1989. Transposable elements in prokaryotes. In S. B. Levy and R. V. Miller (eds.), *Gene Transfer in the Environment*, McGraw-Hill, New York, pp. 99–137.

Berg, D. E., and Howe, M. M. (eds.). 1989. *Mobile DNA*. American Society for Microbiology, Washington D.C.

Berg, P., Baltimore, D., Brenner, S., Roblin, R. O., III, and Singer, M. F. 1975. Summary statement of the Asilomar Conference on recombinant DNA molecules. *Proc. Natl. Acad. Sci. U.S.A.* 72:1981–1984.

Biel, S. W., and Hartl, D. L. 1983. Evolution of transposons: natural selection for Tn5 in *Escherichia coli* K12. *Genetics* 103:581–592.

Bolivar, F., Rodriguez, R. L., Betlach, M. C., and Boyer, H. W. 1977*a*. Construction and characterization of new cloning vehicles. I. Ampicillin-resistant derivatives of the plasmid pMB9. *Gene* 2:75–93.

Bolivar, F., Rodriguez, R. L., Greene, P. J., Betlach, M. C., Heyneker, H. L., Boyer, H. W., Crosa, J. H., and Falkow, S. 1977*b*. Construction and characterization of new cloning vehicles. II. A multipurpose cloning system. *Gene* 2:95–113.

Bouma, J. E., and Lenski, R. E. 1988. Evolution of a bacteria/plasmid association. *Nature* 335:351–352.

Brill, W. J. 1985. Safety concern and genetic engineering in agriculture. *Science* 227: 381–384.

Broughton, W. J., Samrey, V., and Stanley, J. 1987. Ecological genetics of *Rhizobium meliloti*: symbiotic plasmid transfer in the *Medicago sativa* rhizosphere. *FEMS Microbiol. Lett.* 40:251–255.

Buchanan-Wollaston, V., Passiatore, J. E., and Cannon, F. 1987. The *mob* and *oriT* mobilization functions of a bacterial plasmid promote its transfer to plants. *Nature* 328: 172–175.

Carlson, C. A., Pierson, L. S., Rosen, J. J., and Ingraham, J. L. 1983. *Pseudomonas stutzeri* and related species undergo natural transformation. *J. Bacteriol.* 153:93–99.

Chakrabarty, A. M., and Gunsalus, I. C. 1969. Autonomous replication of a defective transducing phage in *Pseudomonas putida*. *Virology* 38:92–104.

Chao, L., Vargas, C., Spear, B. S., and Cox, E. C. 1983. Transposable elements as mutator genes in evolution. *Nature* 303:633–635.

Clewell, D. B., and Gawron-Burke, C. 1986. Conjugative transposons and the dissemination of antibiotic resistance in *Streptococci*. *Annu. Rev. Microbiol.* 40:635–659.

Cohen, S. N., Chang, A. C. Y., Boyer, H. W., and Helling, R. B. 1973. Construction of biologically functional bacterial plasmids *in vitro*. *Proc. Natl. Acad. Sci. U.S.A.* 70: 3240–3244.

Courvalin, P., and Carlier, C. 1987. Tn1545: a conjugative shuttle transposon. *Mol. Gen. Genet.* 206:259–264.

Covey, C., Richardson, D., and Carbon, J. 1976. A method for the deletion of restriction sites in bacterial plasmid deoxyribonucleic acid. *Mol. Gen. Genet.* 145:155–158.

Craine, B. L. 1982. Novel selection for tetracycline- or chloramphenicol-sensitive *Escherichia coli*. *J. Bacteriol.* 151:487–490.

Curtiss, R. 1976. Genetic manipulation of micro hazards. *Annu. Rev. Microbiol.* 30:507–533.

DeFlaun, M. F., Paul J. H., and Davis, D. 1986. Simplified method for dissolved DNA determination in aquatic environments. *Appl. Environ. Microbiol.* 52:654–659.

Ditta, G., Schmidhauser, T., Yakobson, E., Lu, P., Liang, X. W., Finlay, D. R., Guiney, D., and Helinski, D. R. 1985. Plasmids related to the broad host range vector pRK290, useful for gene cloning and for monitoring gene expression. *Plasmid* 13: 149–153.

Duncan, C. H., Wilson, G. A., and Young, F. E. 1978. Mechanism of integrating foreign

DNA during transformation of *Bacillus subtilis. Proc. Natl. Acad. Sci. U.S.A.* 75: 3664–3668.

El Solh, N., Allignet, J., Bismuth, R., Buret, B., and Fouace, J. 1986. Conjugative transfer of staphylococcal antibiotic resistance markers in the absence of detectable plasmid DNA. *Antimicrob. Agents Chemother.* 30:161–169.

Elton, C. S. 1958. *The Ecology of Invasion of Animals.* Wiley, New York.

Falconer, D. S. 1981. *Introduction to Quantitative Genetics*, 2d ed. Longman, London.

Franke, A. E., and Clewell, D. B. 1981. Evidence for a chromosome-borne resistance transposon (Tn 916) on *Streptococcus faecalis* that is capable of "conjugal" transfer in the absence of a conjugative plasmid. *J. Bacteriol.* 145:494–502.

Fulthorpe, R. R., and Wyndham, R. C. 1989. Survival and activity of a 3-chlorobenzoate catabolic genotype in a natural system. *Appl. Environ. Microbiol.* 55:1584–1590.

Gauthier, M. J., Cauvin, F., and Breittmayer, J. P. 1985. Influence of salts and temperature on the transfer of mercury resistance from a marine pseudomonad to *Escherichia coli. Appl. Environ. Microbiol.* 50:38–40.

Gealt, M. A. 1988. Recombinant DNA plasmid transmission to indigenous organisms during waste treatment. *Proceedings of the International Conference on Water and Wastewater Microbiol.*, Newport Beach, California, p. 29.

Genthner, F. J., Chatterjee, P., Barkay, T., and Bourquin, A. W. 1988. Capacity of aquatic bacteria to act as recipients of plasmid DNA. *Appl. Environ. Microbiol.* 54: 115–117.

Gerdes, K., Rasmussen, P. B., and Molin, S. 1986. Unique type of plasmid maintenance function: postsegrational killing of plasmid-free cells. *Proc. Natl. Acad. Sci. U.S.A.* 83:3116–3120.

Gimelfarb, A. 1986. Multiplicative genotype-environment interaction as a cause of reversed response to directional selection. *Genetics* 116:333–343.

Gowland, P. C., and Slater, J. H. 1984. Transfer and stability of drug resistance plasmids in *Escherichia coli* K12. *Microb. Ecol.* 10:1–13.

Grabow, W. O. K., Prozesky, O. W., and Burger, J. S. 1975. Behaviour in a river and dam of coliform bacteria with transferable or non-transferable drug resistance. *Water Res.* 9:777–782.

Graham, J. B., and Istock, C. A. 1978. Genetic exchange in *Bacillus subtilis* in soil. *Mol. Gen. Genet.* 166:287–290.

Griffith, F. 1928. The significance of pneumococcal types. *J. Hyg.* 27:113–159.

Guiney, D. G., and Lanka, E. 1989. Conjugative transfer of IncP plasmids. In C. M. Thomas (ed.), *Promiscuous Plasmids of Gram-Negative Bacteria.* Academic, London.

Hamer, D. H. 1977. Interbacterial transfer of *Escherichia coli–Drosophila melanogaster* recombinant plasmids. *Science* 196:220–221.

Hanahan, D. 1983. Studies on transformation of *Escherichia coli* with plasmids. *J. Mol. Biol.* 166:557–580.

Harayama, S., and Timmis, K. N. 1989. Catabolism of aromatic hydrocarbons by *Pseudomonas.* In D. A. Hopwood and K. F. Chater (eds.), *Genetics of Bacterial Diversity.* Academic, London, pp. 151–174.

Hartl, D. L. 1985. Engineered organisms in the environment. Inferences from population genetics. In H. O. Halvorson, D. Pramer, and M. Rogul (eds.), *Engineered Organisms in the Environment: Scientific Issues.* American Society for Microbiology, Washington, D.C., pp. 83–88.

Hauer, B., and Shapiro, J. A. 1984. Control of Tn7 transposition. *Mol. Gen. Genet.* 194: 158.

Heinemann, J. A., and Sprague, G. F. 1989. Bacterial conjugative plasmids mobilize DNA transfer between bacteria and yeast. *Nature* 340:205–209.

Hirsch, P. R., and Spokes, J. R. 1988. *Rhizobium leguminosarum* as a model for investigating gene transfer in soil. In W. Klingmüller (ed.), *Risk Assessment for Deliberate Releases.* Springer-Verlag, Berlin, pp. 10–17.

Hershfield, V., Boyer, H. W., Yanofsky, C., Lovett, M. A., and Helinski, D. R. 1974. Plasmid ColEl as a molecular vehicle for cloning and amplification of DNA. *Proc. Natl. Acad. Sci. U.S.A.* 71:3455–3459.

Holder-Franklin, M. A., Franklin, M., Cashion, P., Cormier, C., and Wuest, L. 1978. Population shifts in heterotrophic bacteria in a tributary of the Saint John River as measured by taxometrics. In M. W. Loutit and J. A. R. Miles (eds.), *Microb. Ecol.* Springer-Verlag, Berlin, pp. 44–50.

Hughes, V. M., and Datta, N. 1983. Conjugative plasmids in bacteria of the "preantibiotic" era. *Nature* 302:725–726.

Ingraham, J. L., Maaloe, O., and Neidhardt, F. C. 1983. *Growth of the Bacterial Cell.* Sinauer Associates Inc., Sunderland, 435 pp.

Istock, C. A. 1989. Genetic exchange and genetic stability in bacterial popultations. L. R. Ginzberg (ed.) *Ecological Risk of Biotechnology,* Butterworth, New York.

Jacob, F. 1955. Transduction of lysogeny in *E. coli. Virology* 1:207–220.

Jacobs, W. R., Jr., Tuckman, M., and Bloom, B. R. 1987. Introduction of foreign DNA into mycobacteria using a shuttle plasmid. *Nature* 327:532–535.

Jacoby, G. A. 1986. Resistance plasmids of *Pseudomonas.* In *The Bacteria*, vol. 10, J. R. Sokatch (ed.). Academic, Orlando, Florida, pp. 265–293.

Jain, R. K., Burlage, R. S., and Sayler, G. S. 1988. Methods of detecting recombinant DNA in the environment. *CRC Crit. Rev. Biotechnol.* 8:33–84.

Jobling, M. G., Peters, S. E., and Ritchie, D. A. 1988. Plasmid borne mercury resistance in aquatic bacteria. *FEMS Microbiol. Lett.* 49:31–37.

Jones, J. G., Gardener, S., Simon, B. M., and Pickup, R. W. 1986. Antibiotic resistant bacteria in Windermere and two remote upland tarns in the English Lake District. *J. Appl. Bacteriol.* 60:443–453.

Kaijalainen, S., and Lindstrom, K. 1989. Restriction fragment length polymorphism analysis of *Rhizobium galegae* strains. *J. Bacteriol.* 171:5561–5566.

Kelly, W. J., and Reanney, D. C. 1984. Mercury resistance among soil bacteria. Ecology and transferability of genes encoding resistance. *Soil Biol. Biochem.* 16:1–8.

Kleckner, N. 1989. Transposon 10. In D. E. Berg and M. M. Howe (eds.), *Mobile DNA*, American Society for Microbiology, Washington D.C., pp. 227–268.

Knudsen, G. R., Walter, M. V., Porteous, L. A., Prince, V. J., Armstrong, J. L., and Seidler, R. J. 1988. Predictive model of conjugative plasmid transfer in the rhizosphere and phyllosphere. *Appl. Environ. Microbiol.* 54:343–347.

Kretschmer, P. J., and Cohen, S. N. 1979. Effect of temperature on translocation frequency of the Tn3 element. *J. Bacteriol.* 139:515–519.

Kumari, A. A., and Vidaver, A. K. 1988. Transfer and maintenance of IncP and IncW group plasmids into and between extra-slow-growing *Bradyrhizobium japonicum* strains. *FEMS Microbiol. Lett.* 49:423–427.

Leary, J. V., Thomas, M. D., and Allingham, E. 1984. Conjugal transfer of *E. coli* F'lac from *Erwinia chrysanthemi* to *Pseudomonas syringea* pv. *glycinea* and the apparent stable incorporation of the plasmid into the pv. *glycinea* chromosome. *Mol. Gen. Genet.* 198:125–127.

Lenski, R. E., and Levin, B. R. 1985. Constraints on the coevolution of bacteria and virulent phage: A model, some experiments, and prediction for natural communities. *American Naturalist.* 125:585–602.

Lenski, R. E., and Nguyen, T. T. 1988. Stability of recombinant DNA and its effects on fitness. *Trends in Biotechnology.* 6:518–520.

Levin, B. R., Stewart, F. M., and Rice, V. A. 1979. The kinetics of conjugative plasmid transmission: fit of a simple mass action model. *Plasmid.* 2:247–260.

Levine, M. M., Kaper, J. B., Lockman, H., Black, R. E., Clements, M. L., and Falkow, S. 1983. Recombinant DNA risk assessment studies in man: efficacy of poorly mobilizable plasmids in biologic containment. *Recomb. DNA Tech. Bull.* 6:89–97.

Levy, S. B., and Marshall, B. M. 1988. Genetic transfer in the natural environment. In M. Sussman, C. H. Collins, F. A. Skinner, and D. E. Stewart-Tull (eds.) *The Release of Genetically-Engineered Microorganisms.* Acad. Press, London. pp. 61–76.

Liang, L. N., Sinclair, J. L., Mallory, L. M., and Alexander, M. 1982. Fate in model ecosystems of microbial species of potential use in genetic engineering. *Appl. Environ. Microbiol.* 44:708–714.

Lorenz, M. G., and Wackernagel, W. 1987. Adsorption of DNA to sand and variable degradation rates of adsorbed DNA. *Appl. Environ. Microbiol.* 52:654–659.

Lorenz, M. G., Aardema, B. A., and Wackernagel, W. 1988. Highly efficient genetic transformation of *Bacillus subtilis* attached to sand grains. *J. Gen. Microbiol.* 134: 107–112.

Love, P. E., and Yasbin, R. E. 1984. Genetic characterization of the inducible SOS-like system of *Bacillus subtilis*. *J. Bacteriol.* 160:910–920.
Mach, P. A., and Grimes, D. J. 1982. R-plasmid transfer in a wastewater treatment plant. *Appl. Environ. Microbiol.* 44:1395–1403.
Maloy, S. R., and Nunn, W. D. 1981. Selection for loss of tetracycline resistance by *Escherichia coli*. *J. Bacteriol.* 145:1110–1112.
Mancini, P., Fertels, S., Nave, D., and Gealt, M. A. 1987. Mobilization of plasmid pHSV106 from *Escherichia coli* HB101 in a laboratory scale waste treatment facility. *Appl. Environ. Microbiol.* 53:665–671.
McClure, N. C., Weightman, A. J., and Fry, J. C. 1989. Survival of *Pseudomonas putida* UWC1 containing cloned catabolic genes in a model activated sludge unit. *Appl. Environ. Microbiol.* 55:2627–2634.
Melton, D. A., Krieg, P. A., Rebagliati, M. R., Maniatis, T., Zinn, K., and Green, M. R. 1984. Efficient *in vitro* synthesis of biologically active RNA and RNA hybridization probes from plasmids containing a bacteriophage SP6 promoter. *Nucleic Acids Res.* 12:7035–7056.
Mergeay, M., and Gerits, J. 1978. F′-plasmid transfer from *Escherichia coli* to *Pseudomonas fluorescens*. *J. Bacteriol.* 135:18–28.
Morrison, W. D., Miller, R. V., and Sayler, G. S. 1978. Frequency of F116-mediated transduction of *Pseudomonas aeruginosa* in a freshwater environment. *Appl. Environ. Microbiol.* 36:724–730.
Nguyen, T. N. N., Phan, Q. G., Duong, L. P., Bertrand, K. P., and Lenski, R. E. 1989. Effects of carriage and expression of Tn10 tetracycline-resistance operon on the fitness of *Escherichia coli* K12.
Odelson, D. A., Rasmussen, J. L., Smith, J., and Macrina, F. L. 1987. Extrachromosomal systems and gene transmission in anaerobic bacteria. *Plasmid.* 17:87–109.
Ogunseitan, O. A. 1988. Ph.D. Dissertation. The University of Tennessee, Knoxville.
Ogunseitan, O. A., Sayler, G. S., and Miller, R. V. 1990. Dynamic interaction between *Pseudomonas aeruginosa* and viruses in lakewater. *Microb. Ecol.* 19:171–185.
O'Morchoe, S., Ogunseitan, O. A., Sayler, G. S., and Miller, R. V. 1988. Conjugal transfer of R68.45 and FP5 between *Pseudomonas aeruginosa* in a natural freshwater environment. *Appl. Environ. Microbiol.* 54:1923–1929.
Page, W. J. 1982. Optimal conditions for competence development in nitrogen-fixing *Azotobacter vinelandii*. *Can. J. Microbiol.* 28:389–397.
Priefer, U. B., Simon, R., and Pühler, A. 1985. Extension of the host range of *Escherichia coli* vectors by incorporation of RSF1010 replication and mobilization functions. *J. Bacteriol.* 163:324–330.
Rafii, F., and Crawford, D. L. 1988. Transfer of conjugative plasmids and mobilization of a non-conjugative plasmid between streptomyces strains on agar and in soil. *Appl. Environ. Microbiol.* 54:1334–1340.
Rafii, F., and Crawford, D. L. 1989. Gene transfer among streptomyces. In S. B. Levy and R. V. Miller (eds.), *Gene Transfer in the Environment*. McGraw-Hill, New York, pp. 309–345.
Reanney, D. C., Gowland, P. C., and Slater, J. H. 1983. Genetic interactions among microbial communities. In J. H. Slater, R. Whittenbury, and J. W. T. Winpenny (eds.), *Microbes in Their Natural Environments*, Thirty-Fourth Symposium of the Society for General Microbiology, Cambridge University Press, pp. 379–421.
Redfield, R. J. 1988. Evolution of bacterial transformation: Is sex with dead cells better than no sex at all? *Genetics* 119:213–221.
Richaume, A., Angle, J. S., and Sadowsky, M. J. 1989. Influence of soil variables on *in situ* plasmid transfer from *Escherichia coli* to *Rhizobium fredii*. *Appl. Environ. Microbiol.* 55:1730–1734.
Rissler, J. F. 1984. Research needs for biotic environmental effects of genetically engineered microorganism. *Recomb. DNA Tech. Bull.* 7:20–30.
Roberts, M., and Falkow, S. 1979. *In vivo* conjugal transfer of R plasmids in *Neisseria gonorrhoeae*. *Infect. Immun.* 24:982–984.
Roberts, I., Holmes, W. M., and Hylemon, P. B. 1988. Development of a new shuttle plasmid system for *Escherichia coli* and *Clostridium perfringens*. *Appl. Environ. Microbiol.* 54:268–270.

Rochelle, P. A., Fry, J. C., and Day, M. J. 1989*a*. Factors affecting conjugal transfer of plasmids encoding mercury resistance from pure cultures and mixed natural suspensions of epilithic bacteria. *J. Gen. Microbiol.* 135:409–424.
Rochelle, P. A., Fry, J. C., and Day, M. J. 1989*b*. Plasmid transfer between *Pseudomonas spp.* within epilithic films in a rotating disc microcosm. *FEMS Microbiol. Ecol.* 62:127–136.
Roszak, D. B., and Colwell, R. R. 1987. Survival strategies of bacteria in the natural environment. *Microbiol. Rev.* 51:365–379.
Sambrook, J., Fritsch, E. F., and Maniatis, T. 1989. In *Molecular Cloning. A Laboratory Manual*, 2d ed. Cold Spring Harbor Laboratory Press, Cold Spring Harbor, New York.
Saye, D. J., Ogunseitan, O., Sayler, G. S., and Miller, R. V. 1987. Potential for transduction of plasmids in a natural freshwater environment: Effect of plasmid donor concentration and a natural microbial community on transduction in *Pseudomonas aeruginosa. Appl. Environ. Microbiol.* 53:987–995.
Saye, D. J. and Miller, R. U. 1989. The aquatic environment: Consideration of horizontal gene transmission in a diversified habitat. In S. B. Levy and R. U. Miller (eds.) *Gene Transfer in the Environment*, McGraw-Hill, New York, pp. 223–259.
Saye, D. J., Ogunseitan, O. A., Sayler, G. S., and Miller, R. V. 1990. Transduction of linked chromosomal genes between *Pseudomonas aeruginosa* during incubation *in situ* in a freshwater habitat. *Appl. Environ. Microbiol.* 56:140–145.
Schaberg, D. R., and Zervos, M. J. 1986. Intergeneric and interspecies gene exchange in gram-positive cocci. *Antimicrob. Agents Chemother.* 30:817–822.
Schilf, W., and Klingmüller, W. 1983. Experiments with *Escherichia coli* on the dispersal of plasmids in environmental samples. *Recomb. DNA Tech. Bull.* 6:101–102.
Schofield, P. R., Gibson, A. H., Dudman, W. F., and Watson, J. M. 1987. Evidence for genetic exchange and recombination of *Rhizobium* symbiotic plasmids in a soil population. *Appl. Environ. Microbiol.* 53:2942–2947.
Shoemaker, N. B., Smith, M. D., and Guild, W. R. 1980. DNase resistant transfer of chromosomal *cat* and *tet* insertions by filter mating in *Pneumococcus. Plasmid* 3:80–87.
Shultz, J. E., Latter, G. I., and Matin, A. 1988. Differential regulation by cyclic AMP of starvation protein synthesis in *Escherichia coli. J. Bacteriol.* 170:3903–3909.
Simon, R., Priefer, U., and Pühler, A. 1983. A broad host range mobilization system for *in vitro* genetic engineering: transposon mutagenesis in gram negative bacteria. *Bio/technology* 1:784–791.
Slater, J. H., Weightman, A. J., and Hall, B. G. 1985. Dehalogenase genes of *Pseudomonas putida* PP3 on chromosomally located transposable elements. *Mol. Biol. Evol.* 2:557–567.
Soberon, X., Covarrubias, L., and Bolivar, F. 1980. Construction and characterization of new cloning vehicles. IV. Deletion derivatives of pBR322 and pBR325. *Gene* 9:287–305.
Sonea, S. 1988. The global organism: a new view of bacteria. *The Sciences* 28:38–45.
Sparling, P. E. 1966. Genetic transformation of *Neisseria gonorrhoeae* to streptomycin resistance. *J. Bacteriol.* 92:1364–1371.
Squires, C. H., Heefner, D. L., Evans, R. J., Kopp, B. J., and Yarus, M. J. 1984. Shuttle plasmids for *Escherichia coli* and *Clostridium perfringens. J. Bacteriol.* 159:465–471.
Stewart, G. J. 1989. The mechanism of natural transformation. In S. B. Levy and R. V. Miller (eds.), *Gene Transfer in the Environment*. McGraw-Hill, New York, pp. 139–164.
Stewart, G. J., and Carlson, C. A. 1986. The biology of natural transformation. *Annu. Rev. Microbiol.* 40:211–235.
Stotzky, G. 1989. Gene transfer among bacteria in soil. In S. B. Levy and R. V. Miller (eds.), *Gene Transfer in the Environment*. McGraw-Hill, New York, pp. 165–222.
Stotzky, G., and Babich, H. 1984. Fate of genetically engineered microbes in natural environments. *Recomb. DNA Tech. Bull.* 7:163–188.
Tardiff, G., and Grant, R. B. 1983. Transfer of plasmids from *Escherichia coli* to *Pseudomonas aeruginosa*: Characterization of a *P. aeruginosa* mutant with enhanced

recipient ability for enterobacterial plasmids. *Antimicrob. Agents Chemother.* 24: 201–208.
Thiry, G. 1984. Plasmids of the epiphytic bacterium *Erwinia uredovora. J. Gen. Microbiol.* 130:1623–1631.
Tiedje, J. M., Colwell, R. K., Grossman, Y. L., Hodson, R. E., Lenski, R. E., Mack, R. N., and Regal, P. J. 1989. The planned introduction of genetically engineered organisms: Ecological considerations and recommendations. *Ecology* 70:298–315.
Toranzo, A. E., Combarro, P., Lemos, M. L., and Borja, J. L. 1984. Plasmid coding for transferable drug resistance in bacteria isolated from cultured rainbow trout. *Appl. Environ. Microbiol.* 48:872–877.
Trevors, J. T., Barkay, T., and Bourquin, A. W. 1987. Gene transfer among bacteria in soil and aquatic environments: a review. *Can. J. Microbiol.* 33:191–198.
Trevors, J. T., and Oddie, K. M. 1986. R-plasmid transfer in soil and water. *Can. J. Microbiol.* 32:610–613.
Trieu-Cuot, P., and Courvalin, P. 1986. Evolution and transfer of aminoglycoside resistance genes under natural conditions. *J. Antimicrob. Chemother.* 18(suppl.)C:93–102.
Tsuda, M., and Iino, T. 1987. Genetic analysis of a transposon carrying toluene degrading genes on a TOL plasmid pWWO. *Mol. Gen. Genet.* 210:270–276.
Tsuda, M., and Iino, T. 1988. Identification and characterization of Tn4653, a transposon covering the toluene transposon Tn4551 on Tol plasmid pWWO. *Mol. Gen. Genet.* 213:72–77.
Twigg, A. J., and Sherratt, D. 1980. Transcomplementable copy-number mutants of plasmid ColE1. *Nature* 283:216–218.
Van Elsas, J. D., Govaert, J. M., and Van Veen, J. A. 1987. Transfer of plasmid pFT30 between bacilli in soil as influenced by bacterial population dynamics and soil conditions. *Soil Biol. Biochem.* 19:639–647.
Van Elsas, J. D., Trevors, J. T., and Starodub, M. E. 1988. Plasmid transfer in soil and rhizosphere. In W. Klingmüller (ed.), *Risk Assessment for Deliberate Release.* Springer-Verlag, Berlin, pp. 89–99.
Vanstockem, M., Michiels, K., Vanderleyden, J., and Van Gool, A. P. 1987. Transposon mutagenesis of *Azospirillum brasilense* and *Azospirillum lopoferum*: physical Analysis of Tn5 and Tn5-mob insertion mutants. *Appl. Environ. Microbiol.* 53:410–415.
Weinberg, S. R., and Stotzky, G. 1972. Conjugation and genetic recombination of *Escherichia coli* in soil. *Soil Biol. Biochem.* 4:171–180.
Wiggins, B. A., and Alexander, M. 1985. Minimum bacterial density for bacteriophage replication. Implications for significance of bacteriophages in natural ecosystems. *Appl. Environ. Microbiol.* 49:19–23.
Willets, N., and Wilkins, B. 1984. Processing of plasmid DNA during bacterial conjugation. *Microbiol. Rev.* 48:24–41.
Zinder, N. D., and Lederberg, J. 1952. Genetic exchange in *Salmonella. J. Bacteriol.* 64: 697–699.

Chapter

9

Management of Transgenic Plants in the Environment

Kathleen H. Keeler

School of Biological Sciences
University of Nebraska–Lincoln
212 Lyman Hall
Lincoln, Nebraska 68588-0343

Charles E. Turner

Biological Control of Weeds
U.S. Department of Agriculture
Agricultural Research Service
Western Regional Research Center
800 Buchanan Street
Albany, California 94710

Goals of Genetic Engineering of Plants

The uses proposed for genetically engineered plants encompass all the goals of agriculture: more productive and nutritional crops with fewer losses to pests, including weeds (Gasser and Fraley 1989, Gotsch and Rieder 1989, Ratner 1989). Certainly, genetically engineered plants—henceforth referred to as "transgenic plants"—and their transferred genes will be used, as plants were in the past, to produce food, fiber, and industrial raw materials, and for ornamental purposes. Thus we can expect to find transgenic plants growing in fields as crops, in forests for timber, in pasturelands for forage, in urban and suburban settings as ornamentals, and on roadsides and slopes for erosion control. Whether transgenic plants will have a role in natural or restored wild areas is difficult to predict, but it seems likely that they will (Table 9.1).

TABLE 9.1 The Goals and Applications of Genetic Engineering of Plants

Goals
More productive food crops
Reduced losses to pests and weeds
Higher nutritional value
Better postharvest qualities
Applications
Human and animal food
Fiber
Raw materials for the chemical industries
Building materials
Ornamental and shade plants
Plants for erosion control
Environmental Setting (Uses)
Fields (crops)
Forests (timber)
Pasture (forage)
Indoor and outdoor, urban and suburban (ornamentals)
Roadsides and slopes (erosion control)
Natural or restored wild areas (especially to counter human-generated stresses such as introduced diseases or smog damage)

SOURCES: Keeler 1988, Mellon 1988, OTA 1988, Gasser and Fraley 1989, Gotsch and Reider 1989, Ratner 1989.

The actual genes transferred, referred to as "transgenes," should run the gamut of genes to carry out these functions, including genes that enhance plant productivity; increase pest, herbicide, and disease resistance; raise nutritional value; and improve drought, cold, and salt tolerance. Likewise an array of genes improving handling at harvest, and storage and shipping properties, can be expected. While engineering genes in some plant groups is more technically challenging than in others, it is probable that by one means or another foreign genes will be introduced into most if not all economically important plants within this century (e.g., Gasser and Fraley 1989). This means that introduction of plants expressing genes from sources as diverse as mammals and bacteria can be expected to be a routine part of agriculture, horticulture, range management, and forestry soon after the turn of the century (e.g., Gotsch and Rieder 1989). This places a terrific burden on the scientists developing this technology, and on the regulators overseeing it, to take advantage of the benefits genetic engineering can provide while protecting the public and the environment from injurious side effects.

In this chapter we discuss weed problems that might arise from the use of transgenic plants, methods to monitor transgenic plants, and ways to reduce undesirable effects should weeds arise. While current releases are confined to small-scale field tests, our emphasis is on the future, when transgenic plants will be in large-scale commercial use. We define weeds and weed problems, consider the practicality of anticipating novel or unique problems, and discuss the importance of familiar problems that might be created by transgenic plants. With that in mind, we review methods of release, movement, and establishment of plants, especially weeds, and of genes by themselves, a concern novel to genetic engineering. Then we review methods for controlling transgenic plants and transgenes. In particular, the problems of monitoring introduced plants, assessing risks, and intervening if necessary, are considered.

Environmental Concerns

Weeds and their consequences

Weeds are defined as plants in the wrong place or plants interfering with human activity (Salisbury 1961, Buchholtz 1967, Holzner 1982). Thus, weed problems include any consequences of transgenic plants "growing where they are not desired." We deal chiefly with transgenic crops and transgenes interfering with human activity. Disruption of biotic communities and ecosystem processes are discussed in Chap. 1. Other possible problems resulting from transgenic plants are considered in Chaps. 4 and 7.

A very diverse group of plants are considered weeds because plants that differ in terms of taxonomic grouping, growth habit, and habitat interfere with different human activities: toxic plants in pastures, nontimber trees in forest plantations, noncrop plants in croplands, aquatic weeds in aquatic systems (e.g., Holm et al. 1977, Holzner 1982, Mack 1989). Nevertheless, many weeds share a suite of characteristics that enhance their success in habitats disturbed by humans. These include rapid growth, rapid seed production, large numbers of small seeds, tolerance of a wide variety of environmental conditions, and mechanisms for resisting eradication such as rooting from fragments (Baker 1965, 1974, Keeler 1985, 1989).

Whether they possess the typical traits of weeds or not, some plants pose consistent and widespread problems for human activities. Holm et al. (1977) determined the "world's worst weeds" by summing weed-control priorities around the world. Their worst weed, purple nutsedge (*Cyperus rotundus* L.), is a weed of 52 crops in 92 countries, and their top 17 weeds are on the average serious problems in 33 crops in 54 countries. The U.S. Department of Agriculture (USDA) long ago took this seriously enough to establish the federal Seed Act of 1939 and the federal Noxious Weed Act of 1974 to prevent the spread of serious weeds (Crooks et al. 1983, Foy et al. 1983). Preventing increased weed problems due to incorporation of transgenes into weedy species is recognized as important by all workers in the field (e.g., Brill 1985, Hauptli et al. 1985, Tiedje et al. 1989).

Weeds cost U.S. agriculture over $18 billion per year, and millions more are spent by homeowners, and by managers of golf courses, parks, and outdoor areas (Foy et al. 1983). Weed problems have been known to drive crops out of some regions (e.g., Gauthier et al. 1981), and they regularly determine the choice of crop by farmers (Dover and Croft 1986, Keeler et al., in preparation). Thus, increased weed problems from transgenic plants or transgenes could be very costly.

Potential novel and unique problems of transgenic plants

Regulators are, understandably, especially concerned about novel or unique problems from transgenic organisms (e.g., Brill 1985, NAS 1987). However, these two types of problems are the ones we are least able to predict, for the logical reason that they have not been encountered previously. Specifically, by definition, novel problems cannot have previously occurred and so are quite likely to be unanticipated. Unique environmental problems of transgenic plants, on the other hand, simply cannot occur until transgenic plants are released into the environment. Thus, while we can assert that unique or novel problems from transgenic plants seem to us highly unlikely, we have, by definition, no actual information (yet) upon which to base this opinion.

Thus, as we address the subject of weed problems that might result from transgenic plants, we may be considering unique problems, but probably not novel ones. Indeed, by considering "weed problems," we classify our topic as familiar, not novel. Although no weed problem could be completely novel, areas of concern about unexpectedly bad weed problems with transgenic plants might be looked for in at least three areas: (1) if an introduced gene made an otherwise nonweedy plant into a problem weed, (2) if an introduced gene made a weed more difficult to control, and (3) if an introduced gene allowed a species to invade new habitats. (The latter two differ more in their practical aspects than theoretically: human responses to increased difficulty in controlling familiar weeds differ from human responses to species discovered in new regions).

Conversion of a nonweedy crop into a weed. Potentially, a transgene could act pleiotropically to make a surprising and tenacious weed of a crop not otherwise weedy—for example, an invasive maize or strangler soybean. It would be just another new weed, but it might require innovative control methods, if only because there has been little study of how to control maize or soybeans.

This type of response to a new gene is theoretically possible, but seems to us quite unlikely. Weediness is generally the result of a number of traits (e.g., Baker 1965, Keeler 1989), and an introduced gene or group of genes are not expected to have profound effects (Brill 1985, Hauptli et al. 1985, Tiedje et al. 1989). Moreover, it should be possible to detect any drastic changes in the plant during small-scale field tests, and curtail production before serious problems occur (Szybalski 1985, Center for Science Information 1988).

Increased difficulty of weed control. If the presence of a transgene produces a weed that is particularly hard to control, this might constitute a novel weed problem resulting from genetic engineering. Weedy races of crop plants have always been difficult problems for weed control (e.g., Thomas and Jones 1976, Holm et al. 1977). Some introduced genes are expected to reduce the control methods available. Herbicide resistance, for example, among the weeds of a crop would eliminate that herbicide from the array of control methods available. This is clearly seen in the case of oilseed rape (*Brassica* sp.) in Canada, where serious weed problems with weedy brassicas nearly eliminated the crop, until naturally occurring genes for triazine resistance were introduced into the crop, allowing the weedy brassicas to be controlled by triazine (LeBaron 1984, Beversdorf 1987). Obviously, *introgression,* the movement of genes from one species into another, of the transgenes for triazine resistance from the crop to the wild brassicas would be a major setback to this industry.

Naturally occurring herbicide resistance is known for all the major herbicides (e.g., LeBaron 1983, Tangley 1985), so the development of herbicide resistance in weeds is not novel to genetic engineering. However, naturally occurring mutants for resistance have appeared only in scattered plant groups and isolated locations: in many areas the important weeds are not resistant to the herbicide of choice. One consequence of widespread adoption of herbicide-resistant crops will be the introduction of genes for herbicide resistance into new areas. If there are compatible weedy relatives present, then herbicide resistance is likely to become a more widespread problem.

In addition, it is very much to the advantage of weeds to have multiple herbicide resistances. Introduction of herbicide resistance genes into outcrossing plants with weedy relatives, for example sunflower (*Helianthus annuus* L.) or sorghum [*Sorghum bicolor* (L.) Moench], could lead to their escape into wild populations. We suggest a possible, but perhaps improbable, consequence: by repeated introgression, a successful weed might acquire the various resistance genes introduced by different seed companies. With neighboring fields of glyphosate-resistant rape and triazine-resistant rape, a weedy brassica resistant to both glyphosate and triazine could arise from the area. Such a weed would put tremendous stress on agriculture. Under current agricultural practice, spread of herbicide resistance is slow and multiple resistances highly unlikely. Steps should be taken to prevent the novel problem of multiple resistances from ever occurring. One simple method might be to prohibit use of several crop varieties with different resistances in the same area.

Novel problems might also arise from weeds with transgenes for insect and disease resistance. Ecologists are only just starting to address the problem of when and how insects or diseases limit natural plant populations (e.g., Louda 1982, 1983, Paul and Ayers 1986, Crawley 1989). Success with biological control of weeds using both insects and diseases demonstrates that losses to insects and diseases can hold down plant populations (Dahlsten 1986, Julien 1987, Turner 1988, Crawley 1989). It follows that release from losses to insects or disease could, in some cases, allow great expansion of plant numbers (Orians 1986). Indeed, the choice of insect resistance and disease resistance genes to introduce into crops implies that losses to these pests are believed to have an appreciable impact on crop yield. If that is the case, one might expect close relatives of these crops to also suffer severely from phytophagous organisms and to benefit noticeably from resistance genes. From this scenario, we raise the caution that transfer of pest resistance genes to weed populations might allow those weeds dramatically greater success, increasing their seriousness as weeds and certainly reducing our ability to control them through the introduction of natural enemies for biological control.

While introgression of transgenes resulting in enhanced weediness is unlikely to happen in very many cases, both theory and the limited experience available suggest it is possible. If it occurs, it would be a novel method of weed production.

Increased invasiveness of weeds. A third important concern is the possibility of enhancing the ability of weeds to invade new habitats. Any successful weed is likely to be invasive. However, if the gene or gene complex introduced enhances crop performance parameters normally limited by physical factors, these genes might similarly prove particularly useful to weedy populations and allow them to expand to new habitats (Tiedje et al. 1989). For example, better drought tolerance or salt tolerance, if transferred from crops to weedy races, should allow the weeds to grow better not only in dry or saline fields but also in surrounding natural areas. These would be unique problems because the [illegible]rance complex arrived only through the introduction of that higher-t[illegible] gene complex into transgenic crops.

It m[illegible] in some cases, to anticipate other problems unique to transg[illegible]ample, Tiedje et al. (1989) suggested that transgenic plants [illegible] against herbivores might disrupt natural ecosystems. T[illegible]lnut (*Juglans regia* L.) engineered to be quite toxic to i[illegible]genic walnuts (*Juglans* sp.) or maples (*Acer* sp.) or oak[illegible]important ornamental and lumber trees, as well as [illegible]astern forests) are planted in large numbers in [illegible] might cause crashes of populations of her[illegible] lead to crashes of their predators, such [illegible]assive changes in the composition of [illegible]nario is worth drawing because, although better [illegible]ses are continuously evolving in plants, under natural co[illegible] occur first as a single mutant in an isolated plant. Even when [illegible]ing a great increase in fitness, that defense nevertheless takes many generations to spread throughout the population, giving the insect community ample opportunity to adapt. Thus, widespread destabilization of herbivore populations would be an outcome unique to transgenic plants, one that might occur if genetic engineering introduced a gene for a highly toxic alien compound from an unrelated organism into walnuts or maples. Similar impacts might result from planting large numbers of toxic nonnative trees, which would affect primarily native insect species with a sufficiently broad host range to include the introduced trees. Transgenic native trees could affect an array of native insect species with narrow as well as broad host ranges (i.e., both generalists and specialists).

Scenarios like the previous ones tend to strain credibility. It is possible, but is it probable? The paradox is that although one can imagine novel and unique problems, there is no evidence to support such effects. Thus, suggestions of such consequences understandably raise great skepticism among ecologists and molecular biologists alike. Potential problems that have never occurred and may never occur are often criticized as "worst case scenarios" and "doomsaying." The doomsaying of Rachael Carson (Carson 1962) was decried at the time, but has on the whole been substantiated (e.g., Graham 1970). Similarly, that introduced species could cause the near extinction of very numerous and successful native species is too improbable for most experts to anticipate, but occurred when chestnut blight drove American chestnut [*Castanea dentata* (Marshall) Borkh.] to near extinction and also when Dutch elm disease attacked American elm (*Ulmus americana* L.) (Turner 1988). We conclude that even improbable events deserve consideration.

Until the transgenic plants are released into the environment, all interactions with transgenic plants will be hypothetical. We see no solution to this paradox. If it hasn't happened yet, it cannot be shown to be sure to occur, and preparations may be proven unnecessary. The best approach seems to be that of gathering data on the subject by making well-designed, well-monitored tests of increasing scale and complexity, so that undesirable impacts are observed while there is still an opportunity to correct them.

The potential problems from transgenic plants for which we can marshall convincing evidence are relatively familiar problems, such as those falling comfortably within the well-known category of weed problems. It is on those we concentrate, believing that however familiar, potential weed problems from transgenic plants or transgenes are important enough to merit prevention.

Evolutionary consequences: pollution of native gene pools

An additional concern, springing from considerations of evolutionary biology, is the "pollution" of native gene pools by transgenes. This problem is older than genetic engineering (e.g., Millar and Libby 1989), but will be exacerbated by transfer of genes from distantly related species. Where the engineered genes are from a taxon that could not have interbred with the recipient in nature, introgression of exotic genes into native populations presents perplexing problems. Modern evolutionary ecology suggests that the paucity of variation (availability of mutant genes) often limits the ability of species to adapt to changing conditions (e.g., Regal 1986, Tiedje et al. 1989). If humans provide populations with new genes, this could dramatically alter the evolution of the species. Modern evolutionary theory considers the genotypes of surrounding individuals as an integral part of the environment to which individuals and species adapt (e.g., Futuyuma 1986). Therefore a change in the genes of one species can lead to very complex, permanent changes in all of the community. For native species and natural areas this may not be acceptable. Where practical considerations suggest that the introduction of an alien gene that might introgress into native populations is called for (e.g., a gene to protect New England's maples from the gypsy moth), serious prerelease analysis and in-depth public discussion must occur before release, since recall is probably impossible (Colwell et al. 1988, Mellon 1988, Millar and Libby 1989, Tiedje et al. 1989).

Overview of Plant Release, Movement, and Establishment

Release

In the simplest analysis there are two categories of environmental release: (1) accidental release and (2) intentional release. The accidental release of plants includes cases in which the plant was simply not intended to get out into the environment and, more commonly, cases in which the plant is moved from one region to another into which introduction was not intended. Intentional releases cover all planting of transgenic plants outside of containment facilities, greenhouses, and buildings.

TABLE 9.2 Sites of Potential Problems from Movement of Transgenic Plants and Introduced Genes

	Plant movement	Gene movement	
	Weed problems	Weed problems	Gene pool pollution
Agricultural field	X	X	
Suburban and urban areas	X	X	
Roadsides and waste areas	X	X	
Rangelands and pastures	X	X	X[a]
Commercial forests	X	X	
Recreational areas	X	X	
Natural areas and preserves	X	X	X[a]

[a]Within wild populations of native species.

Weed problems from released transgenic crops will vary with each site. In croplands, the intensive management of crops means that a variety of methods of control are available if new weeds appear. In pasturelands or rangelands, which have considerably lower rates of economic return than croplands, lower tolerable costs mean that the same degree of weed problem will be more difficult to control in terms of economic feasibility and probably also accessibility. If weeds invade natural areas, the complexity of the natural areas and the need to preserve sensitive species present will make it particularly difficult to remove them. At the same time, the characteristics which make weeds problematic change with the area: a herbicide-tolerant weed will have no advantage in natural areas, but could become a very expensive problem in croplands (Center for Science Information 1988) (Table 9.2).

There are two kinds of escape to consider: (1) escape of the plant itself and (2) of the introduced genes, which can move independently of the parent plant. The plant can move as propagules or diaspores (seeds, fruits, or vegetative units capable of reproduction). (Botanical terminology is generally based on morphological characteristics, so seeds and spores require different terms even though they may perform the same function. "Propagule" and "diaspore" are catch-all terms for any plant part adapted for dispersal, and as such can include everything from seeds to branches that root.) The transgenes can move if pollen is carried to distant members of the same species. In addition, the transgenes may be transferred to other species via hybridization. Theoretically, transgenes can also move between plant species when carried on a virus, bacterium, or other infective agent, although this has not been shown in nature for higher plants (see Chapter 7).

Movement of plants

We consider movement of whole plants first. Gene movement is discussed in the following section. If the plant grows only on the site where it was originally planted, it would not be expected to cause a problem. Or perhaps we should say that if it grows only where it was originally planted and it causes a problem, it is a relatively simple matter to remove the offending individual. Thus, the key to preventing serious problems is monitoring the movement of plants. Plant movement has three components: dispersal, persistence, and establishment.

Dispersal. Active spreading of plants not mediated by animals or other dispersal agents, including humans, can cover moderate distances: root sprouts of trees such as *Eucalyptus* can emerge 50 m from the parent plant, while single clones of velvet grass (*Holcus mollis* L.) and huckleberry (*Gaylussacia brachycerilea*) cover hundreds of meters, and aspen clones (*Populus tremuloides* Michx.) have spread to occupy 43 ha (Cook 1983). The time involved to move such distances, however, is measured in years, not days or weeks, so from a containment standpoint, this is relatively easy to control.

All extended plant movement is passive and depends upon dispersal agents. Plants can nevertheless cover long distances. Plants are dispersed by animals (including humans), wind, and water. Floating coconuts cross oceans. Propagules attached to fur or feathers of an animal may be carried between continents. While most wind dispersal is on the order of meters to a few kilometers (Daubenmire 1974, Pijl 1982), bracken fern [*Pteridium aquilinum* (L.) Kuhn] spores must have reached Hawaii on the wind, a minimum distance of 720 km (Carlquist 1974). Most plant movement does not span much distance, but long-distance migration is always possible, particularly when propagules are transported by humans, either intentionally or unintentionally (Turner 1988).

Even short-distance dispersal can move plants long distances or expand their range to large areas if there is adequate time. For example, paleontological evidence suggests that white spruce [*Picea glauca* (Moench) Voss] moved 2000 km in 1000 years at the end of the last glaciation (Ritchie and MacDonald 1986). In the modern world, when weed expansions have occasionally been documented, the data indicate that plants can expand their ranges dramatically and increase their area of infestation exponentially (e.g., to hundreds of thousands of hectares in less than a century) if conditions are favorable (Mack 1981, Forcella 1985, Maddox and Mayfield 1985, Pemberton 1986, Thompson et al. 1987, and see Fig. 9.1).

At present, the most important means of plant dispersal is human activity (Baker 1986, Turner 1988). Humans carry plants intentionally when traveling, but also allow the spread of weeds as seed contaminants of crop seed and accidental inclusions in movements of soil or ballast and attached to domestic animals, farm machinery, vehicles, and travelers. This is particularly true of weedy species and will probably be true of transgenic weeds as well.

Persistence. "Persistence" is the ability of a plant to live at a site for some period after arrival; persistence is necessary for any weed problem to occur. The persistence of annual and perennial plants is very different. Annual plants die at the end of the growing season (i.e., with the onset of frost in temperate regions), and so they persist only through their seeds. Among long-lived perennials, the individual planted may survive on a site for decades whether or not it reproduces. If the annual does not reproduce within the first growing season, it will not establish a population or create a weed problem. The longer persistence of perennials increases the chance that mutation or the arrival of a compatible plant may produce conditions for spread. If the plant neither reproduces sexually nor spreads vegetatively, there can be no weed problems.

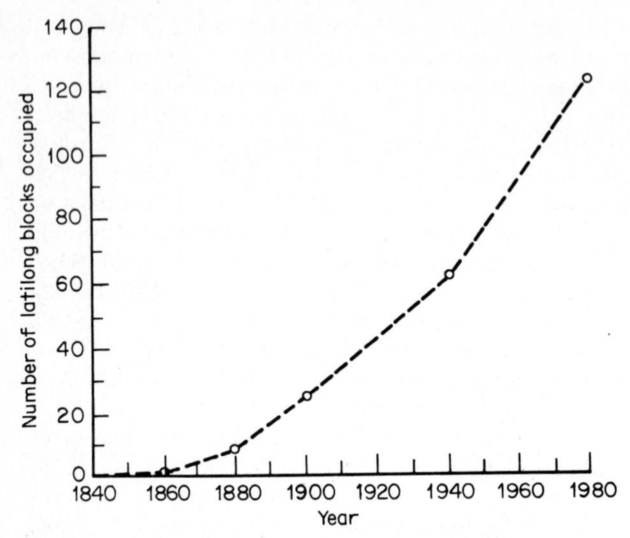

Figure 9.1 Number of 15-ft by 15-ft latilong blocks occupied in westward spread of purple loosestrife (*Lythrum salicaria*) through New York, Michigan, Wisconsin, and Minnesota from 1840 to 1980. *N* total blocks = 1171; area of each block = 771.7 km^2. (*Thompson et al. 1987*)

Establishment. "Establishment" is the production of a breeding population and requires reproduction. For annuals, this means seed production. Most perennials also reproduce by seed formation. However, some perennials also have a mechanism of establishment that does not occur in annuals: vegetative spread. Many perennials reproduce vegetatively, spreading by rhizomes, bulbils, etc. In these species a population can be established from a single individual, and can spread in the absence of pollination or seed set. Examples include the spread of the aquatic weed *Elodea canadensis* Michx. in England and *Hydrilla verticillata* (c.f.) Royle in the United States by strictly vegetative growth (Elton 1958, Anderson 1987).

With vegetative reproduction, what spreads is frequently a clone of identical shoots, which may become disconnected from each other. Thus, the established "population" may consist of only a single genetic individual, but still occupy a great deal of space. Consequently, establishment needs to be recognized as maintenance or expansion of the area controlled by the introduced plants or in terms of the number of shoots or some such measure, rather than in terms of number of genetic individuals or ability to set viable seed.

Movement of genes independently of the parent plant

The genes of plants can move independently of the parent by cross-pollination. Thus, transgenes from crops can contribute to weed problems by hybridization with weeds. Most commonly, compatible plants will be weedy races of crops or wild relatives (e.g., deWet and Harlan 1975, Oka and Morishima 1982). The probability of this sort of weed evolution depends greatly on the crop plant: for tomatoes, the risk is apparently minimal due to low rates of outcrossing (Rick in lit. 1989). For oats, on the other hand, derived weeds and seriously weedy close relatives are numerous (Thomas and Jones 1976, Holm et al. 1977).

Compared to evolution of weedy races derived from crops, introgression of genes from crops is a much faster method of producing effective weeds, since it combines adaptive genes from the crops with the genes of plants that are already successful weeds. The escape of transgenes from crop plants into wild populations may in general be an undesirable effect, but, speaking practically, it is not important to prevent it unless we expect negative consequences should wild species carry those genes. Therefore, concern depends upon the effect (phenotype) conferred by the transgenes. The most obvious problem would be the creation of hard-to-control agricultural weeds.

For transgenes to move into weeds via hybridization, weedy plants compatible with the crops must be present. Although virtually all crops have wild relatives, many crops are grown in areas where their wild relatives do not occur. However, wild relatives may have been introduced along with the crop. Certainly a substantial proportion of the serious weeds in the United States are nonnative (Holm et al. 1977, Foy et al. 1983).

Presence is not itself enough for hybridization: the relatives have to be cross-compatible. Plant breeders have tried crosses with the wild relatives of major crop plants, so the information on ability to hybridize is often available (e.g., Simmonds 1976, Keeler and Turner, in preparation). Usually, only members of the same genus, and usually not all of them, are compatible. For example, of more than 70 species of sunflower (*Helianthus*, Asteraceae), only about 10 species, chiefly the annuals most closely related to the common sunflower (*H. annuus* L.), will hybridize with it (Heiser et al. 1969, Whelan 1978). In addition, some crops, although capable of crossing with other species, rarely do so. This can be because they self-fertilize most of the time (e.g., beans, *Phaseolus*), are largely infertile [sweet potatoes, *Ipomoea batatas* (L.) Lam.], or because the commercial product is harvested before flowers open (e.g., cabbage, *Brassica oleracea* L.) (Keeler and Turner, in preparation). The deliberate creation of highly productive sterile hybrids, available only from the seed companies, has also genetically isolated some crops.

There is a gradient among crops in terms of the frequency of gene exchange with weedy relatives, so that different levels of stringency of containment are required to contain genes of different crops. Furthermore, even if a wild relative is present and compatible, it need not be weedy. Only a small subset of plants occur as weeds; and of all the wild relatives of any particular crop plant, only a few are likely to be considered weeds. There are 300 or so species of *Ipomoea* (morning glory, sweet potato) (Cronquist 1971), of which 57 are listed as weeds by Holm et al. (1979)—a large number, but still the minority of the species in the genus. For yams (*Dioscorea* spp.), only 2 of some 600 species are considered weedy (Coursey 1978, Holm et al. 1979). If the wild relatives, although present, are not weedy, then even if introgression occurs, it is unlikely to cause an agricultural problem.

However, one cannot assume that there will be no problems with wild relatives because the wild relatives are not weedy in their native areas. There are two possible causes of problems. First, plants that are innocuous or minor weeds in their native areas have sometimes become major weeds in new areas, in part because they leave their native natural enemies behind (Foy et al. 1983, Zimdahl 1983, Baker 1984, Turner 1988). Second, changing agricultural conditions or changes in plant genetic structure can change the ability of plants to act as weeds. For example, Cavers and Bough (1985) report the development of seriously weedy races of proso millet (*Panicum miliaceum* L.) after a century of cultivation without serious weed problems. What caused the changes in the millet remains unclear (Cavers and Bough 1985). Furthermore, the emergence of annual weedy grasses such as *Panicum, Digitaria,* and *Sorghum* as significant weeds is due in part to the shift from mechanical cultivation to reliance on effective broadleaf herbicides which remove the competitors of the grasses (Jensen and Bandeen 1979).

Substantial amounts of information will be needed to evaluate the role of compatible wild relatives in posing weed problems, and the situation for each crop is different. While few of the problems considered are novel or unique, the possibility of serious transgenic weeds is quite realistic, and control of these plants is likely to be expensive. The production of such weeds is only a possibility, not a certainty; however, given the expense of weed control, precautions to reduce the chance of transgenic weeds emerging are merited. Suggestions are discussed below.

Immediate and Future Weed Priorities

Weed problems that arise may be due to escape of weedy crops and ornamentals, or to the introgression of transgenes into weedy populations. The former is the chief cause of current weed problems, but is unlikely to be a problem for transgenic plants until they are abundant. The introgression of transgenes, leading to increased weed success, can occur even as a result of small-scale field trials.

Transgenic plants as weeds

Purposeful introduction of transgenic plants into areas where they become a problem—analogous to the intentional establishment of Johnsongrass [*Sorghum halepense* (L.) Pers.] and kudzu [*Pueraria lobata* (Willd.) Ohwi] in the United States, where each has become a major weed problem (Zimdahl 1983, Turner 1988)—although possible, is not a major concern of ours because we expect the elaborate process necessary for introducing transgenic plants to screen out such species. Eventually there will be risks of weed problems from transgenic plants that are unintentionally or illegally moved to areas into which introduction was not intended. It is also possible that epistatic interactions between genes may cause some plant to spread beyond the predictions of its maker. These are important considerations, but good monitoring (see below) and well-designed, well-enforced procedures for eliminating problems as they emerge (see below) should be effective in preventing the spread of transgenic plants as weeds.

Transgenes in weedy species

The immediate necessity if one wishes to prevent the emergence of weeds as a result of genetic engineering is to prevent the escape of transgenes by hybridization between crops and weedy relatives. This problem occurs as soon as any transgenic plant is planted in proximity to compatible weeds, regardless of the number of plants released. However, unless very specific precautions are taken, the risks of problems from transgenes that escape into weedy populations will increase as increased numbers of transgenic crops are planted—i.e., with commercialization.

The potential risk of evolution of weeds by introgression of genes from crop plants into wild relatives varies greatly between plants likely to be genetically engineered. In particular, two biological aspects are considered crucial: (1) the nature of the gene introduced and (2) the potential for crossing between crops and weedy relatives. These are discussed in the remainder of this section—and also see Table 9.3, which outlines a risk assessment protocol based on the parameters discussed below, and Table 9.4, which summarizes levels of risk for a variety of possible transgenic plant characteristics.

The introduced gene. Not all introduced genes are of similar concern if transferred to weedy relatives. For example, if the gene improves nutritional quality or prevents early ripening, then it is unlikely to assist weedy relatives in surviving in an agricultural field. However, genes that may confer a selective survival or reproductive advantage in weeds have the potential to produce serious problems if weedy relatives of the crop plant receive them. Herbicide tolerance is perhaps the most obvious of these genes. However, insect and virus resistance should also be considered since they may allow the resistant plant to outcompete other weeds or other crops where the susceptible plants could not. Resistance to physical stress such as drought or salt would also enhance fitness of the weeds. As previously suggested (e.g., Milewski and Tolin 1984, Keeler 1988, Tiedje et al. 1989), a scaled order of risk of introduced genes can be produced and should be applied. Table 9.3 lists points to consider in preventing transgene escape, and Table 9.4 offers suggestions as to the relative risks of different traits.

Some introduced genes will be beneficial only in restricted environments. For example, triazine resistance increases plant fitness only where triazine is applied, and the fitness of triazine-resistant weeds such as common groundsel (*Senecio vulgaris* L.) is low in the absence of the herbicide (Holt 1988), so that in unsprayed areas the resistance gene is unimportant to weed evolution. However, in fields where triazine is applied, triazine-resistant weeds could become serious problems. Genes which are expected to cause problems only under specific and definable conditions deserve detailed a priori analysis because for these genes a little forethought will allow trouble-free use of the transgenic plant, while lack of planning may lead to serious weed problems.

Experience with weeds shows that the same species can be a trivial weed in one area and a serious weed in another, based on only slight differences in

TABLE 9.3 Information Needed to Assess Risk of Transgenes Contributing to Weed Problems:

An Affirmative Answer Indicates the Potential for Weed Problems and the Need for Further Analysis.

1. Nature of genes inserted:
   A. Do the inserted genes have the potential to raise the fitness of a weed in any environment?
   B. Will this be a novel trait in recipient weed species or populations, raising the chances for unexpected responses?
2. Ability to exchange genes with weedy populations:
   A. Is the crop outcrossing?
   B. Is the crop produced by open pollination with seed set necessary for production of the product?
   C. Can the crop naturally cross with weedy relatives?
   D. Does the crop frequently cross with weedy relatives?
3. Weediness of wild relatives:
   A. Is the crop itself, or one of its races, a major weed somewhere in the world?
   B. Are the crop's relatives major weeds?
   C. Are weedy wild relatives found in areas where the crop is cultivated?
   D. Are weedy relatives currently problems in the areas where the transgenic crop is to be released?
   E. Are weedy relatives currently problems in areas of similar climate?
   F. Is the improved crop likely to be attractive for growing in areas where its relatives are already a weed problem?
4. Potential for establishment of new weeds:
   A. Could large hybrid populations be produced within a year or two?
   B. Can hybrids flower their first year?
   C. Could new populations grow rapidly by producing a large number of seeds?
   D. Are the weedy relatives easily spread so that transgenic hybrids could easily travel well beyond the site of origin?
   E. Could hybrid plants spread vegetatively despite failure of seeds?
5. Availability of monitoring and mitigation:
   A. Are plants with transgenes difficult to distinguish from nontransgenic plants?
   B. Are populations of the hybrid likely to be difficult or expensive to control in fields of the transgenic crop?
   C. Are populations of the hybrid likely to be difficult or expensive to control in any environment?

environment (e.g., Elton 1958, Baker 1965, 1974, Holm et al. 1977, Orians 1986). Transgenic plants might be similarly contained. The other side of this approach is the warning that species considered innocuous in one area may pose a problem elsewhere, as is the case with yellow starthistle (*Centaurea solstitialis* L.), an insignificant weed within its native range (e.g., in Greece), but a major weed in northern California (Holm et al. 1979, Maddox and Mayfield 1985).

The relationships of plants with other organisms are an important factor in their ability to act as weeds. There is considerable evidence that competitive interactions determine weed success (Harper 1977, Crawley 1989, Louda et al. 1990), and that herbivore pressure can reduce plants from major weeds to minor nuisances—as indicated by successful biological control of prickly pear cactus (*Opuntia* spp.) in Australia, and Klamath weed (*Hypericum*

TABLE 9.4 Characteristics Affecting Risk of Weed Problems from Transgenic Plants

Characteristic of transgenic plant	High risk	Uncertain risk	Low risk
Nature of genes inserted	Herbicide resistance	Pest resistance, pathogen resistance, environmental tolerance (to frost, drought, etc.)	Nutritional quality, ease of harvest, shelf life
Containment of gene	Hybridizes with weedy relatives	Crop hybridizes with weedy relatives but commercial product can be produced from sterile or nonreproducing plants, so only breeders and seed producers must work with plants capable of passing transgenes to weed populations	Incompatible with wild relatives; sterile
Weediness of crop or wild relatives	Crop itself or compatible relatives are major weed here;[a] crop itself or compatible relatives are major weeds in similar climates	Crop or relatives are minor weeds here or in similar climates; crop or relatives are major weeds in areas of quite different climate	Wild relatives not weedy; very low risk if no wild relatives
Distribution of weedy relatives	Native or introduced here	Found in areas with climates similar to here; native or introduced to areas near here	Native or introduced in regions of very different climate from here; found only in distant regions
Breeding system	Sexual and outcrossing with abundant seed set	Usually selfing with good seed set	Usually selfing with poor seed set or nearly sterile; very low risk if truly sterile
Compatibility with weedy relatives	Crosses freely with weedy relatives	Capable of crossing with weedy relatives, but rarely does so	Cannot cross with weedy relatives
Rate of reproduction	Reproduces within months of germination	Reproduces within a year of germination	Requires several years for sexual maturity
Ease of seed production	Very numerous tiny seeds produced by single individual; production of very abundant tiny seeds, but two compatible individuals need to be present	Produces few, large seeds from single individual or pair of compatible individuals	Produces no seed; can be raised commercially without allowing seed production
Ability to spread vegetatively	Aggressive vegetative reproduction	Capable of vegetative reproduction	Perennial without vegetative reproduction; annual
Presence of offensive characteristics	Extremely aggressive and hard to eradicate; spines, thorns, etc.; contact toxic for humans or poisonous to livestock	Low palatability for livestock; bad-smelling or otherwise repugnant to humans	No offensive characteristics
Availability of effective control methods	Herbicide-resistant; tolerant of mowing, plowing, grazing, trampling, etc.; effective control method prohibitively expensive	Some available methods somewhat expensive	Susceptible to inexpensive, easy control method

[a]As used in this table, "here" means "in the region of interest."

perfoliatum L.) and tansy ragwort (*Senecio jacobaea* L.) in northern California (Huffaker and Kennett 1959, Baker 1986, Pemberton and Turner 1990). Thus, releases of engineered plants whose biotic interactions are likely to be changed by the engineered genes should receive particular scrutiny as potential weed problems. Our concern is that poor competitive ability or losses to insects may have kept the plants in check, and that, with the new gene, better competitive ability or increased resistance to insects will permit vastly improved reproduction, allowing the plant to expand as a weed. Fairly simple field experiments should indicate whether changes in the biotic interactions of the new release are likely to present a problem: species significantly limited by herbivores should show dramatic improvement in survival or seed production if herbivores are removed (for example, by spraying with an insecticide), while plants whose numbers are regulated by other factors will not show substantial change in fitness if herbivores are removed. Native plants seem to run the gamut of responses (e.g., Crawley 1989).

Predictions of the potential impact of some introduced genes may be very difficult. For example, in one agricultural situation the presence of weeds with insect resistance genes might have a net beneficial impact by reducing total insect populations supported in the area, while in other situations the advantage of lower numbers of pest insects could be outweighed by the increased success of the resistant weed (but see analysis in Tiedje et al. 1989). Critical analysis of the ecology and evolutionary biology of the plants and genes involved and well-designed small-scale field tests will be necessary for successful anticipation of the results of introductions of genes where enhanced fitness in weedy species depends on complex details of interspecific interactions.

The potential for crossing between crops and weedy relatives. The escape of a transgene can create a weed problem only if weedy species receive that gene. Consequently there are three conditions, all of which must exist for a transgene to produce weed problems: (1) the relatives must occur where the crop is grown, (2) weed and crop must be compatible and must cross, and (3) the relatives must be weedy enough to cause problems.

Weedy relatives must occur where the transgenic crop is grown. Introgression of transgenes into weedy populations is possible only in the presence of compatible weedy relatives. Although a substantial body of data exists (e.g., Holm et al. 1979 and included references), detailed information on the distribution of the relatives of a crop can be surprisingly difficult to find. The need for assembling these data for use in deliberate release of transgenic plants has repeatedly been noted (e.g., Colwell et al. 1987, Center for Science Information, 1988), and we support that effort.

Although few wild relatives of crops are native to North America, in many cases the wild relatives have been introduced along with the crop. Thus it does not give sufficient basis to note that, for example, oats (*Avena sativa* L.) are native to Eurasia for concluding that there are no problems with weedy wild oat relatives in the United States. In the case of oats, the weedy relatives (*Avena barbata* Brot., *A. fatua* L., and others) have also been widely introduced, and are established throughout much of the United States (e.g., Holm et al. 1979). Like crops, weeds tend to be more successful outside their natural range because they too have escaped their insect pests and diseases (Turner 1988).

Consequently, the actual rather than the original distribution of weedy relatives is important. Since gene exchange between crop plants occurs over distances of meters, not kilometers (except for some highly specialized cases), the scale of distributional information needed to prevent escape of transgenes is very local indeed. Again, such local data would be necessary if one wished to accomplish containment of transgenes by spatial isolation. Since regulators tend to err in the direction of caution, efforts to accumulate and maintain current small-scale maps of the distribution of weedy relatives would benefit both those releasing the transgenic plants and those protecting the environment.

References for determining the distribution of weedy relatives include regional floras (noting that some floras exclude nonnative species), published lists of weeds (e.g., Holm et al. 1977, 1979, University of Illinois 1981), and regional checklists. The rapid movement of weed species within and between regions, and the importance of such movement to containment of transgenes, means that accuracy of published lists decreases rather rapidly over time, a problem that will have to be directly addressed in monitoring plans for transgenic plants.

Weed and crop must be compatible and must cross. Serious weeds produced by hybridization with weedy wild relatives will only occur if crop and weed exchange genes. Consequently, many pairs of species pose no threat. First, despite being members of the same genus, many pairs of species are simply not compatible. Second, the compatible relatives may not be weedy. Third, the crop may not outcross freely or may not do so under conditions of cultivation. This last is considered in detail because of its importance here.

Crops which are vegetatively produced and seldom or never make viable pollen or seeds are not going to create hybrid races no matter what genes they carry or where their weedy relatives occur. Such plants include rhubarb (*Rheum* sp.—because it is sterile) and potatoes (*Solanum tuberosum* L.—because they are propagated from tuber cuttings) (e.g., Simmonds 1976).

Other species, such as radishes (*Raphanus sativus* L.) or lettuce (*Lactuca sativa* L.), are raised commercially for vegetative tissues, so that in commercial production they do not flower. As commercial plantings, they should never hybridize. However, seeds are raised by seed companies to generate each year's crop. In addition, some crop plants may set seed if they are missed at harvest or neglected. Thus, while not free of potential to hybridize, the number of hectares of crops such as radishes and lettuce in which pollen is shed and where hybridization may be a problem is much smaller than the total area planted to these crops.

Some crop plants, although open-pollinated to produce the crop, are in fact, largely selfing, so the frequency of gene transfer to wild relatives, should they be present, is low. Tomatoes (*Lycopersicon esculentum* Miller) and peas (*Pisum sativum* L.) fall in this category (Simmonds 1976). Historically, this group has not been considered important in weed production. However, the situation changes radically for transgenic varieties of these crops. Some transgenes in the crop might confer high fitness to weedy relatives. Even a very low rate of gene flow (e.g., one pollinator visit per 1000 flowers) might be sufficient to produce a hybrid swarm of weeds if the hybrid gains resistance to the herbicide applied to the field. In addition, scale affects the risk of hybridization problems in infrequent outcrossers: an outcrossing rate of 0.1% becomes 10,000 events in a field where 10 million flowers are produced over the course of a summer. Thus, transgenic varieties of predominantly selfing species may need careful containment if weedy relatives are present (Manasse and Kareiva 1989).

Similar considerations apply to crops with low fertility such as sweet potatoes (*I. batatas*) and sugarcane (*Saccharum officinarum* L.). These plants are largely infertile due to chromosome abnormalities (Simmonds 1976). This does not, however, preclude rare normal gametes and the possibility of interspecific hybrids with weedy relatives. Again, these are cases in which the escape of even 1 herbicide-resistant gene in 10 million could create serious problems, given the importance of *Ipomoea* species and *Saccharum spontaneum* L. as weeds (e.g., Holm et al. 1979).

Plants raised for open-pollinated seeds form the group most likely to hybridize with wild relatives. Sorghum, oilseed rape, and sunflowers are in this category. Pollen moves freely on insects or in the wind in order to produce a crop. Consequently, if there are weedy relatives around the field, hybrids are likely to form. For these species, transgenes that could potentially enhance the fitness of wild relatives are both important and particularly difficult to control. Realizing the benefits of genetic engineering in open-pollinated plants without creating new weed problems will be a challenge to plant biologists. Some methods for biotic isolation are indicated on page 212 (see under "Containment Methods").

Weedy relatives must be weedy enough to cause problems. Not all weeds pose equal problems. Some are serious threats to agriculture because of aggressive growth or toxicity to livestock. Other weeds are abundant, but easily controlled and thus relatively innocuous. An assessment of the wild relatives of a potential transgenic plant should also consider the seriousness of its relatives as weeds. In cases where the weedy relatives are particularly hard to control, extremely stringent measures are needed to prevent introduced genes from reaching wild populations: an example would be weedy brassicas (e.g., LeBaron 1984, Beversdorf 1987). Where the wild relative is not problematical even when abundant, the risk of a weed problem from hybridization is real, but not as serious as if the wild relative is a proven problem. Cultivated and wild radishes (*R. sativus* and *R. raphanistrum* L.) form an example of compatible relatives that, although abundant and weedy, are not considered noxious in the United States (Holm et al. 1979).

Amount and Type of Data Needed to Make Reliable Risk Estimates

Specifying the amount of data needed to assess the risks of creating weed problems through transgenic plant introduction requires specifying the degree of certainty desired. It will never be possible to amass enough data to release an engineered plant without any chance of a problem. However, widespread use of transgenic plants could occur without adding to current weed problems if a few precautions are heeded (e.g., Foy et al. 1983, Kim 1983, OTA 1988). Below we describe information needed for preventing weed problems.

Monitoring Introduced Plants and Genes

Plants will be introduced as accidental releases or intentional plantings. It is important to know if plants or genes escape—i.e., move beyond the original site—in order to detect problems. For this, monitoring is crucial.

Recognition

In general, the fact that plants are both macroscopic in size and immobile means that workers with some training can easily follow plant movement if it occurs. However, monitoring of transgenic plants will be impossible if observers cannot distinguish the transgenic plants from nonengineered members of the same species. Distinctive characteristics are needed for monitoring to be effective. For example, a color-producing gene that is tightly linked to the economically important transgene would make it possible to observe any spread of transgenic plants. Ideally, identifying markers should be constitutively expressed in preflowering tissues for optimal field recognition.

Monitoring

The next question is: How important is it to find all plants carrying genetically engineered genes, everywhere they may have escaped? As the number of introductions increases, this problem will be magnified. Currently, monitoring of small-scale field trials is facilitated by sizeable bare zones and very careful human monitoring. When transgenic crops reach commercialization, such intensive methods will be much more difficult.

The problems posed by detection of escaped transgenic plants or genes has some similarity to those posed by protection of crops from introduced weedy species. However, adequate monitoring of transgenic plants, even with effective morphological markers, will be more of a problem than stopping introduced alien plants because (1) the transgenic plants are being generated and released within the United States and (2) most transgenic plants will look like useful plants that are generally considered innocuous. (This is not true for weedy hybrids: they may look like weeds.)

Historically, threats from introduced plants began at points of contact with other continents, particularly ports of entry. The federal Seed Act of 1939 and the federal Noxious Weed Act of 1974 are geared to this pattern (e.g., Crooks et al. 1983, Foy et al. 1983, Kim 1983). While legal jurisdiction to control weeds and transgenic plants within the United States has been established (Foy et al. 1983, USDA 1987), due to shortage of personnel, the complexity of interior movement, and lack of public concern, movement within the continental United States remains poorly monitored, allowing accidental transport of plants to different areas. In addition, there are large areas with small human populations, where great changes could occur unnoticed. The spread of yellow starthistle (*C. solstitialis*) and cheatgrass (*Bromus tectorum* L.) over large areas of rangeland within the United States form models of the population growth that can occur and the difficulty of obtaining good information on changes until weed populations are very large (Mack 1981, Maddox and Mayfield 1985). Controlling the internal spread of weeds has been much more difficult than preventing exotic plants entering from other continents.

An additional problem is that people do not see crops as much of a threat. Carryover of crops into subsequent years is widespread, and the escape of a few individuals into marginal habitats is quite common, but few major crop plants are sufficiently weedy that the appearance of escaped individuals prompts control (Fig. 9.2, 9.3). The historical failure of crops such as maize (*Zea mays* L.) and wheat (*Triticum aestivum* L.) to naturalize supports this attitude. People who would be sure to note the novel occurrence of a roadside orchid are unlikely to comment on a soybean. However, to avoid future weed problems, it will be critical to note transgenic crop plants wherever they occur. Even if transgenic crops prove no more weedy than nontransgenic crops, several decades of experience will be needed to establish that. During the interim period, the historical complacency about the occasional escape of crops poses a problem for adequate monitoring.

A serious difficulty with monitoring escapes to prevent the emergence of new weeds is that in virtually every case a number of years will have to elapse before problems develop, and that small populations of even very serious weeds generally appear harmless (Elton 1958, Mack 1985). Occasional plants of noxious weeds such as thistles are seen as attractive roadside flowers or interesting novelties. Careful long-term data collection will be critical (Mack 1985, Tiedje et al. 1989). In many cases, experts may be able, if provided with sufficiently detailed information on the plant populations, to define changes in populations that signal the emergence of a problem and therefore the need for initiation of weed control.

Figure 9.2 Carryover corn in soybeans. Carryover is a common problem, but most crops are easily controlled.

Figure 9.3 Johnsongrass hybrids in field of sorghum. Genes from weeds cause tall Johnsongrass hybrids in sorghum field.

Weeds generally show exponential growth: sparse populations gradually expand until large areas are colonized annually and control is extremely difficult (Elton 1958, Mack 1985, Fig. 9.1). Often, by the time the weed is a recognizable problem, it occupies too much land to be easily controlled. Past weeds have been drawn from an array of introduced species, making monitoring all of them very difficult. With transgenic plants, the number of species that are potential weeds is much smaller, so in that sense monitoring will be easier. The large amount of data available on each transgenic species to be introduced also improves our ability to detect problems earlier.

Since plants can be very abundant without being weed problems if they possess no characteristics which prompt humans to regret their presence, the advent of weed problems with transgenic plants should be heralded by local weed problems (Szybalski 1985). A plant that is unwanted on a small scale will almost certainly be unwanted on a large scale. Spiny plants such as thistles, for example, are obviously going to be problems if they become abundant, so if any landowner reports problems with a spiny species, it suggests that they pose a problem. Consequently one would expect major weed problems to be foreshadowed by local problems with the same species. Functionally, this means that transgenic plants for which no particular weed problem is anticipated can be left uncontrolled, even if they escape, until such time as someone reports problems controlling these plants. Once some report of local problems is noted, however, serious inquiry and application of eradication measures are indicated.

Mitigation and Control Measures

Both federal and state governments recognize "noxious weeds," a category of plants that are sufficiently undesirable that their destruction is a high priority and can be required of property owners. Such lists are under continuous scrutiny, and there are effective mechanisms for addition (or removal) of species (e.g., Kim 1983, Foy et al. 1983). The federal Plant Protection Act provides strong statutory basis for action should weed problems develop. Consequently once a problem is perceived, well-established procedures and responsible agencies are in place for responding appropriately.

It should be noted, however, that our ability to control the action of the public is very limited. For example, allowing plants to flower in the presence of wild relatives, however clearly forbidden on the seed package, is likely to occur, just as vegetative propagation of protected genotypes or illegal importation of useful seeds happens commonly on a noncommercial scale. These practices make problems that are possible into virtual certainties.

The escape of crop plants and their naturalization as weeds have a very long history (deWet and Harlan 1975), while the introduction of useful single genes into weed populations, if it occurs, will be a new problem, a product of genetic engineering. We discuss the former first.

Escaped transgenic plants

Granted that transgenic plants are effectively monitored, the next question is what to do about escaped plants. Should all escaped transgenic crops and plants carrying transgenes be destroyed at once?

The decision as to whether to eradicate escaped plants depends upon the potential for problems. This is a function of (1) the genes carried, (2) the weediness of the plant species carrying those genes, and (3) the environment. If any of these are high-risk—for example, if there is a gene conveying an advantage in the particular environment, a crop with numerous weedy races, or a fragile environment—even a few escaped transgenic plants should be eliminated, without waiting for a problem to occur. Plans should be drawn up before environmental release designating the level of control to be applied to escapees of each release, by habitat. To prevent weed problems, transgenic plants found in habitats not mentioned in original release documents should be destroyed.

However, if neither the plant, the transgenes, nor the environment suggest a high risk, it should suffice to simply note the location of the escaped plants and watch carefully, eliminating the plants only if problems develop.

Emergence of a transgenic plant release as a weed problem in one area does not necessarily mean that the transgenic plant will be a problem everywhere. The success of even the worst weed depends on environmental conditions (e.g., Baker 1986). Thus, control measures initiated because of a local problem with a transgenic plant need only apply to areas sharing the environmental conditions of the problem area. Given the nature of plant breeding that will culminate in deliberate release, the environmental tolerances of the weed are likely to be well known (Brill 1985, Hauptli et al. 1985). However, one report that a plant has emerged as a weed should raise the level of care with which that species is monitored wherever it occurs.

Some crop plants have weedy tendencies or seriously weedy close relatives. Experience suggests that these are more likely to establish naturalized populations or to further expand and colonize as widespread weeds. Escape should condemn members of these groups (for example, sorghum or sunflowers) to immediate eradication. For members of other groups with less of a record as weeds (for example tomatoes and wheat), careful observation should suffice. Classifying transgenic plants into these categories can be done fairly readily (e.g., Milewski and Tolin 1984, OTA 1988, Tiedje et al. 1989). There are also good lists of species that are weed problems elsewhere in the world (e.g., Reed 1977, Holm et al. 1977, 1979, Zimdahl 1983). These should be considered seriously in determining how to respond to the escape of transgenic plants.

Escaped transgenes

For many crops, genes are not released via cross-pollination or there are no relatives with which to cross, so these considerations are unimportant. If the transgene escapes into populations of troublesome weeds, the populations should be eradicated at once. If the gene escapes into plant species that have not previously been a problem, then careful monitoring is called for but eradication can wait until a problem is detected (see above).

Control measures

Eradication of a plant means that no parts capable of reproduction are left at the site. While herbicides are quite effective in killing plants, attention needs to be paid to the potential for seed escape or recovery by perennial rootstocks. Given the information required for each introduction at present, we believe that this can be done very effectively. However, plans should be drawn up in advance: the herbicide (or other method) to be used should be specified, as well as whether the escaped plant is to be eradicated when vegetative, when flowering, by destruction of seeds, etc.

Over a period of time, rare events are likely to occur. Even if seeds are rarely alive when carried by birds between continents, the rare survivor can start the introduced population. This sort of event defies probabilistic predictions but is a crucial part of evolutionary change (Regal 1986, Colwell 1988). Pragmatic regulators will generally have to ignore highly improbable events. But all the participants in this process must recognize that, given sufficient time and area of use, rare events will occur.

A practical response to the probability that the rare event will eventually occur is that if a problem of the scale of water hyacinth [*Eichhornia crassipes* (Mart.) Solms] or yellow starthistle is predicted as a worst case scenario by experienced weed scientists, then the introduction should probably be prevented altogether. A less cautious approach is to develop cost-benefit models: how much economic benefit will the release provide; how much economic cost would an escape entail? Everyone should remember that eradication of the Mediterranean fruit fly from California in 1982 was very expensive ($150 million, Dahlsten 1986), and that the cost of cleanup of pesticide contamination of U.S. waters is far higher. Given the high annual weed-control costs, the expense of controlling a transgenic weed could be substantial. Preventing problems will undoubtedly be more cost-effective.

Pollution of native gene pools is another problem that those contemplating release must consider. If the wild relatives are native, then introgression of any genes is undesirable because incorporating new alleles into the population will permanently change the future evolution of the species in ways we lack the experience to predict. Given the rapid loss of biological diversity in our present world (e.g., Wilson 1988), for transgenes to contribute to loss of natural variation is unwarranted. Methods to prevent this should be in place before release. Eradication of hybrid seedlings will generally be warranted. Since we find it is difficult to conceive of a practical and effective monitoring protocol for this, prevention is essential.

Containment Methods

Field testing and academic research

At the scale of experimental field tests, control of the escape of transgenic plants and transgenes can be virtually absolute. A bare area free of compatible relatives, of a size determined by the breeding system of the genetically engineered plants, combined with extensive human monitoring, can reduce the chance of escape (and therefore of future problems), to virtually zero. As the scale of use of transgenic plants increases, however, the ability to be certain that no plants or genes escape decreases. Few weed problems are likely for small-scale testing or academic research, provided containment methods are properly applied.

Commercialization

The movement of transgenic plants is far more likely to lead to new weed problems, if ever it does, when transgenic plants are routinely used in agriculture, have been planted in large numbers as ornamentals, or are established as populations in lawns or golf courses. At that time, it will be possible for propagules to move out of their intended location into areas where ecological differences, or differences in human perceptions, make them weeds. The vast majority of current weed problems are the result of the dispersal of weedy plants themselves, rather than the movement of genes useful to weeds via introgression into weedy populations. Generally intentional or unintentional transport by humans is the major cause of long-distance plant introductions (e.g., Foy et al. 1983, Turner 1988). This form of dispersal will be unimportant during the stages of field testing and introduction of genetically engineered crops into the marketplace. That should not, however, obscure the immense potential for problems from human dispersal of transgenic plants in the next century.

When transgenic plants reach commercialization and are adopted by large numbers of agriculturalists, the combination of (1) rare events becoming probable when numbers are large and (2) some users' disregard for safety procedures suggest that, where problems are possible, some will occur. Good regulatory procedures, carefully applied, cannot totally prevent problems. However, they can drastically reduce the number and magnitude of the problems.

Physical containment. The scale of the release significantly affects the feasibility of methods to mitigate. Killing all plants in a radius of 100 m is feasible for problems arising from small-scale field tests but may be impractical after commercialization. Thus, it is essential to develop strategies for control of weed problems should they develop 5 to 10 years after commercial release of the transgenic crop. To prevent problems, practical methods for controlling or eradicating an introduced plant or its hybrids, even after it establishes countywide populations, need to be in place before the plants are released. Functionally, this means determining in detail susceptibilities to herbicides *before* commercialization and describing methods effective in controlling the plant in an array of habitats, from fields of closely related crops to parks and natural areas.

For crops in which seeds are not the commercial product (e.g., lettuce and beets, *Beta vulgaris* L.), accidental seed production or escape from breeders' fields are the chief risks. Seed producers already operate under relatively stringent conditions, requiring, for example, substantial distances between the seed field and other compatible plants in order to produce certified seed. For these species the risk of escape of transgenes by hybridization is both low and subject to stringent regulation. Current procedures may require review, however (e.g., Manasse and Kareiva 1989).

Biological containment. There will remain problems with control of naturally outcrossing crop plants. For some species, there is the potential for genetic modification to prevent gene exchange with wild relatives. Methods of genetic isolation include male sterility, chromosomal incompatibility, polyploidy, and aneuploidy. For example, male-sterile plants receive, but do not make, pollen. Consequently, they are much less likely to contribute to introgression of transgenes into wild populations. If fruit-producing, transgenic, male-sterile plants receive nontransgenic pollen, transgenic seed can be produced without the risk of movement of the transgenes via pollen release. Many species have known male-sterile mutants that could be incorporated into commercial varieties (Kaul 1988). Seed escape would have to be prevented, however, as would the occurrence of the occasional mutant male-sterile plant which produces healthy pollen.

Chromosomal translocations, polyploidy, and aneuploidy are all genetic rearrangements that reduce gene flow and fertility (Stebbins 1971). Combined with genes that are lethal when homozygous, chromosomal translocations can severely restrict or eliminate genetic recombination. The result is that genes cannot escape from the parental genome (Stebbins 1971). Polyploids are generally incompatible with their nonpolyploid parents (e.g., Stebbins 1971). If crops are polyploid, they may be able to receive genes from wild relatives but not contribute genes to diploid wild populations. Since agriculture finds polyploids much less tractable than diploids or haploids, this method of biological containment is not particularly attractive. Furthermore, despite theoretical incompatibility, there are economically important plant groups in which species of differing ploidy cross freely—for example, cultivated oats (*A. sativa*) and wild oats (*A. fatua*) cross freely, despite the fact that the former is diploid and the latter is hexaploid (Thomas and Jones 1976).

Aneuploidy, the presence of one or more unpaired chromosomes in the genome, can reduce interfertility without necessarily reducing vigor. Individuals with extra or missing chromosomes may grow vigorously but be sterile or nearly so with plants not sharing their chromosome complements. Polyploidy, aneuploidy, and other methods of biological containment should be seriously considered for transgenic crops if weedy relatives present a threat of transgene escape.

Other containment methods. If the transgenic plant and its genes can be effectively and economically controlled should they escape, this will alleviate much concern. Biological containment (above) is one method. So is susceptibility to some control agent—whether it is a chemical, a mechanical barrier, or a biological control organism—that could readily be applied without endangering nontarget plants or ecosystems. If the plant is susceptible to a host-specialist insect or to frost damage, then there should be less concern. Sensitivity to growth hormones, narrow-spectrum herbicides, and pathogens might also permit practical control mechanisms.

Summary

Weed problems from transgenic plants do not constitute novel risks, but require attention nevertheless. At present, they are most likely from the movement of useful transgenes into weed species. Since this requires compatibility between crop and weed, it can be prevented by (1) planting where no compatible weedy species exist or (2) preventing gene exchange. When transgenic plants are in widespread use, transgenic plants themselves may move into new areas in which they become weeds.

Prevention of potential problems can make use of existing information about which groups of plants are weedy and about the traits that favor successful weeds. Careful planning, long-term monitoring (for decades), and well-enforced eradication procedures will be necessary if we really want to prevent transgenes from contributing to weed problems.

Acknowledgments

We thank L. W. J. Anderson, H. G. Baker, J. Beyea, C. T. Bryson, F. L. Gould, S. C. Hake, and M. E. Hogan for helpful comments on the manuscript. We thank K. L. Chan for help with manuscript preparation.

Kathleen Keeler gratefully acknowledges a research leave from the School of Biological Science (University of Nebraska–Lincoln) at the Western Regional Research Center of the USDA Agricultural Research Service. The support of both institutions is acknowledged, also.

References

Anderson, E. 1952. *Plants, Man and Life*. University of California Press, Los Angeles.
Anderson, L. W. J. 1987. Exotic Pest Profile no. 11. Hydrilla (*Hydrilla verticillata*). California Department of Food and Agriculture, Sacramento, California.
Baker, H. G. 1965. Characteristics and modes of origin of weeds. In H. G. Baker and G. L. Stebbins (eds.), *The Genetics of Colonizing Species*. Academic, New York, pp. 147–168.

Baker, H. G. 1974. The evolution of weeds. *Annu. Rev. Ecol. Syst.* 5:1–24.
Baker, H. G. 1986. Patterns of plant invasion in North America. In H. A. Mooney and J. A. Drake (eds.), *Ecology of Biological Invasions of North America and Hawaii.* Springer-Verlag, New York, pp. 44–57.
Baker, H. G., and Cox, P. A. 1984. Further thoughts on islands and dioecism. *Ann. Missouri Bot. Gard.* 71:230–239.
Beversdorf, W. D. 1987. Classical approaches to the development of herbicide tolerance in crop cultivars. In H. M. LeBaron, R. O. Mumma, R. C. Honeycutt, and J. H. Duesing (eds.), *Biotechnology in Agricultural Chemistry*. American Chemical Society, Washington, D.C., pp. 108–114.
Beyea, J., and Keeler, K. H. Environmental implications of genetic engineering applied to biomass energy and chemical production. *CRC Crit. Rev. Biotechnol.* (in press).
Bowring, J. D. C., Evans, A. W., and Sneddon, J. L. 1980. Objectives and methods in seed production. In P. D. Hebblethwaite (ed.), *Seed Production*. Butterworth, London, pp. 1–14.
Brill, W. J. 1985. Safety concerns and genetic engineering in agriculture. *Science* 227: 381–384.
Buchholtz, K. P. 1967. Report of the terminology committee. *Weed* 15:388–389.
Carlquist, S. 1974. *Island Biogeography*. Columbia Univ. Press, New York.
Carson, R. L. 1962. *Silent Spring*. Houghton Mifflin, New York.
Cavers, P. A., and Bough, M. A. 1985. Proso millet (*Panicum miliaceum* L.): a crop and a weed. In J. White (ed.), *Studies on Plant Demography*. Academic, New York, pp. 143–156.
Center for Science Information. 1988. *Regulatory Considerations: Genetically Engineered Plants*. Summary of a workshop at the Boyce Thompson Institute, Cornell University, Ithaca, New York. Center for Science Information, San Francisco.
Colwell, R. K. 1988. Ecology and biotechnology: expectations and outliers. In J. Fiksel and V. T. Covello (eds.), *Safety Assurance for Environmental Introduction of Genetically-Engineered Organisms*. Springer-Verlag, Berlin, pp. 163–180.
Colwell, R. K., Norse, E. A., Pimentel, D., Sharples, F. E., and Simberloff, D. 1985. Genetic engineering in agriculture. *Science* 229:111–112.
Cook, R. E. 1983. Clonal plant populations. *Am. Scientist* 71:244–253.
Coursey, D. G. 1978. Yams. In N. W. Simmonds, *Evolution of Crop Plants*. Longman, London, pp. 70–74.
Crawley, M. J. 1989. Insect herbivores and plant population dynamics. *Annu. Rev. Entomol.* 34:531–564.
Cronquist, A. 1971. *A Synthetic Classification of Vascular Plants*. Columbia Univ. Press, New York.
Crooks, E., Havel, K., Shannon, M., Snyder, G., and Wallenmaier, T. 1983. Stopping pest introductions. In C. L. Wilson and C. L. Graham (eds.), *Exotic Plant Pests and North American Agriculture*. Academic, New York, pp. 240–261.
Dahlsten, D. L. 1986. Control of invaders. In H. A. Mooney and J. A. Drake (eds.), *Ecology of Biological Invasions of North America and Hawaii*. Springer-Verlag, New York, pp. 275–301.
Daubenmire, R. F. 1974. *Plants and Environment*, 3d ed. Wiley, New York.
deWet, J. M. J., and Harlan, J. R. 1975. Weeds and domesticates: evolution in the man-made habitat. *Econ. Bot.* 29:99–107.
Dover, M. J., and Croft, B. A. 1986. Pesticide resistance and public policy. *Bioscience* 36:78–85.
Drake, J. A., Kenny, D. A., and Voskuil, T. 1988. Environmental biotechnology. *Bioscience* 38:420– 422.
Elton, C. S. 1958. *The Ecology of Invasions by Animals and Plants*. Chapman and Hall, London.
Forcella, F. 1985. Spread of *Kochia* in the northwestern United States. *Weeds Today* 16:4–6.
Foy, C. D., Forney, D. R., and Cooley, W. E. 1983. History of weed introductions. In C. L. Wilson and C. L. Graham (eds.), *Exotic Plant Pests and North American Agriculture*. Academic, New York, pp. 65–92.
Futuyuma, D. J. 1986. *Evolutionary Biology*, 2d ed. Sinauer, Sunderland, Massachusetts.

Gasser, C. S., and Fraley, R. T. 1989. Genetically engineered plants for crop improvement. *Science* 244:1293–1299.
Gauthier, N. L., Hofmaster, R. N., and Semel, M. 1981. History of Colorado potato beetle control. In J. H. Lashomb and R. Casagrande (eds.), *Advances in Potato Pest Management*. Hutchinson Ross, Stroudsberg, Pennsylvania, pp. 13–31.
Gotsch, N., and Rieder, P. 1989. Future importance of biotechnology in arable farming. *Trends Biotechnol.* 7:29–34.
Graham, F. 1970. *Since* Silent Spring. Houghton Mifflin, Boston.
Harper, J. L. 1977. *Population Biology of Plants*. Academic, London.
Hauptli, H., Newell, N., and Goodman, R. M. 1985. Genetically engineered plants: environmental issues. *Bio/technology* 3:437–442.
Hebblethwaite, P. D. (ed.). 1980. *Seed Production*. Butterworth, London.
Heiser, C. B., Jr., Smith, D. M., Clevenger, S., and Martin, W. C. 1969. The North American sunflowers (*Helianthus*). *Mem. Torrey Bot. Club* 22:1–218.
Holm, L., Pancho, J. V., Herberger, J. P., and Plunknett, D. L. 1979. *A Geographical Atlas of World Weeds*. Wiley, New York.
Holm, L. G., Plunknett, D. L., Pancho, J. V., and Herberger, J. P. 1977. *The World's Worst Weeds*. University Press, Honolulu.
Holt, J. S. 1988*a*. Reduced growth, competitiveness, and photosynthetic efficiency of triazine-resistant *Senecio vulgaris* from California. *J. Appl. Ecol.* 25:307–318.
Holt, J. S. 1988*b*. Ecological and physical characteristics of weeds. In M. A. Altieri and M. Liebman, *Weed Management in Agroecosystems: Ecological Approaches*. CRC Press, Boca Raton, Florida, pp. 7–23.
Holzner, W. 1982. The nature of weeds. In W. Holzner and N. Numata (eds.), *Biology and Ecology of Weeds*. W. Junk, The Hague, pp. 3–15.
Huffaker, C. B., and Kennett, C. E. 1959. A ten-year study of vegetational changes associated with biological control of Klamath weed. *J. Range Mgmnt.* 12:69–82.
Jensen, K. I. N., and Bandeen, J. D. 1979. Triazine resistance in annual weeds. In E. Häflinger (ed.), *Maize*. Ceiba Geigy Tech. Monogr., Ceiba Geigy, New York, pp. 55–57.
Julien, M. H. 1987. *Biological Control of Weeds: A World Catalogue of Agents and Their Target Weeds*, 2d ed. CAB International, Wallingford, U.K.
Kaul, M. L. H. 1988. *Male Sterility in Higher Plants*. Springer-Verlag, Berlin.
Keeler, K. H. 1985. Implications of weed genetics and ecology for the deliberate release of genetically-engineered crop plants. *Recomb. DNA Tech. Bull.* 8:165–172.
Keeler, K. H. 1987. Survivorship and fecundity of the polycarpic perennial *Mentzelia nuda* (Loasaceae) in Nebraska sandhills prairie. *Am. J. Bot* 12:785–791.
Keeler, K. H. 1988. Can we guarantee the safety of genetically-engineered plants in the environment? *CRC Crit. Rev. Biotechnol.* 8:85–97.
Keeler, K. H. 1989. Can genetically engineered crops become weeds? *Bio/Technology* 7: 1134–1139.
Keim, C. R., and Venkatansubramanian, K. 1989. Economics of current biotechnological methods of producing ethanol. *Trends Biotechnol.* 7:22–29.
Kim, K. C. 1983. How to detect and combat exotic pests. In C. L. Wilson and C. L. Graham (eds.), *Exotic Plant Pests and North American Agriculture*. Academic, New York, pp. 262–320.
LeBaron, H. M. 1983. Herbicide resistance in plants—an overview. *Weeds Today* 14(2): 4–6.
LeBaron, H. M. 1984. Herbicide resistance in plants—future research needs. *Weeds Today* 15(4):2–6.
Louda, S. M. 1982. Limitation of the recruitment of the shrub *Happlopappus squarosus* (Asteraceae) by flower and seed-feeding insects. *J. Ecol.* 70:43–53.
Louda, S. M. 1983. Seed predation and seedling mortality in the recruitment of a shrub *Happlopappus venetus* (Asteraceae) along a climatic gradient. *Ecology* 64:511–521.
Louda, S. M., Potvin, M. A., and Collinge, S. K. 1990. Predispersal seed predation, postdispersal seed predation and competition in the recruitment of seedlings of a native thistle in sandhills prairie. *American Midland Naturalist* 124:104–113.
Mack, R. N. 1981. Invasion of *Bromus tectorum* L. into western North America: an ecological chronicle. *Agro-Ecosystems* 7:145–165.

Mack, R. N. 1985. Invading plants: their potential contribution to population biology. In J. White (ed.), *Studies in Plant Demography*. Academic, New York, pp. 127–142.
Mack, R. N. 1989. Environmental issues. In *Transgenic Plant Conf. Proc.*, Keystone Center, Keystone Colorado, pp. 127–136.
Maddox, D. M., and Mayfield, A. 1985. Yellow starthistle infestations are on the increase. *Calif. Agric.* 39:10–12.
Manasse, R., and Kareiva, P. 1989. Quantitative approaches to questions about the spread of recombinant genes or recombinant organisms. In L. Ginzburg (ed.), *Risk Assessment in Biotechnology*. Wiley, New York.
Mellon, M. 1988. *Biotechnology and the Environment*. National Biotechnology Policy Center, National Wildlife Federation, Washington, D.C.
Milewski, D., and Tolin, S. A. 1984. Development of guidelines for field testing plants modified by recombinant DNA techniques. *Recomb. DNA Tech. Bull.* 7:114–124.
Millar, C. I., and Libby, W. J. 1989. Restoration: Disneyland or native ecosystems? The genetic question. *Restor. Mgmnt. Notes*. In press.
National Academy of Sciences (NAS). 1987. *Introduction of Recombinant DNA-Engineered Organisms into the Environment: Key Issues*. National Academy Press, Washington, D.C.
Office of Technology Assessment (OTA) of the U.S. Congress. 1988. *New Developments in Biotechnology: Field Testing Engineered Organisms: Genetic and Environmental Issues*. OTS-BA-350, U.S. Government Printing Office, Washington, D.C.
Oka, H-I., and Morishima, H. 1982. Ecological genetics and the evolution of weed species. In W. Holzner and N. Numata (eds.), *Biology and Ecology of Weeds*. W. Junk, The Hague, pp. 73–90.
Orians, G. H. 1986. Site characteristics favoring invasion. In H. A. Mooney and J. A. Drake (eds.), *Ecology of Biological Invasions of North America and Hawaii*. Springer-Verlag, New York, pp. 133–148.
Paul, N. D, and Ayres, P. G. 1986. The impact of a pathogen (*Puccinia lagenophorae*) on populations of groundsel (Senecio vulgaris) overwintering in the field. *J. Ecol.* 74: 1085–1094.
Pemberton, R. W. 1986. The distribution of halogeton in North America. *J. Range Mgmnt.* 39:281–282.
Pemberton, R. W., and Turner, C. E. 1990. Biological control of *Senecio jacobaea* L. in northern California, an enduring success. *Entomophaga* 35:71–77.
van der Pijl, L. 1982. *Principles of Dispersal in Higher Plants*, 3d ed. Springer-Verlag, New York.
Ratner, M. 1989. Crop biotech '89: research efforts are market driven. *Bio/Technology* 7:337–341.
Reed, C. F. 1977. *Economically Important Foreign Plants*. Agricultural Hndbk. no. 498. U.S. Department of Agriculture, U.S. Government Printing Office, Washington, D.C.
Regal, P. J. 1986. Models of genetically engineered organisms and their ecological impact. In H. A. Mooney and J. A. Drake (eds.), *Ecology of Biological Invasions of North America and Hawaii*. Springer-Verlag, New York, pp. 111–129.
Ritchie, J. C., and MacDonald, G. M. 1986. The patterns of post-glacial spread of white spruce. *J. Biogeogr.* 13:527–540.
Salisbury, E. 1961. *Weeds and Aliens*. Collins, London.
Schery, R. W. 1972. *Plants for Man*. Prentice-Hall, New York.
Simmonds, W. H. 1976. *Evolution of Crop Plants*. Longman, London.
Stebbins, G. L. 1971. *Chromosomal Evolution in Higher Plants*. Addison-Wesley, Reading, Massachusetts.
Szybalski, W. 1985. Genetic engineering in agriculture. *Science* 229:112–115.
Tangley, L. 1985. Managing pesticide resistance. *Bioscience* 35:216–218.
Thomas, H., and Jones, I. J. 1976. Origins and identification of weed species of *Avena*. In D. P. Jones (ed.), *Wild Oats in World Agriculture*. Agricultural Research Council, London, pp. 1–19.
Thompson, D. G., Stuckey, R. L., and Thompson, E. B. 1987. Spread, Impact and Control of purple loosestrife (*Lythrum salicaria*) in North American Wetlands. U.S. Fish and Wildlife Service, Region 2.
Tiedje, J. M., Colwell, R. K., Grossman, Y. L., Hodson, R. E., Lenski, R. E., Mack, R. N.,

and Regal, P. J. 1989. The planned introduction of genetically engineered organisms: ecological considerations and recommendations. *Ecology* 70:297–315.

Turner, C. E. 1985. Conflicting interests and biological control of weeds. In E. S. Delfosse (ed.), *Proc. VI Int. Symp. Biol. Contr. Weeds, 19–25 Aug. 1984, Vancouver, Canada*, Agriculture Canada, Ottawa, pp. 203–225.

Turner, C. E. 1988. Ecology of invasions by weeds. In M. A. Altieri and M. Liebman (eds.), *Weed Management in Agroecosystems: Ecological Approaches*. CRC Press, Boca Raton, Florida, pp. 41–55.

University of Illinois at Urbana-Champaign. 1981. Weeds of the north central states. Agric. Exp. Stn. Bull. no. 772.

U.S. Department of Agriculture (USDA)/Animal and Plant Health Inspection Service. 1987. Introduction of organisms and products altered or produced through genetic engineering which are plant pests or which there is reason to believe are plant pests. *Fed. Reg.* 52:22892–22915, June 16.

Vitousek, P. M. 1986. Biological invasions and ecosystem properties: can species make a difference? In H. A. Mooney and J. A. Drake (eds.), *Ecology of Biological Invasions of North America and Hawaii*. Springer-Verlag, New York, pp. 163–177.

Whelan, E. D. P. 1978. Cytology and interspecific hybridization. In J. F. Carter (ed.), *Sunflower Science and Technology*. American Society for Agronomy, Madison, Wisconsin, pp. 339–369.

Wilson, E. O. (ed.). 1988. *Biological Diversity*. National Academy Press, Washington, D.C.

Yarwood, C. E. 1983. History of plant pathogen introductions. In C. L. Wilson and C. L. Graham (eds.), *Exotic Plant Pests and North American Agriculture*. Academic, New York, pp. 40–64.

Zimdahl, R. L. 1983. Where are the principal exotic weed pests? In C. L. Wilson and C. L. Graham (eds.), *Exotic Plant Pests and North American Agriculture*. Academic, New York, pp. 183–217.

Chapter

10

Statistical Techniques for Field Testing of Genetically Engineered Microorganisms

Marla S. McIntosh

Department of Agronomy
University of Maryland
0111 H. J. Patterson Hall
College Park, Maryland 20742

Introduction

Statistics is a tool that can be used by scientists to aid in accurately describing populations and in testing hypotheses. In order to be able to use statistics correctly, appropriate sampling and experimental designs must be utilized. This can be done effectively by understanding the theories underlying experimental and sampling design, and then applying this understanding to well-conceived objectives. Without the proper foundation, "statistics abuse" can easily occur. The result of improper statistical design and analysis is often inaccurate or imprecise answers to questions, or answers to questions that never needed to be asked.

The following chapter is not meant as a "cookbook" on how to plan, sample, and analyze field research of genetically engineered organisms. Instead, it is intended to help scientists work with statisticians to effectively design experiments to test the right questions and to understand the pitfalls and objectives of field research. Much of what follows describes standard statistical methodology that can be found in more detail in a number of statistical references on agricultural and biological experimentation (Steel and Torrie 1980, Little and Hills 1978, Gomez and Gomez 1984, Green 1979, Cochran 1963).

The chapter is divided into two sections. The first section explains how to obtain useful and appropriate estimates of statistics used to describe populations. The second section discusses statistics used to test hypotheses about population or treatment means. Statistics used for hypothesis testing must satisfy more stringent requirements than simple descriptive statistics. This second section of the chapter also describes the design of field experiments to test relevant hypotheses.

Descriptive Statistics

Defining populations

After the release of an engineered microorganism into the environment, it is necessary to monitor the fate of the microorganism as well as the effects of the release on competing and target organisms. To describe the effects of the release, measurements are taken in the field on random, representative samples of the populations of interest. Field measurements differ from laboratory measurements in that they are more variable and less repeatable. In laboratories, the environment can be controlled, but each time the organism is released in a field, the environmental conditions will be different. Therefore, for field studies, a population should be defined in terms of the environment as well as the microorganism. For example, to describe the field survival rate of *Pseudomonas* sp. after release, because of variable field conditions, the *Pseudomonas* sp. of interest should be released and measured in different seasons, years, and locations. However, for initial testing, the population could be more narrowly defined, and *Pseudomonas* sp. survival described for a particular year and site. Additional field tests would then be needed to determine whether the survival of the *Pseudomonas* sp. was sensitive to changes in environment.

Useful Statistics

For a preliminary study to characterize the effects of released microorganisms, statistics can be used to describe such parameters as the number, survival, and mortality of the microorganism, as well as the response of the target organism. In many cases, description of the genetically engineered population is the primary objective of the study, and in the initial investigation it is not necessary to test whether populations are different from each other at a given significance level. Often, it is more relevant and experimentally simpler to estimate the means of the populations and how confident you are in these estimates.

The statistics of interest usually include a measure of central tendency (mean, median, or mode) and a measure of dispersion or variability (standard deviation, standard error, or range) of the population. The statistics that best describe the population will depend on the population's distribution.

Normally distributed populations

For a normally distributed population, the mean, the median, and the mode are identical, and the population frequency distribution follows a bell-shaped curve. In this case, the most commonly used measure of central tendency would be the mean.

A measure of variability should be presented with the mean. Two statistics that measure the variability of a population that are often used and confused with each other are the "standard deviation" and the "standard error." To add to the confusion, the "standard error" is also called the "standard deviation of the mean." The two statistics are compared in Table 10.1.

The "standard deviation" is the square root of the error variance and is a measure of the variability of a population of *individual observations*. The "standard error" is the square root of the variance of a treatment mean and is equal to the standard deviation divided by the square root of n (the number of samples used to estimate the mean). The standard error is the measure of variability of a population of *means based on* n *samples*. Sixty-six percent of the population of sample means falls within one standard error of the mean and 99% within two standard errors of the mean. The standard deviation estimates the variability of individuals, while the standard error estimates the variability of means. Therefore, means should be presented with their standard error to indicate the variability associated with the estimate of this mean. The standard deviation should be used to describe the variability associated with the individuals in the population.

The standard error indicates the precision of the estimate of the mean. The precision used in estimating the mean is measured as the reciprocal of the variance of a treatment mean. As the sample size used to estimate the mean increases or as the variance of the populations decreases, precision increases.

The importance of the standard error is often overlooked. Many researchers tend to put primary importance on the mean, and to look secondarily, if at all, at the standard error. However, the standard error is important because it is a measure of the precision of the methods of detection, the inherent variability of the population, and the adequacy of the sampling procedure and design.

Another useful statistic that is often presented with a mean is a confidence interval. A confidence interval at the 95% probability, for example, sets the limits that have a 95% probability of containing the true population mean. Therefore, if the confidence intervals for means of two treatments do not overlap, the means are considered to be different with a 5% probability of an error.

TABLE 10.1 Comparison of Standard Error and Standard Deviation

Character	Standard deviation	Standard error
Population	Individuals	Means
Population symbol	σ_y or σ	$\sigma_{\bar{y}}$
Sample symbol	S_y or S	$S_{\bar{Y}}$
Abbreviation	SD	SE
Calculation (for n samples)	$\sqrt{\Sigma(Y - \bar{Y})/(N - 1)}$	$\sqrt{\Sigma[(Y - \bar{Y})/(N - 1)]/n}$
Calculation for ANOVA (analysis of variance for r replications)	$\sqrt{\text{Error mean square}}$	$\sqrt{\text{Error mean square}/r}$

A 95% confidence interval for a mean is calculated as the mean $\pm t_{.05}$ (standard error), where $t_{.05}$ is the student t value with $n-1$ degrees of freedom.

Log-normally distributed populations

Many populations of microorganisms cover a wide range of values (varying by an order of magnitude or more) and are log-normally rather than normally distributed. The log-normal distribution is skewed, and the mean is larger than the median or the mode. For a population that is not highly skewed, the sample mean and variance can adequately estimate the population mean and variance (Parkin et al. 1988). However, for highly skewed distributions, the mode and the range are probably more useful than the mean and variance in describing the population. Alternatively, the data can be log-transformed, and the transformed data will be normally distributed. Then the population parameters can be estimated by the sample mean and variance of the transformed data.

Sample number

The number of samples used to estimate the mean depends on both practical and statistical considerations. A general statistical guideline for sampling is: The more samples, the better the estimate. However, other considerations such as the cost of collecting and assaying each sample are also important. For a given sample size, a decision needs to be made about how to allocate samples. The choice between collecting few large samples or many small samples depends on the distribution of the organism in the field. If the organism is distributed randomly throughout the field, either alternative will give similar results. However, if the organism is distributed in aggregates rather than randomly throughout the field, then it is better to measure a larger number of small samples (Green 1979). Samples can be composited for analysis without affecting the estimate of the mean. However, to estimate the variability of the population, it is essential to have data for replicate samples. The estimate of the standard error can be based on measurements taken individually or composited and will usually be smaller if based on composite samples.

Because of resource limitations, it is desirable to determine the minimum number of samples necessary to obtain an adequate estimate of the population mean. The formula given as Eq. 1 can be used to estimate the minimum sample number n (Stein 1945).

$$n = \frac{(Z^2)S^2}{d^2} \qquad (1)$$

where n = number of samples
Z = Z with probability ($Z_{.05}$ = 1.96)
S^2 = error variance of samples
d^2 = margin of error for the plot

For example, in order to estimate the mean abundance of the test organism to within 5×10^3 colony-forming units (d) at the .05 level of confidence

(Z = 1.96), the sampling error variance is estimated as 25×10^3, based on samples collected from the plot. The minimum sample number is estimated to be $1.96^2 (25 \times 10^3)/(5 \times 10^3)^2 = 3.8$, rounded to 4. Although this formula is one of the most commonly used to estimate sample size, it underestimates n (Kupper and Hafner 1989). To correct for the underestimate, increase the number of samples by 4 if α = .01 or by 2 if α = .05 or .10. (For an explanation of α, see the section titled "Probabilities of Incorrect Conclusions" below.) Another alternative is to use tables provided by Beal (1989).

Testing Hypotheses

Many field studies of microorganisms released into the environment will be planned to test specific hypotheses rather than just describe populations. The first step in hypothesis testing is to define the hypotheses to be tested and then determine how to test them. As with descriptive statistics, the populations to be compared must be carefully defined to ensure that the results will satisfy the objectives of the experiment. Examples of hypotheses that might be tested are:

1. The engineered organism is not present.
2. The abundance of engineered and nonengineered organisms is the same.
3. The number of engineered organisms does not decrease over time.
4. The number of engineered organisms decreases in time at the same rate as the number of nonengineered organisms.
5. The engineered organism is not moving.
6. The engineered organism is moving at the same rate as the nonengineered organism.
7. The release of the engineered organism does not affect the competing species.
8. The target species is not affected by the release of the engineered organism.

The first six hypotheses concern monitoring the engineered organism, whereas the last two hypotheses are related to efficacy and ecological effects. All of these hypotheses can be tested in one planned-release experiment using different sampling designs that are appropriate for the parameter being measured. Outlined in the next section are considerations necessary for testing these hypotheses using analysis-of-variance techniques.

Analysis of Variance

"Analysis of variance" is an arithmetic technique used to test hypotheses. Analysis of variance quantifies sources of variation such as treatment, time, or distance effects to determine whether these means are different due to these sources or solely to random-error variation. With analysis of variance, the variation due to each explained source can be estimated as a mean square.

The residual, or unexplained, variation is attributed to random error, which is estimated as the error mean square. The tests of significance of treatments are called "*F* tests." An *F* test is conducted by calculating an *F* value, which is a ratio of two mean squares. To test the hypothesis of whether treatment means are different, the *F* value is calculated as the ratio of the treatment mean square to the error mean square. *F* tables based on the degrees of freedom of the treatment and error mean squares contain critical *F* values for given probabilities. If the calculated F value exceeds the critical *F* value, then the treatment means are considered to be significantly different at the chosen level of probability. Critical *F* values are usually chosen to be at the .05 or .01 level, which means that there is only a .05 or .01 probability, respectively, that the means are equal if the critical value is exceeded. If the calculated *F* ratio is less than the table *F* ratio, then the hypothesis that the means are equal is not rejected and the means are probably not different.

For a test of a hypothesis involving only two treatments, the *F* test gives results identical to a *t* test. However, a *t* test can be used only to test for differences between two populations at any particular time, whereas *F* tests can be used to determine whether there are differences among any of several populations or treatments. Also, analysis of variance, which employs *F* tests, can determine the significance of more than one factor and can be used to determine whether the effects of these factors are independent of each other.

Probabilities of Incorrect Conclusions

There are two types of errors that can be made when testing a hypothesis as to whether treatment means (populations) are significantly different from each other. These errors are called "type I" and "type II" errors.

A type I error occurs if one concludes that the population means are different from each other when the true population means are not different. The chance of making a type I error is called α, which is the probability associated with the critical *F* value. The "α level," or "probability level," is the level of significance. A common misconception is that the chance of making a type I error is affected by the experimental error or the number of replications. It is not. The α level is typically set by researchers at .01 or .05. These two levels are the traditional significance levels and are popular because they are acceptably low. *F* tables for these values are readily available. However, since computer printouts from statistical packages provide the actual probability of significance, there is no longer a need to be constrained by the availability of tables, and probability levels different from .05 or .01 that are acceptably low (such as .06) may be chosen (Freund and Littell 1981, Carmer 1976).

A type II error occurs if it is concluded that the means of the populations are *not* different when the true population means are different. The chance of making a type II error is β. β is not chosen or known by the researcher, but can be controlled. The probability of a type II error decreases as (1) α increases, (2) the true differences between means increases, (3) the experimental error decreases, (4) the number of replications or samples increases.

Many researchers consider type I errors to be much more serious than type II errors. However, for initial field trials, a type II is probably as serious as a type I. Therefore, decreasing the type II error rate by increasing the power of the test should be a major consideration in these tests. The "power" of a test is the ability of the test to find two means to be significantly different from each other. The factors which decrease probability of a type II error will also increase the power of the test.

Error Variation in the Field

"Error variation," estimated by the error mean square in analysis of variance, is the random variation that occurs among experimental units receiving the same treatment. A basic difference between laboratory or greenhouse experiments and field experiments is the degree of control over random error variation. In the laboratory and the greenhouse, as compared to the field, conditions are fairly uniform within a particular experiment. This lack of control over error in the field results in relatively higher error variation and a less powerful test.

There are many unexplained or uncontrolled factors that contribute to the error variation in biological experiments. Common sources of error variation in the field include (1) differences in the properties of soil, water, or air; (2) microclimatic variations; (3) plant heterogeneity; (4) variation in populations of competing microorganisms; (5) differences in substrate availability; (6) variation in distance and direction from source; and (7) time of sampling. These sources of variability can be reduced or removed as sources of error through careful technique and by using the optimum experimental and sampling designs. It is desirable to minimize the error variation because the larger the error variation, the more difficult it is to find significant differences.

Requirements for Analysis of Variance

Three requirements that must be satisfied for the F tests from analysis of variance to be valid are:

1. Treatment and environmental effects must be independent.
2. The experimental errors of each treatment must be normally distributed and independent of each other.
3. Error variances must be homogeneous—that is, the variance associated with each treatment must be similar to all of the others.

In order to satisfy the requirements of analysis of variance, treatments must be randomized and replicated. However, even with appropriate experimental design, data still may not satisfy the requirements of analysis of variance. A data transformation or other remedial measure may be used to correct problem data.

Replication of treatments

Replication of treatments is essential to analysis of variance. Without replication of treatments, there is no estimate of experimental error to be used to test the significance of treatment effects. For an experiment to have replications, there must be more than one experimental unit of each treatment. A common mistake in designing field experiments is to use "pseudoreplication" rather than true replication (Hurlbert 1984). Pseudoreplication occurs when the sampling unit is replicated instead of the experimental unit. An "experimental unit" is the unit to which the treatment is *applied*. The "sampling unit" is the unit that is *measured*. An experiment that contains replications of sampling units rather than experimental units will usually suffer an increase in the probability of a type I error (Millard et al. 1985). In an experiment designed to monitor the engineered organism as well as to determine the efficacy of the microorganism, the experimental units will be the same for all the variables measured, but the sampling units will be different. Experimental units for field testing of organisms will usually be field plots of a size that makes treatment application practical and accurate. For variables such as plant yield, the entire plot (excluding border rows) would be measured, so the sampling unit and the experimental unit will be the same. For variables related to the microorganism, it would be impractical to measure the entire plot, and several samples would be taken from each plot. Figure 10.1 shows an experiment with eight experimental units (experimental plots) and four replications of each treatment. Sampling units would usually be one or more subsets of the field plot. If the organism of interest is soilborne, a gram of soil could be the sampling unit. If 12 soil samples (sampling units) were collected per plot, these 12 samples would constitute one experimental unit.

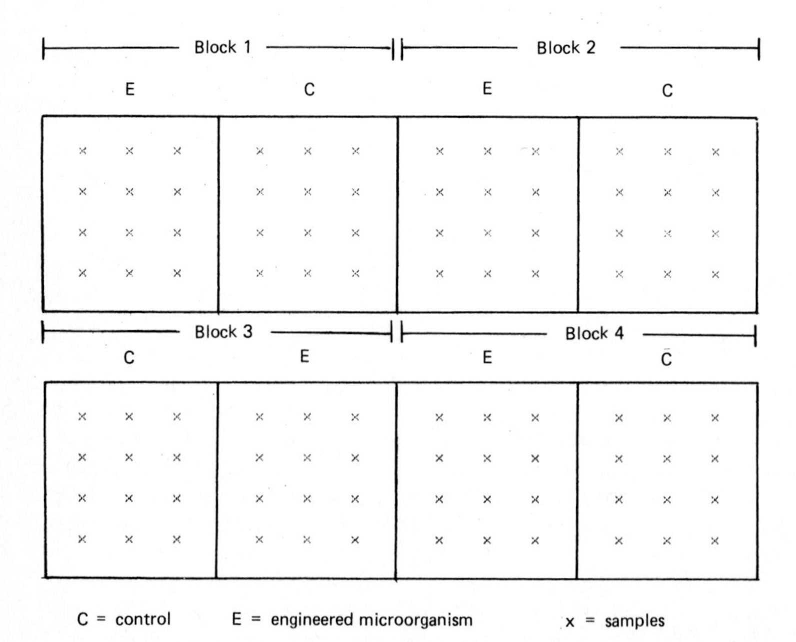

Figure 10.1 Schematic drawing of a randomized-complete-block design.

Randomization of experimental units

For observations to be independent and means to be unbiased, treatments must be randomized. Randomization is especially important for field studies because of the frequent occurrence of gradients of nontreatment factors such as soil moisture and fertility. The purpose of randomization is to eliminate bias in estimation of the treatment means and ensure that environmental and treatment effects are independent of each other. Systematic arrangement of treatments can result in confounding of the treatment effects with nontreatment effects. For example, in a study comparing the number of colony-forming units of engineered and nonengineered microorganisms, if the engineered microorganisms were released on one side of the field and the nonengineered microorganisms were released on the other side of the field, the differences or lack of differences in number of colony-forming units for the two types of microorganism could be due to either the engineering or the environment. These two effects could not be separated from each other. However, if the release of the organism were randomized in the field, the treatment effects would be independent of the environment.

Data transformation

Data transformations are often used for data that are not normally distributed and/or do not have homogeneous variances. Microbial population counts are often log-normally rather than normally distributed (unless the observed values cover a narrow range of concentrations). Also, the variances of each treatment mean are usually not homogeneous—because the higher the mean value, the higher the variance. Therefore, microbial count data often do not meet two of the conditions for analysis of variance: normality and homogeneity of variance. Fortunately, after a log transformation, the data should be normally distributed with homogeneous variances. An easy check to see if data should be log-transformed would be to calculate the mean and variance of each treatment. If the variances differ by more than fivefold and if the variance increases as the mean increases, a log transformation is probably needed. After taking the log of the data [or $\log(x + 1)$ if there are zeroes in your data], recalculate the means and variances. The variances should no longer be correlated to the means, and the ratio of the high variance to the low variance should have decreased.

Data measured as percentages (e.g., percent infected tissue) will also often require transformations to make variances homogeneous. If the percentages fall within a range of 30 to 70% or are calculated as percentage of control, the data usually do not need a transformation (Steel and Torrie 1980). However, if the percentages range from 0 to 100% (or even from 0 to 50% or 50 to 100%), an arcsine transformation is often needed to make variances homogeneous. To determine whether an arcsine transformation is necessary, calculate the mean and variance of each treatment. If the treatments with means near 50% have large variances and the treatments with means near 0% or 100% have small variances and the variances differ by more than fivefold, an arcsine transformation is indicated. Recalculate the treatment means and variances using data that have been transformed. (The arcsine transformation is the arcsine of the square root of the observation.) If the transformation was effective, the ratio of the high and low variances should have decreased.

"Outliers" are another cause of heterogeneity of variance. Values that fall more than three standard deviations from the mean and result from poor measurement, sample contamination, or incorrect transcription of data are considered outliers. It is very important that the raw data be carefully scrutinized to avoid using any outliers in your analysis because outliers can have a considerable effect on the estimation of the mean and the variance. However, care should be taken not to omit data from the analysis just because it doesn't appear to fit. If the actual cause of the suspected outlier cannot be ascertained, it is possible that the cause is a skewed distribution rather than an outlier. Computer packages such as Statistical Analysis System (SAS) can be used to identify outliers using regression techniques (Freund and Littell 1981), but they should not be used unless the distribution is normal. Log-transformed data are less affected by outliers, and a log transformation may be the most appropriate when there are outliers of unexplained cause (Berry 1987).

Two commonly used statistical procedures that test for homogeneity of variance are Bartlett's χ-square and F-max tests. These tests can determine if variances are not homogeneous, but not whether the lack of homogeneity is serious enough to invalidate the F test. The consequence of violating the condition of homogeneity of variance is that the significance level for the F tests will not be correct. However, because F tests are robust, they will generally be valid if the ratio of the variances is 20 or less (Harris 1975). Nonparametric tests are distribution-free and can be used when there is a gross violation of the conditions of normality and homogeneous variance, but these tests lack the power of F tests.

Design of Preliminary Field Experiments

Randomized-complete-block design

The "randomized-complete-block design" (RCB design) is the experimental design most commonly used in field experiments. In an RCB, the field is divided into "blocks" (also called "reps" or "replicates"). Each block contains one experimental unit of each treatment. This design is used to reduce nonrandom variation due to error by minimizing variation within a block and maximizing variation among blocks. With use of statistical techniques, the variation among blocks can be removed from random error and instead attributed to the "block effect." Because adjacent plots in the field are usually more alike than nonadjacent plots, randomizing treatments so that each treatment occurs once in each block will remove some of the variation due to difficult-to-control effects, such as soil and moisture variability, from the random error.

Choosing a field

When a field experiment is designed, the history of the field should be known. Learning this history is probably the most cost-effective step in designing an experiment. Some examples of previous history of the field having severe consequences for testing are: a field is chosen that had been used to test a fungicide that has residual activity and affects the microorganism to be studied; a study in which plant yield is to be measured is conducted in a field that was previously used for a fertility study; a field is selected that has been habitually infested with a noxious weed like Johnsongrass. Most field researchers make this kind of mistake only once.

Measurements taken before the microorganism is released or from previous experiments in the same field can be an invaluable tool in determining how to "block" the field. The effectiveness of blocking can be determined from a previous analysis. If the blocks were not significantly different, a different blocking strategy should probably be considered. Also, the coefficient of variation from previous experiments in that field provides information about the field. The "coefficient of variation" (cv) is a measure of relative variation, and is the ratio of the standard deviation to the mean. If the cv is large (greater than 25%), this indicates that field variability is a potential problem, and the problem should be identified and rectified before starting the experiment.

Blocking a field

The next step in designing the experiment involves actually going to the field to select and block the study area. The study area selected should be as uniform as possible. Also, potential problem areas should be avoided. For example, if one corner of the field is underwater, that area would be a poor location for the experimental plots.

Once the plot area has been selected, the field variability should be observed and the shape and location of the blocks chosen to minimize variation within a block. For example, if the field is sloped, the blocks should run perpendicular to the slope, or if there are two soil types in the field, a block should contain only one soil type. The objective in blocking is to make the plots within a block as similar as possible. An example of a typical plot layout is given in Fig. 10.1.

Unless the experiment is very small in terms of number of treatments and replicates and in terms of space, blocking a field experiment is recommended. Even if there is no visual evidence of heterogeneity in the field, the field area is probably not uniform. In the absence of any obvious variation or prior knowledge of the field, it is probably best to use blocks that are square (Weibe 1935).

Number of blocks

The number of blocks to include in an experiment depends upon several parameters. A formula to determine minimum number of blocks requires selecting or estimating the difference to be detected, the desired type I error rate, and the error variance. The difference that needs to be detected (for example, 10^4 colony-forming units, or 100 kg/ha) is decided upon based on what constitutes a biologically or economically significant difference. The probability of a type I error is chosen to be an acceptably small level. The most commonly used α level is .05. An estimate of the error variance is also needed. This estimate is often the weak link when estimating number of blocks. Although it is difficult to obtain a good estimate for error before the experiment has been conducted, an estimate can be obtained by conducting a preliminary experiment in this same field or by using the analysis from a similar experiment. The estimate of error used in the formula is your best estimate, but not necessarily a particularly good estimate. Although the formula results in one number rather than a range of numbers, the accuracy of the estimate may be limited by the accuracy of the estimates used in the equation and should only be considered a general guideline in planning experiments. Also, this formula ignores the power of the test, and the precautions discussed under "Sample Number" should be considered. This estimate of number of blocks will provide an approximation that can be used in conjunction with information about constraints on space, time, and labor in planning the number of blocks needed. Equation 2 can be used to calculate the minimum number of replications. Equation 2 is similar to Equation 1, but substitutes number of replications for number of samples and experimental error mean square for sampling variance.

$$r = \frac{(Z^2)(\text{error mean square})}{d^2} \tag{2}$$

where r = number of replications or blocks
Z = Z with probability ($Z_{.05} = 1.96$)
d^2 = margin of error for the treatment mean

For example, suppose you wish to estimate the mean abundance of the test organism to within 5×10^3 colony-forming units (d) at the .05 level of confidence ($Z = 1.96$). The error mean square is estimated as 25×10^3, based on an experiment with four blocks. The minimum number of blocks is estimated to be $1.96^2\ (25 \times 10^3)/(5 \times 10^3)^2 = 3.8$, rounded to 4.

Monitoring Microorganisms in the Field

Sampling design

When microorganisms and plants are monitored, the entire plot cannot be measured; therefore, the number, location, and size of samples to be collected must be determined. Sampling designs can be either random or systematic, or a combination of both. For simple random sampling, a plot is divided up into units, and each unit has an equal chance of being sampled. This strategy is good if there is little variation within a plot or the microorganism is randomly dispersed. If the microorganism is found in clumps in the field, its distribution is aggregated rather than random, and stratified random sampling is a better sampling procedure (Green 1979). For stratified random sampling, each plot is divided into strata, and one randomly selected sample is collected from each stratum. Stratification is similar to blocking in that variation within a stratum should be minimized. If the plots are not homogeneous, sampling from each stratum will give a representative estimate of the plot mean. Sampling can also be systematic, with each sample taken at a certain location. This method can also give representative estimates of the plot mean. These three methods all rely on using a plot plan and locating samples in an unbiased manner. When using one of these sampling methods, the plot plan may indicate a plant that is either dead or diseased; in that case it would be best to use the nearest healthy plant to avoid outliers. On the other hand, do not use the "typical sample method" when sampling. The typical sample method can lead to bias because of use of a subjective selection criteria, since "typical" often translates to the healthiest or the most accessible.

Number and size of samples for analysis of variance

For variables such as abundance of microorganisms, it is not practical to measure an entire plot. Instead, one or more samples are collected and measured in each plot. If the organism being measured is uniform throughout the plot, few measurements are needed per plot. However, the more variable the population is within a plot, the more samples are needed per plot. The goal in planning an experiment is to choose the number of replicates and the number of samples within each replicate which will provide the desired levels of precision and power to test the hypothesis. Increasing either the number of replications or samples per plot will increase the precision and power of the experiment. The question then arises as to whether it is better to increase number of samples or number of replications. To make this decision, several factors should be examined. The relative costs of increasing samples or replications should be weighed against the effect of their increases on precision. In most cases, it is cheaper to increase the number of samples per plot than to add an additional replication of the treatments. On the other hand, increasing the number of samples will not improve precision as much as increasing the number of replications. The choice of the number of samples and replications depends on which combination of samples and replications will be the least costly in terms of space, labor, and time for a chosen level of precision.

The relative advantage of increasing replication number rather than sample number depends on the ratio of the sampling error and experimental error. The "experimental error" is the random variability associated with the experimental units, whereas the "sampling error" is the random variability associated with the sampling units. In most cases, the sampling error is less than the experimental error. The larger the experimental error is in relation to the sampling error, the bigger the advantage of increasing the number of replications. Table 10.2 illustrates how the relationships between sample size, replication number, sampling error, and experimental error influence confidence intervals.

Power function charts provided by Pearson and Hartley (1951) can be used in a procedure for determining numbers of experimental and sampling units (Geng and Hills 1978). An alternative method estimates minimum sample number required to obtain a desired confidence interval for the population mean. This method does not consider the power of the test and will tend to underestimate sample and replication number, but it will set a minimum

TABLE 10.2 Confidence Intervals and Standard Errors for Means Based on Different Variances, Numbers of Replicates, and Numbers of Samples

Replicate number	5	2	5	2	5	2
Sample number	2	5	2	5	2	5
Experimental error variance	4	4	100	100	200	200
Sampling error variance	4	4	100	100	2	2
Standard error	1.1	1.5	5.5	7.7	6.7	10.2
Confidence interval	2.4	4.8	12.0	24.5	14.6	32.4

number. This formula (Eq. 3, below) is similar to Eq. 1 except that estimates of both experimental error and sampling error are needed and either the number of reps or samples must be specified.

$$n = \frac{(Z_\alpha^2)(V_s)}{r(d^2) - (Z_\alpha^2)(V_e)} \tag{3}$$

where n = number of samples
Z_α = Z value at the α probability level ($Z_{.05} = 1.96$)
V_s = estimate of sampling variance component = sampling error mean square
r = number of replications
d = difference that is significant
V_e = estimate of experimental variance component = (experimental error mean square − sampling error mean square)/s

The formula given above can be used to determine minimum sample number for different numbers of replications. By determining the number of samples needed for several different numbers of replications, the most cost-effective combination of numbers of replications and samples can be selected.

Monitoring changes over time

Field studies concerning the release of microorganisms will involve monitoring the change in the released populations until the microorganism is no longer present, as well as measuring the effects of the release on target organisms over a growing season. In order to test whether there is a significant difference between the treated plots and the control at any particular time, an analysis of variance could be conducted for each time. This will provide information about when there are significant differences. However, it is also of interest to test whether there are differences due to the treatments averaged over time, whether the differences due to the treatments were consistent over time, and if there are differences among the treatments in the rate of change. The analysis-of-variance procedure used to test these latter hypotheses is called a "repeated-measures" or a "split-plot-in-time" analysis. The standard

TABLE 10.3 Analysis of Variance for an Experiment Analyzed for a Particular Time and as a Repeated Measure over Time

One Time			
Source of variation	Degree of freedom	Mean square	*F* ratio
Block	$r-1$	M_1	M_1/M_3
Population	$p-1$	M_2	M_2/M_3
Error	$(r-1)(p-1)$	M_3	
Combined over Times			
Source of variation	Degree of freedom	Mean square	*F* ratio
Block	$r-1$	M_1	
Population	$p-1$	M_2	M_2/M_3
Error *a*	$(r-1)(p-1)$	M_3	
Time	$t-1$	M_4	M_4/M_5
Error *b*	$(r-1)(t-1)$	M_5	
Population × time	$(p-1)(t-1)$	M_6	M_6/M_7
Error *c*	$(r-1)(t-1)(p-1)$	M_7	

sources of variation for the analysis by time, and then combined over times, are described by Steel and Torrie (1980) and presented in Table 10.3.

Sampling distances and times are often chosen to be at equally spaced intervals, but this should depend on the nature of the information needed. For example, if the microbial population declines rapidly and slowly approaches zero as in Fig. 10.2, samples should be collected daily until the population drops to a prespecified size or has stabilized; sampling should then continue on a weekly basis until the microorganism is no longer detectable.

When planning an experiment in the field, the distance from the source at which the engineered organism should be sampled is not always known. In such cases, a strategy of initially sampling to a distance farther than one would expect to detect the organism could be employed. At the next sampling, sample to the last distance where the organism was detected plus one. If the microorganism is no longer present, sample at least one additional time.

For parameters such as colony-forming units of the microorganism, a hypothesis as to whether the percentage change in rate of growth of the microorganism is the same over time can be tested by taking the log of the data before the analysis. If the data follow a logarithmic curve over time, then the transformation will linearize the response over time. An example of data collected over time on the original and log scale is given in Fig. 10.2. Analysis of covariance can be used to test whether the slopes derived for the log-transformed data for the engineered and nonengineered organisms are different.

Data that are below the detection limit present a problem for log transformation. The values recorded for counts below the detection limit will affect the slope and the fit of the line. There are several approaches that can be taken:

1. The counts below the detection limit can be recorded as 0 and all values log-transformed. In this case, the counts below the detection limit will be missing because there is no log of 0.
2. The counts below the detection limit can be recorded as 0 and the raw data log-transformed as the (count + 1). In this case, the log of the counts below the detection limit will be 0. However, the slope of the line will be slightly affected because the effect of adding 1 to small values is greater than the effect of adding 1 to large numbers.
3. The counts below the detection limit can be set to 1, and then all the data can be log-transformed. In this case, the log of the counts below the detection limit will be 0, and the other values will be unaffected.
4. The counts below the detection limit can be recorded as .01 because this is a reasonable estimate of the value. The log of the count at the detection limit would be −2, which is probably not a biologically reasonable number to use.

For the data shown in Fig. 10.2, alternatives 2 and 3 are the best choices because these alternatives result in the best fitting (highest r^2) models. Because the results will be different depending on how the data below the

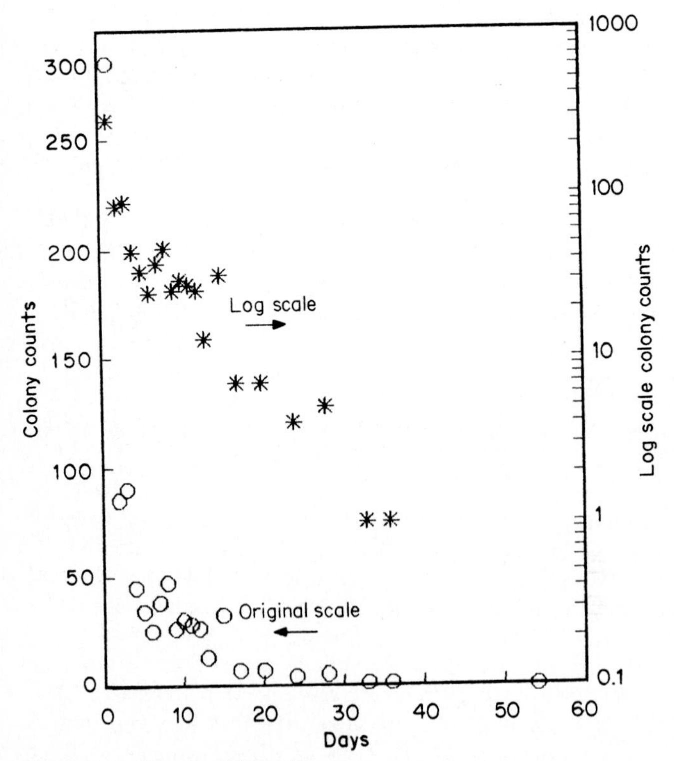

Figure 10.2 Counts of *Pseudomonas*.

detection limit are treated, it is important that the procedure be standardized among scientists.

There is another hypothesis about time or depth that can be tested that would not involve data transformation or coding of data below the detection limit. This hypothesis would test whether there are significant differences between the engineered and the nonengineered microorganism in the number of days until the organism is below the detection limit. This hypothesis can be tested instead of or in addition to the hypothesis about the percent change in the growth rate and would not require a log transformation of the data or coding of the counts below the detection limit.

Advanced Testing for Release of Microorganisms

Testing at more than one location or year

Following initial testing at one location, testing in additional environments will be necessary to estimate the survival and efficacy of the engineered organism in environments representative of the range of variation of its intended commercial release. The same experiment would be conducted in environments at different sites and over more than 1 year. The information gained from analyzing experiments combined over environments will indicate whether the measured characteristics of the engineered organism are consistent over years and sites. Significant differences between years may indicate that weather effects such as rainfall and temperature play a role in determining the microorganism's characteristics. On the other hand, significant differences between locations may be caused by differences in soil, water, or air as well as climatic differences.

The analysis for experiments combined over locations and/or years is given in Table 10.4, where each experiment is considered an environment. The sources of variation and F tests for experiments combined over locations and/or years for additional experimental designs are given by McIntosh (1983).

Number of locations and years

In planning for large-scale testing, decisions need to be made concerning numbers of locations, years, replications, and samples. These decisions are all interrelated. Factors pertaining to replications and samples have been discussed

TABLE 10.4 Analysis of Experiments Combined over Environments

Source of variation	Degree of freedom	Mean square	F ratio
Environments (E)	$e - 1$[a]	M_1	M_1/M_2
Blocks/E	$e(r - 1)$	M_2	
Populations (P)	$p - 1$	M_3	M_3/M_5
$E \times P$	$(e - 1)(p - 1)$	M_4	M_4/M_5
Pooled error	$e(r - 1)(p - r)$	M_5	

[a] e = number of environments, r = number of blocks, p = number of populations.

in the previous sections on sampling and blocking. However, because the variance per mean will be decreased more rapidly by increasing the number of environments rather than the number of replications, it has been recommended that the optimum allocation of resources would consist of one replication in as many different environments as possible (Balaam and Hunter 1962). This suggestion, however, is not practical because increasing the number of environments for testing is usually much more costly than increasing the number of replications in each environment. As a compromise, the number of replications could be reduced to two or three and remain adequate for testing at multiple locations.

Factors to consider in determining the number of locations and years of testing are similar to those for choosing number of replications, except that the number of environments should also be sufficient to represent the range of environments for the intended release of the microorganism. As with replication, the more years and sites an engineered organism is tested at, the better the estimates obtained. However, the number of sites and years tested is limited by cost. The costs involved in increasing the number of sites include the labor involved in preparation, treatment application, maintenance and sampling of the plots, travel expenses, and land acquisition. The costs involved in testing over years are even greater because they include these same costs plus the cost of delaying the commercial release of the engineered organism.

In deciding how many years and locations are needed for testing, it is useful to have estimates of interactions of treatments with years and locations. Interactions are caused by the treatment effects not being consistent from year to year or from location to location. The more inconsistent the effect of the treatment, the larger the interaction. For example, a small treatment × year interaction might be a result of the treatment being 3 times better than the control in the first year, but only 2 times better than the control the second year; a large interaction might be the result of the treatment being 3 times better than the control the first year but not as good as the control the second year. In general, the larger the treatment interaction with years or locations, the more years or locations, respectively, are needed for testing purposes. Because it is costlier to increase the number of years of testing, locations can be substituted for years if the divergence in climatic conditions among locations is comparable to differences between years (Fehr 1987).

Testing for the commercial release of a low-risk engineered microorganism is in many ways analogous to testing for the commercial release of a new crop cultivar. It is typical for a private company to test potential cultivars in 5 to 15 locations for 3 to 4 years before release. This could be used as a guideline for determining the number of years and locations required for testing for efficacy before the release of an engineered microorganism. Of course testing of a high-risk microorganism would need to be more extensive because of the safety-related issues and the need to increase the power of the test.

Adaptive research

Because results cannot always be anticipated, research plans should be adaptive. A general plan can be set at the initiation of the testing, but these plans should be subject to modification at any time. This is particularly important when testing is planned for several years and locations. Data acquired from the initial analysis can be useful in modifying treatments, sampling techniques, and sampling design. Analysis of the data will often be complicated by changes in treatment and experimental design, but the main consideration should always be to test meaningful hypotheses.

Conclusion

Field testing before the release of an engineered microorganism is important because the consequences of the release cannot always be predicted from laboratory testing. Due to the large amount of uncontrolled variability in the field and the large expense involved in each field trial, experimental design needs to be carefully worked out for each trial. A statistician who is familiar with field plot technique can be an invaluable team player in planning field experiments. A good statistician can work with the researchers to maximize the amount of useful information obtained from the data. Statistical analysis of the data can be facilitated by using a software package like SAS, but statistical packages cannot determine the most appropriate analyses. Therefore, statistical analysis is also best done with the assistance of an experienced statistician. Many problems have been documented concerning design, data analysis, and policy decisions involving environmental monitoring that could have been avoided with the advice of qualified statisticians (Millard 1987).

Researchers involved in performing field research and trials of engineered organisms can benefit from a basic understanding of statistical and experimental design concepts. This foundation will allow the researcher to effectively communicate with a statistician in order to develop cost-effective, statistically sound designs and analyses.

Glossary

Coefficient of variation a measure of relative variability calculated as the ratio of the standard deviation to the mean

Confidence interval an interval determined from the sample mean and its standard error of the mean. The interval will contain the population mean with a chosen level of certainty.

Experimental error the variability of experimental units within a treatment. This is estimated as the error mean square in analysis of variance.

Experimental unit the unit or plot to which the treatment is applied

Interaction the effect of a factor (type of treatment) being dependent on level of another factor in the experiment

Mean the arithmetic average

Median the value which is greater than half the population and less than the other half of the population

Mode the most commonly occurring value of the population

Population all possible measurements of a variable. The population may be defined to include only measurements of a variable from a specific treatment.

Power of a test the ability of the test to detect significant differences

Precision the sensitivity of the test

Sample a part of a population

Sampling error the variability of sampling units within an experimental unit

Sampling unit the unit which is measured

Significance level the probability of incorrectly finding means (that are the same population value) to have different values

Standard deviation the measure of the average deviation of the observations from the population mean

Standard error the measure of the average deviation of the sample means from the population mean

Treatment the procedure or factor whose effect is measured

References

Balaam, L. N., and Hunter, R. D. 1962. The analysis of a series of wheat varietal trials. *Aust. J. Stat.* 4:61–70.

Beal, S. L. 1989. Sample size determination for confidence intervals on the population mean and on the difference between two population means. *Biometrics* 45:101–105.

Berry, D. A. 1987. Logarithmic transformation in ANOVA. *Biometrics* 43:439–456.

Carmer, S. G. 1976. Optimal significance levels for application of the least significant difference in crop performance trials. *Crop Sci.* 16:95–99.

Cochran, W. G. 1963. *Sampling Techniques*, 2d ed. Wiley, New York.

Fehr, W. R. 1987. *Principles of Cultivar Development.* Macmillan, New York, 525 pp.

Freund, R. J., and Littell, R. C. 1981. *SAS for Linear Models.* SAS Institute, Cary, North Carolina.

Freund, R. J., and Littell, R. C. 1986. *SAS System for Regression.* SAS Institute, Cary, North Carolina.

Geng, S., and Hills, F. J. 1978. A procedure for determining numbers of experimental and sampling units. *Agron. J.* 70:441–444.

Gomez, K. A., and Gomez, A. A. 1984. *Statistical Procedures for Agricultural Research.* Wiley, New York, 680 pp.

Green, R. H. 1979. *Sampling Design and Statistical Methods for Environmental Biologists.* Wiley, New York, 257 pp.

Harris, R. J. 1975. *A Primer of Multivariate Statistics.* Academic, New York.

Hurlbert, S. H. 1984. Pseudoreplication and the design of ecological field experiments. *Ecol. Monogr.* 54:187–211.

Kupper, L. L., and Hafner, K. B. 1989. How appropriate are sample size formulas? *Am. Statistician* 43:101–105.

Little, T. M., and Hills, F. J. 1978. *Agricultural Experimentation.* Wiley, New York, 350 pp.

McIntosh, M. S. 1983. Analysis of combined experiments. *Agron. J.* 75:153–155.

Millard, S. P. 1987. Environmental monitoring, statistics, and the law: room for improvement. *Am. Statistician* 41:249–253.

Millard, S. P., Yearsley, J. R., and Lettenmaier, D. P. 1985. Space-time correlation and its effects on methods for detecting aquatic ecological change. *Can. J. Fisheries Aquat. Sci.* 42:1391–1400.

Parkin, T. B., Meisinger, J. J., Chester, S. T., Starr, J. L., and Robinson, J. A. 1988. Evaluation of statistical estimation methods for lognormally distributed variables. *Soil Sci. Soc. Am. J.* 52:323–329.

Pearson, E. S., and Hartley, H. O. 1951. Charts of the power function for analysis of variance tests, derived from the noncentral *F*-distribution. *Biometrika* 38:112–130.

Steel, R. G., and Torrie, J. H. 1980. *Principles and Procedures of Statistics*. McGraw-Hill, New York, 633 pp.

Stein, C. 1945. A two-sample test for a linear hypothesis whose power is independent of the variance. *Ann. Math Stat.* 16:243–258.

Weibe, G. A. 1935. Variation and correlation in grain yield among 1500 wheat nursery plots. *J. Agric. Res.* 50:331–357.

Chapter

11

Use of Fate and Transport (Dispersal) Models in Microbial Risk Assessment

Harlee S. Strauss

H. Strauss Associates, Inc.
21 Bay State Road
Natick, Massachusetts 01760

Morris A. Levin

Maryland Biotechnology Institute
University of Maryland
Baltimore, Maryland 21228

Introduction and Scope

This chapter focuses on models that have been developed to predict microbial dispersal, many of which are based on chemical or particulate fate and transport models. These models integrate information necessary to evaluate a key element of a microbial risk assessment: the determination of where microorganisms will go in the environment, and how many viable microbes will arrive at a particular location where a sensitive receptor might be located.

Transport (dispersal) models are primarily concerned with the movement and location of the organism of interest. In principle, transport models can be linked to fate models, which take microbial reproduction and die-off into account, and to dose-effects models. Linking these three models would provide an estimate of the probability associated with the occurrence of the effect being modeled. In contrast, epidemic models are primarily concerned with describing the location and intensity of the effects of the organism of concern, and neglect the location of microorganisms that are not causing the effect being examined. Epidemic models are the subject of Chap. 12.

That some bacteria can move in a variety of environmental media, and arrive in a form capable of reproduction in sufficient numbers to cause an adverse effect, is readily demonstrated in the public health literature. For example, in the United States between 1920 and 1980, 386,000 cases of waterborne disease outbreaks were recorded, resulting in 1083 deaths (Yates and Yates 1988). The importance of this route of disease transmission is still evident. Between 1971 and 1979, 267 outbreaks of waterborne disease, associated with 57,947 cases of disease, were reported (Craun 1984). The use of untreated or inadequately treated groundwater for drinking water was associated with 52% of the outbreaks and 45% of the disease. The remainder was due to drinking contaminated surface water (Craun 1984).

Everyday experience with the flu and the common cold demonstrate that some disease-causing microorganisms can be transmitted by air. In fact, data demonstrating aerial and/or insect vector transport of disease-causing viruses, bacteria, and fungi are available in the public health, epizoology, and plant pathology literature. For example, a small excess of enteric virus infection was associated with population study members who had a high degree of aerosol exposure at a spray irrigation site in Lubbock, Texas (Moore et al. 1988), and a seasonal excess (i.e., during the summer irrigation season) of enteric disease was observed among the 0 to 4 age group in a study of kibbutzim in Israel (Fattal et al. 1986).

This chapter is organized as follows. The next section presents a conceptual model for microbial dispersal and spread, defines the terms used throughout the chapter, and discusses the relationship between genetically engineered microorganisms (GEMs) and naturally occurring microorganisms. The second section summarizes factors affecting dispersal, all of which have been discussed in greater detail in other chapters in this volume. The third section describes practical aspects for using models to predict microbial dispersal, including when to use models, how to select models, and where to find appropriate models and data. The fourth section describes several models for fate and transport in various environmental media that have been applied to microorganisms. The final section describes how fate and transport models can be incorporated into microbial risk assessments.

Definitions and key terms

The movement of microorganisms from one site to another and their subsequent reproduction and colonization of the new site can be thought of as a series of individual processes. Figure 11.1 presents a conceptual model and terminology directed toward microorganisms newly introduced into the environment. However, it does not presuppose whether the release is planned or unplanned, short or prolonged, or it does not presuppose the purpose of the release.

Release, transport, deposition. We begin the conceptual model with the release of microorganisms into an environmental setting. The initial source is

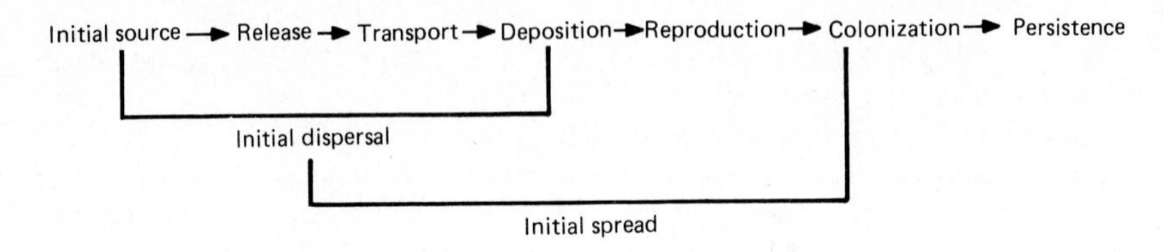

Figure 11.1 Conceptual model to define dispersal and spread.

the collection of laboratory-grown (or production-plant-grown) microorganisms entering the environment for the first time. The "release" is the process by which the microbes constituting the initial source are transferred from their original container (site) to a transport medium or vector. The release could be via many mechanisms, including aerosols generated by nozzle sprayers, aerosols generated during aeration of open fermentation vats, microorganism-laden water discharged from pipes, or the planting of seeds coated with bacteria.

"Transport" is the horizontal or vertical displacement of the microorganism. Although some microorganisms are capable of a limited amount of self-propulsion, usually a transport medium or vector is necessary for the displacement. Transport media or vectors of importance for microorganisms are air, surface water, groundwater, and surfaces. Surfaces can include insects, small animals crossing the site, and the clothing and equipment of workers near the release.

"Deposition" is the transfer of the microorganism from the transport medium to a new site. Deposition does not include the notion that the site has to be suitable for growth and reproduction, only that it provides a "landing strip."

Dispersal. We use the term "initial dispersal" to include the cumulative process of release from the initial source, transport, and deposition of viable microorganisms. By "viable," we mean capable of reproduction under some conditions. It is possible that the viable microorganisms may not be cultureable under certain conditions, but are nonetheless capable of reproduction under other conditions or at a later time after recovery from injury (Hussong et al. 1987).

Spread. Some sites of initial dispersal will provide a suitable (susceptible) habitat for the deposited microorganism(s) to reproduce and thus increase the size of the population. Reproduction may then be followed by initial colonization of the new site by the newly arrived microorganisms and the establishment of a permanent population. Alternately, following reproduction, the new population may die off due to competitive pressures, environmental changes in the habitat, or other adverse effects.

We use the term "initial spread" to describe the cumulative process of initial dispersal, followed by reproduction and colonization of the (susceptible) new site. Thus spread, unlike dispersal, includes characteristics of the new site. We use the term "persistence" to describe whether or not the newly arrived population remains at the new site.

If the microbial population at the site of the initial spread persists at the site, it will be available for release to a transport medium and a repeat of the entire dispersal and spread process. To distinguish between dispersal and spread of microorganisms that were initially released after growth in the laboratory from those that were dispersed and spread after growing under environmental conditions, we use the terms "secondary dispersal" and "secondary spread" for the latter case. This distinction is important because microorganisms grown in the laboratory and those grown in the environment may have quite different physiological characteristics and thus different persistence and growth characteristics. In addition, different numbers of microorganisms are likely to be associated with initial and secondary dispersal, which may alter the predictions for successful spread.

Different authors have used different terms for the individual and composite processes that we have called "dispersal" and "spread." In some cases, technical terms have different meanings, depending upon the authors and their disciplinary training. In this chapter, we have selected terms to describe various processes, and we use the terms consistently. However, the terms used in this chapter may differ from those used by the original author.

Relation between exposure and dispersal

When assessing the risk of introducing a microorganism into the environment, it is necessary to consider both the potential that a susceptible population or individual will be exposed to the microorganism (exposure assessment) and the potential harm that can be caused by that microbe (hazard assessment). For both exposure and hazard assessments, it is also important to quantify, to the extent possible, the number of microorganisms reaching the susceptible receptors and the number required for the response.

Dispersal is one of the factors that contributes to the exposure assessment. Other important contributions to the exposure assessment include the fate of the microorganism and the location and density of a susceptible population or microhabitat. The "fate" of a microorganism refers to the distribution of the microorganism in various environmental media or microhabitats, and the rates of reproduction or die-off in each medium.

Relation between GEMs and naturally occurring microbes

Genetically engineered microorganisms (GEMs) are created by using modern molecular biological techniques to add, delete, or substitute DNA sequences. Most of the DNA, and most of the physiological and ecological functions, are derived from the recipient (parental) microorganism. However, it is possible that some changes introduced by genetic engineering will change the recipient in a manner which will enhance its environmental survival (and thus its potential for dispersal and spread—see below) or ecological competitiveness. It is these changes that are of concern from the risk assessor's point of view.

If a microorganism is engineered in such a way that new DNA is added to a microorganism, then there is a possibility that the introduced DNA may disperse independently of the microorganism into which it was inserted. The mechanisms and frequency of this DNA transfer, known as "horizontal gene transfer," are discussed in detail in Chapters 7 and 8. The potential risks associated with horizontal gene transfer are discussed in Chapters 7, 8, and 13.

Summary of Factors Affecting Dispersal

The dispersal of viruses, bacteria, fungi, and other microorganisms has been demonstrated in a wide range of environments. Information about factors that affect microbial dispersal in various environmental media has been reviewed in other chapters (see Chapters 5 and 6).

Figure 11.2 summarizes the major factors that affect release, transport, and deposition, the processes that cumulatively describe dispersal. Factors important to the description of the initial source are also included. Factors affecting the transport step are subdivided to reflect the different parameters that must be considered in regard to transport of particles through the air by wind, and transport by water flow in water and soil. Factors affecting adventitious transport, such as by insects or the movement of workers on the site, are not included in Figure 11.2.

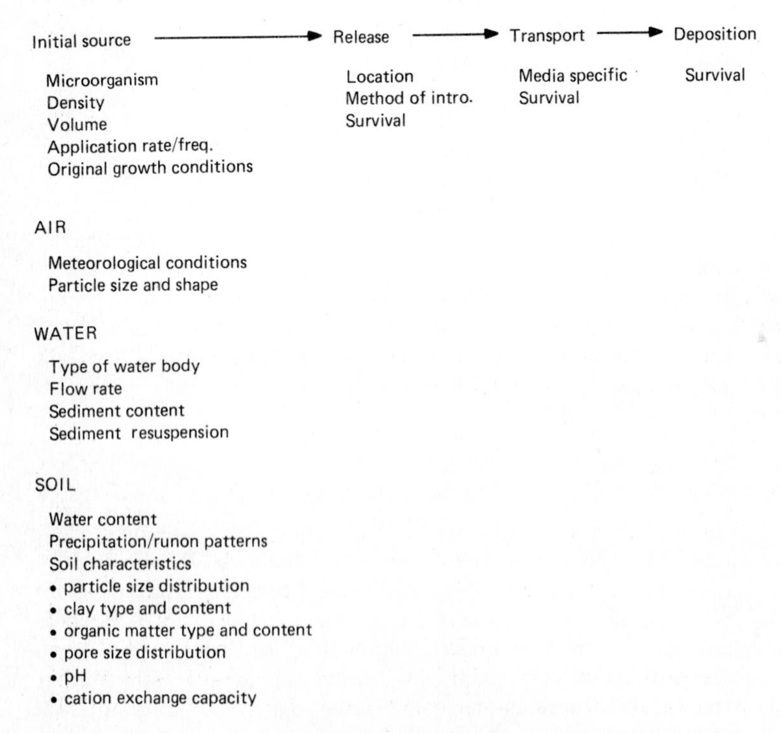

Figure 11.2 Factors that affect dispersal of microorganisms introduced into the environment.

TABLE 11.1 Physical and Chemical Factors That Affect Microbial Survival in Different Environmental Media

Air	Water	Soil
Temperature	Temperature	Temperature
Relative humidity	pH	pH
UV light	Organic matter	Moisture
Particle size	Oxygen content	Moisture holding capacity
Toxicants	UV light	Organic matter
	Toxicants	Nutrient supply
	Nutrients	Oxygen content
	Salinity	UV light
		Toxicants, including antibiotics

Figure 11.2 indicates that survival plays a role in each of the three subprocesses of release, transport, and deposition. The rate of die-off, and the parameter that controls that rate, may differ for each subprocess.

Table 11.1 summarizes the physical and chemical factors that affect microbial survival in air, water, and soil. Temperature, UV light, and the presence of microbial toxicants play a role in each of the three media.

Use of Models for Predicting Dispersal and Spread

Mathematical models have been used to predict the dispersal and spread of microorganisms. These models range from "back of the envelope" calculations to large, user friendly models implemented on mainframe computers. In recent years, there has been a large increase in microcomputer-based models.

Utility, credibility, and limitations

Dispersal models may be appropriate to use for a variety of reasons, including (1) to organize one's thinking about dispersal and spread, (2) to develop input into the exposure assessment term of a risk assessment, or (3) to guide the development of monitoring strategies.

A limitation on the use of any model is the fact that models are simplifications of reality and not necessarily reflections of reality. These simplifications may be a particular problem with fate and transport models of microorganisms in soil, and transport resulting from ground-level, or near-ground-level, aerosol releases. In these cases, patchy, anisotropic (i.e., not the same in all directions) conditions will be the norm, rather than the exception. In addition, there may be temporal patterns, such as diurnal and seasonal cycles, overlaying the spatial patchiness (cf., Wallace 1978). Unfortunately, it is precisely these changing conditions that are the most difficult to model, and where incorporation of parameters based on "average" or "typical" conditions can yield inaccurate results.

Every model is developed with a specific objective in mind. Although there is a tendency to use an "off-the-shelf" model, it may or may not be appropriate to use a particular model to achieve objectives other than the ones originally intended. In addition, even for apparently similar circumstances, a particular model may not apply, or important parameters, such as die-off rate, may require alteration. This lack of universality may be attributed to the complexity of the world being modeled and our lack of complete understanding of the factors which exert an influence.

Types of models and components

A variety of transport and fate models may prove useful for the prediction of the dispersal of microorganisms in water, air, or soil. Many of these models have been adapted from models developed to describe the fate and transport of chemicals or inert particles, such as dust. In general, transport models treat microorganisms as particles, and the fate component introduces the ability of microorganisms to reproduce or die off. These models may be useful in predicting both the spatial and temporal aspects of dispersal of microorganisms. However, they are likely to prove less useful in the prediction of spread, where characteristics of the receptor are also considered.

Two classes of models focus on effects due to microorganisms rather than the microbes themselves: dose-response models and epidemic models (Zanetos 1984). Dose-response models (which may also be called "dose-effects models") focus on the relationship between the administered dose, or route of the dose, and the proportion of hosts that respond under those conditions. Dose-response models emphasize the dynamics of microbial growth and the relationships between the dose, response time, and fraction responding. Epidemic models describe the spread of communicable diseases throughout a community (which could be humans, animals, or plants). They focus on the characteristics of the host population, rather than on the microorganism.

Both dose-response and epidemic models may prove useful in predicting the temporal and spatial aspects of spread if sufficient data are available. That is, these models explicitly account for the characteristics of the new site and the likelihood of the new site providing a hospitable habitat for the microorganism. However, effects-oriented models are likely to be less useful for the determination of dispersal, and certainly of dispersal by environmental media. In addition, effects models may prove impractical for nonpathogens, whose effects may be ill-defined and difficult to measure.

Many terms are used to describe the characteristics of models. For example, the terms "deterministic" and "stochastic" (also called "probabilistic") describe the types of mathematics incorporated into the model. The terms "simulation" (also called "analytical") and "empirical" (also called "statistical" or "synoptic") describe the types of information on which the model is based.

Deterministic models are usually based on transport or rate equations (i.e., the underlying mathematical structure is usually an ordinary or partial differential equation, or an integral equation). Deterministic models implicitly assume that the population as a whole acts in a predetermined fashion. Stochastic models, on the other hand, are based on probability theory. They assume that the different particles or microorganisms act independently (with independent probabilities) and that the population as a whole can be described by a probability distribution function. Stochastic models are often more mathematically complex than deterministic models, but are better suited to small populations where random statistical fluctuations can change the outcome.

Simulation models attempt to simulate each event with an exact functional relationship (which may be based on either deterministic or stochastic equations). For example, in a simulation model, each underlying step in a specified dispersal process (e.g., aerial dispersal from a nozzle sprayer to a strawberry leaf) would be characterized by a separate mathematical description. In principle, the mathematical description for each step would explicitly include the effects of all the physical, chemical, and biological parameters (such as aerosol size, temperature, humidity, windspeed, UV sensitivity of the microbe) that are important for that step, although this is rarely, if ever, practicable. Examples of simulation models used to study plant epidemics are provided in Chap. 12.

Empirical models are formulated directly from the data, rather than from underlying physical or other theoretical principles. Field data are analyzed by various statistical regression techniques to determine the major independent variables influencing the variable (e.g., the microbial distribution) of interest. These techniques generate empirical formulas that are likely to be suitable for predicting the effects of similar organisms under similar conditions. However, they are likely to be applicable (i.e., provide reasonably accurate predictions) to dispersal over only limited distances, times, and environmental and meteorological conditions.

Complex models can have both deterministic and stochastic elements. For some models, such as the Lighthart and Mohr (1987) modified Gaussian-plume model (see the section titled "Descriptions of Selected Models" below), the parameters used in the deterministic simulation model may themselves be empirical models. The values of the parameters are calculated from laboratory or field data similar to the situation being modeled.

In general, a simulation model requires more information, time, and resources to develop and use than an empirical model. On the other hand, simulation models, which are potentially more comprehensive, may allow predictions over a larger range of conditions than is possible for a synoptic, empirical model (Quentin 1979).

Criteria for selecting a model

General considerations. Many considerations should be incorporated into the decision as to whether or not a particular model for dispersal is appropriate for a particular use. These include (1) the purpose for which a model is being selected, (2) the ecology of the microorganisms being modeled, (3) the method of application (i.e., introduction into the environment) of the microorganisms, (4) the amount of data available, (5) the attributes and limitations of the model (see below), and (6) the flexibility of the model to be modified to reflect the particular situation. In general, models (or combinations of models) to be used for risk assessment should be sufficiently comprehensive to treat the entire system that may be affected by the introduced microorganisms. In addition, it is not advisable to use models which are based on simplifying assumptions that will affect the model's predictions under the circumstances of interest (Technical Resources, Inc. 1987).

Prior to the selection of a model, the user must develop an explicit set of model selection criteria that are appropriate to the objective of the study. It is usually helpful if the criteria are either ranked in order of importance, or grouped into categories such as "essential," "preferable," and "wouldn't it be nice." Sometimes a category labeled "inappropriate" may also be of use. Finally, the user should have a reasonable understanding of the model being examined for use, especially in regard to how the model incorporates the parameters which have major impacts on the real world results.

Attributes of models. Every model has specific attributes which can be used to characterize it. In this section, we describe nine model attributes, one or more of which will be important in selecting a model for a particular use. In the section titled "Descriptions of Selected Models," we summarize models that have been developed for microbial dispersal and spread, and include many of these attributes in the description.

Objective. Models are developed with specific objectives in mind, and these shape the form and output of the model. For example, epidemiological models are concerned with the effects of infectious diseases, and thus focus on the results of spread. In contrast, fate and transport models focus on where the microorganisms are likely to be, and how many will be there. As another example, some models focus on dispersal through a particular medium, such as air or water, while others may be multimedia models. When selecting a model for a particular use, it is critical to ensure that the objectives of the original model are similar to, or at least not inconsistent with, the objectives of the user.

Range and limits of applicability. Models will have a range of environmental conditions, time scales, distances, or types of microorganisms over which they will apply in principle. Validated models (see below) will have a range of conditions over which they have been tested for their predictive value. The user of the model should have a clear notion of the limits of applicability of each model being examined for selection and whether the system of concern falls with these limits.

Data requirements. Different models require quite different levels of data to be incorporated by the user. If few data are available to the model user, and only an order of magnitude estimate of dispersal is required, then a model with fewer data requirements is likely to be more appropriate than a model with larger data requirements. The amount of data required often reflects the level of refinement of the model.

Developers of models that require extensive site-specific and/or microorganism-specific data sometimes incorporate "default values" into their models. [MICROBE-SCREEN (see under "MICROBE-SCREEN" in "Descriptions of Selected Models") is an example.] "Default values" are average or typical values for a particular parameter—values which may or may not be appropriate for the application at hand. Users of this type of model should incorporate as much site- and microorganism-specific data as possible into their calculation. In addition, they should fully understand the sensitivity (see below) of the model to the values for which the default value option is permitted.

Computer requirements. Models can have various levels of mathematical complexity. In some cases, the level of complexity dictates the capability of the computer required (if any). If a model has been implemented on a computer, then the type of computer must be known. If it is on a mainframe computer, the user must have access to that mainframe. If it is on a microcomputer, then a compatible microcomputer, with sufficient memory (i.e., RAM), must be available.

Some models have been written primarily for the use of the developer, and little effort has been put into writing documentation or user friendly features so that others can use it with ease. Thus, the availability of documentation is another consideration in the selection of a model developed by others.

Level of development. Models, especially complex models, may be developed in stages. Thus one consideration in selecting a model is whether it has achieved a sufficient level of development to meet a particular need. For example, the multimedia model MICROBE-SCREEN (see under "Descriptions of Selected Models") is not fully developed. Models for a few transport pathways and intermedia transport have been incorporated, but more are planned. Thus, if the underlying model necessary to describe a particular system is already in MICROBE-SCREEN, it may be appropriate to use this system. If not, then the appropriate model should be sought elsewhere.

Quality assurance. Although models are not laboratory operations and thus do not require an assortment of reagents and adherence to protocols that define the reliability of each step, quality assurance is still required. A model is composed of formulae which encode theoretical concepts or summarize empirical data. Quality assurance involves checking that the formulae are correctly stated. If the model is computerized, quality assurance also includes the demonstration that the formulae have been correctly encoded. A user should be comfortable that the model selected has undergone sufficient quality assurance.

Verification and validation. To verify a dispersal model, the developer must demonstrate that the model can successfully predict the results based on hypothetical or actual laboratory or field data. For computer-encoded models, the formulae describing the theories involved (for simulation models) or the mathematical relationships (for empirical models) must be correctly linked to produce an operational program.

Model validation begins with a verified model, which is then challenged with data for a particular microorganism from a specific field-test site. The challenge data must be carefully selected to ensure that it is within the scope of the model's theoretical or empirical basis. The specific objectives of the user of a particular model will dictate the importance of selecting a validated model, or the extent of validation required. A validated model is essential to the assurance that the user isn't simply generating random numbers.

Sensitivity. A sensitivity analysis provides an estimate of the amount of variation in the model output that results from changes in a particular parameter. It can provide a useful index of the uncertainty of the model's output.

A primitive sensitivity analysis can be conducted by varying the values of selected parameters one at a time. A more sophisticated sensitivity analysis, or a Monte Carlo error analysis, can be conducted by using a random number generator to select numbers with a predetermined distribution (e.g., normal, log-normal, uniform) to use as input into each model parameter. The simulation is run numerous times and the results summarized.

Knowledge of which model parameters are most important to the model's predictions can assist researchers in prioritizing parameters to measure. It can also assist the developers of user friendly, multiuser models to determine which parameters can have "typical" values (default values) provided and not substantially degrade the predictive value of the model.

Acceptance. The acceptance of a model by the scientific community is an important attribute for model selection, especially if the results of the model's use may be controversial and incur a large cost. A model can gain acceptance through widespread peer review, publication in a respected, peer-reviewed journal, or modification of a model with a history of successful use. For example, the Gaussian-plume model is a widely accepted model for aerial dispersal of particulates. Modifications of this model for aerial dispersal of microbial aerosols have been developed and used by numerous authors.

Sources of information about the models

Sources of models. Models for the movement of microorganisms, including their transport, fate, and dispersal, come from a variety of disciplines. Hydrologists, hydrogeologists, soil scientists, and atmospheric scientists have developed models for the movement of particles in water, groundwater, soils, and air, respectively. The published scientific literature based in these disciplines is a useful source of models. In addition, U.S. government agencies, including the U.S. Environmental Protection Agency (USEPA), Department of Agriculture (USDA), Department of Defense, and Department of Energy, have sponsored the development of models for specific purposes. Some of these models are described below (see under "Descriptions of Selected Models").

Epidemiologists, epizoologists, and plant pathologists have developed models based on the incidence of disease, or the results of the spread of infectious organisms. In addition, ecologists have developed models to predict biological invasions. These models are presented in the scientific literature of these respective disciplines, and in the mathematical biology and theoretical biology literature. As with the fate and transport models, unpublished models may be found in unpublished government reports, especially reports of the USDA and the agricultural agencies of other countries. Specific examples of epidemic models developed by plant pathologists are discussed in Chap. 12.

USEPA has sponsored several projects to assess and adapt current models, or to develop new models appropriate for assessing the dispersal of introduced microorganisms (Zanetos 1984, Sticksel 1984, Versar 1986), including those that have been genetically engineered. Reports that discuss and summarize models for ecological risk (Technical Resources, Inc. 1987, Barnthouse 1988) also include dispersal models that may be relevant to introduced microorganisms. In addition, many of the models evaluating pathogen risks from various sewage sludge and wastewater disposal options contain relevant models or data (USEPA 1985).

Sources of data. Microorganisms in the environment have been studied in conjunction with numerous disciplines for decades, if not centuries. For example, public health scientists and civil (sanitary) engineers have carefully monitored the dispersal of enteric bacteria and viruses as a result of municipal sludge and wastewater disposal, leaking sewer systems, and the use of septic systems and leachfields. In recent years, USEPA has sponsored numerous field studies and summary reports regarding the movement of pathogenic microorganisms following land spreading or ocean disposal of sewage sludge.

Agricultural scientists have tracked fecal organisms resulting from animal wastes to observe their disease-producing characteristics in farm animals and humans. Many of these studies have been conducted under USDA sponsorship. Plant pathologists in many countries have monitored the spread of plant diseases and the causative organisms. They have also developed models to predict the spread of the disease and the result of various interventions. Data are also available on the growth and dispersal of currently available microbial pesticides, such as *Bacillus thuringienesis* and *Bacillus popilliae*.

Some of the data has been developed over half a century of research (*B. thuringiensis* was first described in 1906) and reported in any of a large number of biological journals. However, there has been an explosion of biotechnology-related data development over the past two decades, reported in a wide variety of sources. The range is great: topics include genetic manipulation techniques, successes in gene transfer, ecological observations regarding adverse effects and efficacy, regulatory information, and product literature; the list is almost endless. As a result, there have been a number of developments in the availability and utility of online systems (user-operated computer systems for searching databases), and the number of information vendors (end-user systems) has increased. Overall, vendor coverage has expanded, and new vendors have become available to aid in searching the literature. A full discussion of databases and data acquisition is beyond the scope of this chapter. The reader is referred to two reviews containing detailed information that have appeared recently: Bruce et al. (1989) and Johnson (1989). Highlights of the reviews are presented in Table 11.2.

The online systems, such as Medline, Biosis, or STN International (formerly *Chemical Abstracts*), can be entered individually by the user or through a librarian. End-user systems provide access to a number of these online databases, which are linked via user friendly, menu-driven software. The cost for this service varies greatly, as does the coverage of the individual vendor. Some (e.g., Agribusiness) charge on a per citation basis ($0.50 per citation) as well as a per hour basis ($96), and some (e.g., Dialog) have a variable hourly charge ($36 to $178).

TABLE 11.2 End-User Systems and Biotechnology Databases

Name	Source	Cost ($/h)	Content
Agricola (D)[a]	National Agriculture Library	39	Agricultural research, regulations, natural resources
Biosis (EU)[b]	Biosis Inc.	45	General: life sciences
Compendex Plus (EU)	Engineering Information Inc.	108	Bioengineering, environmental technology
Medline (D)	National Library of Medicine	36	Medical research
Dialog (EU)		36–178	Medical, life science, agricultural research
Microbial Strain Data Network (D)	Cambridge University, U.K.	17	Microbial strain data

[a]D = database
[b]EU = end user

The same online databases can be entered via any of several end-user systems. Thus, BRS, Dialog, or Paper Chase can be used to search Medline, in addition to direct access through the National Library of Medicine's MEDLAR's system (which is now easily used with Grateful Med software). BRS and Dialog include other databases as part of their service. BRS coverage includes Excerpta Medica, Biosis, PatData, and *Chemical Abstracts*. Dialog provides more extensive coverage, giving the user the ability to search CAB abstracts, Life Sciences Collection, Biocommerce Abstracts, Biosis, *Chemical Abstracts*, BioBusiness, World Patent Index, and two biotechnology abstracting services. Each of the end-user vendors is continually adding databases to its service, and current catalogues should be consulted for the databases that are available.

Finally, the USDA has initiated an electronic Biotechnology Bulletin Board as part of its National Biological Impact Assessment Program (NBIAP), which can be accessed directly using an 800 number (1-800-NBIAPBD). The system contains information about regulations, current literature, field-test approvals, and patents, as well as a listing of other information sources.

Descriptions of Selected Models

In this section, we describe models that may be useful in the prediction of the dispersal or spread of microorganisms introduced into the environment. The models vary in their various attributes, including their objectives, complexity, data requirements, and computer requirements. Some models, such as MICROBE-SCREEN, are multimedia models of fate and transport, while others model transport in only a single medium.

Fate and transport models

Numerous fate and transport models have been developed for chemicals, microorganisms, and/or particulate matter. The theories underlying the models are based on transport mechanisms relevant to the environmental medium under consideration. The source terms allow for (or assume) initial releases of microorganisms that differ in terms of numbers released, frequency of release, and the geometric configuration of the release. In this subsection, we classify models according to the environmental medium they were designed to describe. We use the term "multimedia models" for those models which allow mass transfer among several media during a single model run.

Air dispersal models. Air dispersal models can either be deterministic (e.g., those based on Gaussian-plume dispersion), or they can be stochastic (e.g., a random walk model). Several variants of the Gaussian-plume model that have been adapted to better predict the behavior of microorganisms are described in this section. A random walk model to predict the dispersal of microorganisms has also been published (Lighthart and Kim 1989). Models of microbial releases from large-area sources can utilize a different model formulation, such as that described in the box model. Recent experiments to prepare for the field testing and use of genetically engineered microorganisms in agriculture have led to the development of empirical models of air dispersal, as described under "Empirical Model for Aerosol Dispersal from an Agricultural Sprayer," later in this section.

Modified Gaussian plume. The objective of the modified Gaussian-plume model developed by Lighthart and Frisch (1976) is to provide crude estimates of ground-level concentrations of airborne microorganisms dispersed from a cooling tower. It is intended to be a heuristic model to help nonmeteorologists to understand the factors affecting downwind concentrations.

The Lighthart and Frisch (1976) modified Gaussian-plume model is based on the Pasquill inert-particle-dispersion model that has been modified by the addition of terms for droplet settling and microbial death rate. It assumes that the mean concentrations of the microbial aerosols are normally distributed about the plume axis. It allows for the superposition of many Gaussian plumes: one for each droplet size category. In a sensitivity analysis, Bovallius et al. (1980) showed that the main factor influencing dispersal is the meteorological situation. "Decimal reduction times" (DRTs, or the time it takes for the viable microbial population to decrease 10-fold) of 1 minute or less will significantly shorten the dispersal distance.

Limitations. As with all models, the modified Gaussian-plume model has a range of applicability; the model may not be useful outside this range. For example, the model is applicable to dispersal from small point sources for distances as large as 1 to 10 km from the source. It is only applicable in regions of constant wind and diffusivity. It will be inaccurate near the atmospheric boundary layer. It does not account for terrain, and thus the modified Gaussian-plume model is not suitable to use where the terrain is highly uneven. Finally, even if the model is used within the limits of its applicability, only order of magnitude agreement with experimental results is anticipated.

Data requirements. Data requirements for the modified Gaussian-plume model include source strength (including the emission rate), mean wind speed (usually 1-hour averages), atmospheric stability, particle size(s), microbial die-off rate, stack characteristics, and plume spread. The plume spread rates provided in the original paper describing the model (Lighthart and Frisch 1976) were determined empirically and are not universally applicable.

Validation. The Lighthart and Frisch model has undergone validation for some conditions. For example, the model predictions are consistent with data of bacteria dispersed from sewage treatment plants and via spray irrigation (Bovallius et al. 1980). This model is also readily adapted to a variety of computers and has been implemented on computers of all sizes.

Microbial survival in modified Gaussian-plume dispersion. The modified Gaussian-plume model just described (Lighthart and Frisch 1976) has been further modified to include an airborne microbial survival term that takes weather variables into account (Lighthart and Mohr 1987). The accuracy of the survival term is critical to the prediction of dispersal for long travel times, such as would occur at low wind speeds and with long-distance transport. At long times, the die-off rate controls the dispersal distance of viable microorganisms.

The modified Gaussian-plume equation for aerial dispersal (i.e., the Gaussian-plume equation was modified to include terms for microbial die-off rate and microbial settling) was further modified to include the effect of environmental variables on the microbial die-off rate. Specifically, the model was changed to include an algorithm with a polynomial function modulating airborne microbial death due to temperature, relative humidity, solar radiation, and time in the aerosol. The parameters used in the polynomial function are based on the prevailing conditions when the microorganisms are released; it is assumed that the environmental conditions in the plume remain constant after the initial aerosolization. The coefficients of the polynomial were determined for a "composite virus" based on laboratory survival data.

Limitations. The limitations of the underlying modified Gaussian-plume model are the same as those described in the previous section, although, under steady weather conditions, its applicability may be extended by the modifications introduced by Lighthart and Mohr (1987). The inclusion of the environmental variables in the die-off rate requires values that are microbe-specific and release-specific (i.e., specific to the site and the time of year and day of the release), making this adaptation of the Gaussian-plume model more data intensive than previous versions. For example, the coefficients of the polynomial should be developed or validated for each microorganism being tested. Another limitation of the survival term is that the functional form assumes no interactions among the various environmental factors, although it is known from experimental data that interactions are important.

Data requirements. The data requirements for the Lighthart and Mohr (1987) modification of the Gaussian-plume model include those described in the previous section in addition to temporal weather input data. If a microorganism other than the composite virus utilized by the authors is modeled, then microbe-specific and release-specific values for the effects of environmental variables on microbial die-off rates should be obtained. Laboratory data may or may not reflect the realities of the field.

Computer requirements. The algorithm, and the modified Gaussian-plume model, were written in Applesoft BASIC to run on an Apple IIe with 128K memory.

Model validation. No data available. However, the model was used to simulate virus dispersal at different times of the day or year in Eugene, Oregon, suggesting that the model has been appropriately verified.

Deterministic (Gaussian-plume) aerial-dispersal model. The objective of the modified Gaussian-plume aerial-dispersal model developed by Camann (1980) is to predict the dispersion of microorganisms in wastewater aerosols produced by spray irrigation, cooling towers, and aeration surge basins.

The Camann model has the general form

$$C_d = Q \times D_d \exp(ka_d) + \mathrm{B}$$

where C_d is the concentration of microorganisms at a distance d downwind from the source
D_d is the diffusion factor at d
Q is the source strength,
k is the die-off rate
a_d is the age of the aerosol at distance d
B is the background air concentration of microorganisms

The diffusion factor can be expressed in Gaussian form. The general form of the model is shown in Fig. 11.3.

The source strength term can be determined either empirically or by modeling, depending, in part, on the complexity of the source. For spray irrigation,

$$Q = W \times F \times E \times I$$

where W is the microbial concentration in the water
F is the flow rate
E is the aerosolization efficiency factor
I is the microorganism impact factor

Limitations. The uncertainty and variability of the die-off rate, which is a key parameter at larger distances and times, limits the model's predictiveness beyond 200 to 500 m.

Data requirements. The model requires input of the microbial die-off rates, the microbial impact factor (both of which are expected to be sensitive to environmental conditions and the microorganisms used), the number of microorganisms in the wastewater, the flow rate, and meterological data. The die-off and impact factors can be determined from empirical models based on multivariate regression analysis of I and k measured over a range of environmental variables such as relative humidity, UV light, temperature, wind speed, air pollutant indices, and total suspended solids.

Computer requirements. This model has probably been implemented via several PC programs. The spray irrigation version is the point-source air model component of MICROBE-SCREEN, and thus is available on the GEMS system (see below under "Multimedia Models").

Validation. The model has been validated for the spray irrigation model where rotating impact sprinklers are used to apply wastewater with no residual chlorine. The predictions deteriorate with distance from the source.

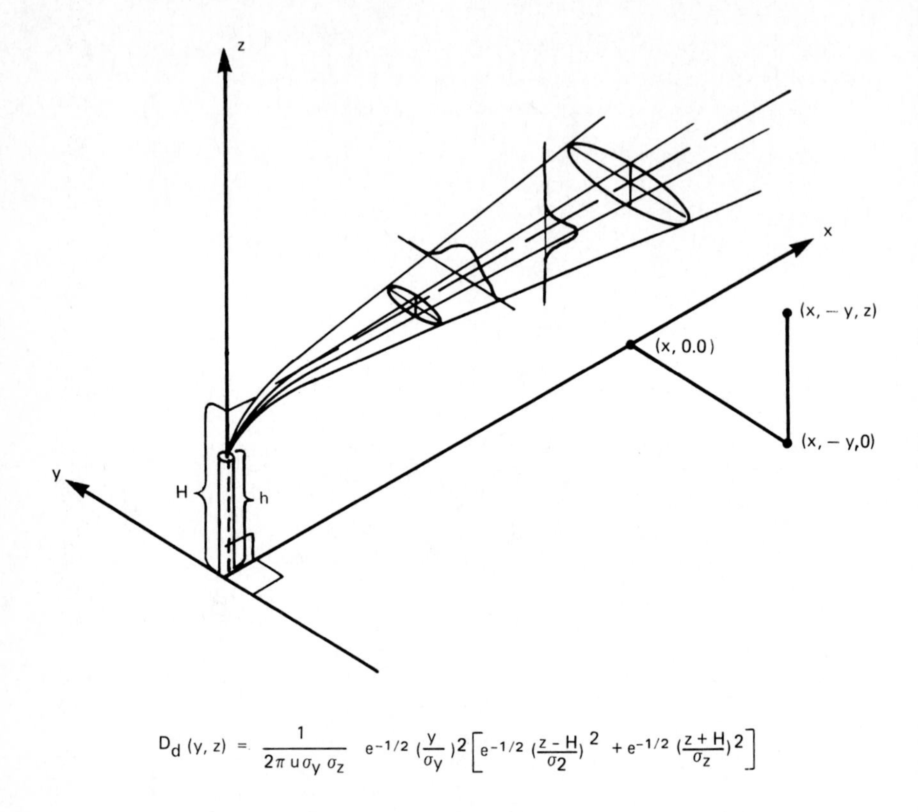

$$D_d\,(y, z) = \frac{1}{2\pi\, u\, \sigma_y\, \sigma_z}\; e^{-1/2\,\left(\frac{y}{\sigma_y}\right)^2}\left[e^{-1/2\,\left(\frac{z-H}{\sigma_2}\right)^2} + e^{-1/2\,\left(\frac{z+H}{\sigma_z}\right)^2}\right]$$

where $D_d\,(y, z)$ = diffusion factor at *d* (sec/m^2)
x = distance in downwind direction, (m)
y = distance in crosswind direction, (m)
t = distance in vertical direction, (m)
u = mean wind speed, (m/sec)
σ_y = standard deviation of plume concentration distribution in the crosswind direction, (m)
σ_z = standard deviation of plume concentration distribution in the vertical direction, (m)
H = height of plume center-line (m)

Figure 11.3 Gaussian form of the diffusion factor D_d. (*Camann 1980*).

Random walk model of aerial dispersal. The dispersion of individual droplets of water containing microbes has been simulated using a model that incorporates a random walk component (Lighthart and Kim 1989). The model is intended to provide relatively high spatial resolution near to, and downwind from, the microbial source.

The simulation model is composed of five submodels: aerosol generation, evaporation, dispersion, deposition, and microbial death. The aerosol generation term is based on the type of sprayer (i.e., a nozzle under pressure) used in the "Ice$^-$" experiments conducted at Tulelake, California, by Lindow et al. (1988). The dispersion submodel incorporates a random walk probability model. Microbial death is modeled as a single exponential decay, where the rate depends upon time, temperature, and relative humidity.

Data requirements. The data requirements are extensive, and include detailed knowledge of the aerosol-generating equipment, detailed meteorological data, and detailed microbial death data.

Computer requirements. The model is written in FORTRAN and has been run on a Digital Equipment Corporation VAX 785/8605 cluster computer.

Validation. The model has been challenged using data from the Ice^- field trial at Tulelake, California, in May 1988. Similar deposition within 30 m was obtained from experimental data and the simulation model. However, the simulation used only 10^4 droplets, compared to 10^{12} released, thus limiting the ability to evaluate fully the predictiveness of the model.

Aerial-dispersal model of foot and mouth disease. Recently, transport models have been used to predict the spread of epidemics where transmission occurs through the air. The following model for the spread of foot and mouth disease (FMD), developed by Gloster et al. (1981), is an example.

The FMD model is designed to provide an estimate of the area most at risk from airborne spread of the FMD virus within a 10-km radius of a known source. The model is based on the Gaussian-plume equation with provision for the effect of relative humidity on virus survival and the effect of local topography on plume dispersion.

The model computes the dose received by a cow at one of 360 grid points within a 10-km radius. It assumes that a parcel of air will stay intact over this distance and that the effect of environmental factors on die-off rate will be constant. The model has a daily time cycle.

Data requirements. The model uses meteorological data from COSMOS (United Kingdom weather service program) to obtain hourly data on wind speed, wind direction, relative humidity, cloud cover, and precipitation. It requires topographical data that may be obtained from geological survey maps augmented by on-site inspection. The model also requires an estimate of daily virus output from infected animals.

Computer requirements. The program was written for an Olivetti PC in FORTRAN and requires 128K storage, 32K memory, and 6 seconds to perform the calculations for a 10-day emission period.

Validation. The model was tested with historical data from two FMD outbreaks in the United Kingdom and successfully identified locations at which animals became infected. It was applied successfully when an outbreak of FMD occurred soon after the model was completed.

Empirical model for aerosol dispersal from an agricultural sprayer. Lindow et al. (1988) have developed a synoptic empirical model for the ground-level deposition of the bacteria *Pseudomonas syringae* immediately after aerosolization through a large-orifice (Tee-Jet 8004) nozzle. Based on a least-square regression analysis of field-sampling data, they determined the relationship

$$N = A \exp(-XD)$$

where N is the number of cells deposited per square centimeter
D is the distance in meters
A and X are parameters used to define the best fit

The values for A ranged between 0.3 and 0.45. The values for X ranged between 17.4 and 29.2.

Modified box model for aerial dispersal from large-scale sources. The objective of this model, developed by Bovallius et al. (1980), is to estimate long-distance downwind concentrations from large-scale sources. The transport is assumed to take place between ground level and an inversion layer. The model assumes complete mixing, and thus homogeneous concentrations of bacteria. The DRT of the bacteria is the dominant factor in the equation. This model is drawn from the air pollution literature.

Soil and groundwater models

Porous media. Corapcioglu and Haridas (1984, 1985) developed a model to predict the transport and fate of microorganisms through water-saturated porous media, such as saturated soils and groundwater. The model represents microbial transport in three dimensions, and includes the time dependence of transport. The fate portion of the model assumes that microbes are particles with defined decay and growth rates. Growth is based on the Monod equation (see under "Fate Models" below) and decay on a first-order decay rate.

The model allows for the removal of microorganisms from the soil-water suspension by three clogging mechanisms: straining in the contact zones of adjacent pores, sedimentation in the pores, and adsorption. Declogging mechanisms are also incorporated. The transport equation allows for movement due to chemotaxis and random (tumbling) motions of bacteria as well as dispersion and convection associated with particle movements. The growth and chemotaxis processes are coupled with a transport equation for bacterial nutrients. The resulting equations are highly nonlinear and require numerical solutions.

Computer requirements. The equations require numerical solutions. This has been done by the authors, but no computer code appears to be generally available.

Data requirements. The model requires several composite rate constants, such as those for clogging and declogging, and growth rates. It also requires bacterial concentrations and hydrologic and soil parameters.

Model validation. None.

Model for transport through both saturated and unsaturated zones of soil. WORM and Sumatra I are one-dimensional, finite-element models that predict the movement of solutes through the unsaturated zone to the saturated zone of soils and through nonhomogenous soils. The effects of linear adsorption (through the use of a K_d, or dissociation constant, for chemicals) and zero- and first-order decay are included. WORM, and to a lesser extent Sumatra, allow a variety of hydraulic and concentration boundary conditions (Versar 1987). Neither model allows for the calculation of altered effective porosities due to clogging (filtration), an important consideration for the movement of bacteria through soil.

Sumatra I was written for a mainframe computer, but has been adapted to run on an IBM PC. Both the program and documentation are available from the Holcombe Research Institute, International Groundwater Modelling Center, in Indianapolis, Indiana.

WORM is derived from Sumatra I, but it uses a different computer algorithm and runs more efficiently on a PC. Although it is more user friendly than Sumatra I, it is not especially so. The code and unpublished documentation for WORM are available from the developer, M. Th. van Genuchten, at the USDA Salinity Laboratory in Riverdale, California.

It does not appear that WORM has been validated for bacteria. It has been used to simulate the movement of *Pseudomonas* through soil under various conditions (Versar 1987), but these simulation runs were apparently not compared with data to test their accuracy.

Aquatic models

EXAMS II. EXAMS II is a complex, multicompartment model developed to predict the transport and fate of chemicals in aquatic media. The chemicals can be dissolved in the water column, adsorbed to the sediment, complexed, or found in the aquatic biota. The model allows for transformation, and could be adapted to incorporate simple microbial die-off rates. EXAMS II provides for advective and dispersive transport, with a relatively simple hydrodynamic model (Burns and Cline 1985). EXAMS II has been incorporated into MICROBE-SCREEN, a multimedia model intended for use with microorganisms (see below).

Limitations. EXAMS II is time-varying, but it is not truly dynamic.

Data requirements. The model requires 44 site constants and other variables as well as chemical parameters. Default values are available for the site parameters.

Computer requirements. EXAMS II is written in FORTRAN 77. It requires 512K RAM and a 5-megabyte hard disk. It is an interactive, command-driven model. The code and documentation is distributed by the USEPA laboratory at Athens, Georgia.

Validation. EXAMS II has been validated for several chemicals and sites. The model is widely accepted and used for chemicals, but not with microorganisms.

WASP3. WASP3 is an aquatic fate model with a complex hydrodynamic component (DYNHYD3) built into it. It was designed as a general chemical model, especially for large lakes or estuaries (Ambrose et al. 1986).

WASP3 includes compartments for chemicals that are dissolved, sorbed to the sediments, and incorporated into the biota. The processes modeled are discharge, tide and wind-driven flow, advective and dispersive transport, sedimentation, scour, pore-water dispersion and percolation, and sediment-water dispersion.

Data requirements. The site-specific data requirements are extensive. However, a simple, stationary site, such as a pond, can be run with minimal data.

Computer requirements. WASP3 is written in FORTRAN 77. It requires a microcomputer with 640K RAM, a 5-megabyte hard disk, and a 132-column printer. The code and documentation are available from the USEPA laboratory in Athens, Georgia (Ambrose et al. 1986).

Validation. WASP3 is still a relatively new model. It has been verified by the original authors and is being tested by others.

Multimedia models

MICROBE-SCREEN. MICROBE-SCREEN is a screening-level multimedia model developed to assess the dispersal and fate of microorganisms released into the air, surface water, or soil. Figure 11.4 diagrams the processes included in the model. Four types of surface-water bodies are considered: lakes, rivers, estuaries, and oceans. MICROBE-SCREEN allows intermedia transport of microbes from air to soil, rivers, lakes, and estuaries; and from soil to rivers, lakes, and estuaries. Transport of microbes on suspended particles in water bodies is not considered, although it has been shown to be important under field conditions. The model assumes that microorganisms are passively, rather than actively, dispersed (i.e., self-propulsion and direct-contact models are neglected). Additional transport pathways are being developed to enhance the model (General Sciences Corporation 1987, Reichenbach et al. 1987, Chiu 1988).

MICROBE-SCREEN was adapted from TOX-SCREEN, a chemical fate model, by modifying TOX-SCREEN to account for processes important to modeling microorganisms, such as growth and decay, settling, aerosolization efficiency, and filtration. Growth is modeled based on the Monod equation and decay as a first-order rate constant.

MICROBE-SCREEN is based on a set of simulation models and provides microbial concentrations at designated time steps (1-hour calculations for the air model, output at 48 hours for all media) throughout a 1-year simulation period. Underlying models in MICROBE-SCREEN include:

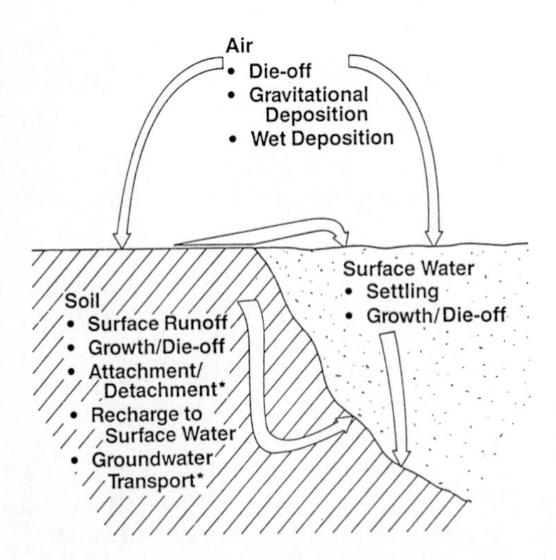

Figure 11.4 Processes modeled in MICROBE-SCREEN.

For air: Modified Gaussian-plume equation for particulates, based on the equations developed by Camann (1980), and described above under "Deterministic (Gaussian-Plume) Aerial-Dispersal Model." Point and area sources are allowed.

For rivers and lakes: Similar to the EXAMS model (see under "EXAMS II" above). It includes growth, decay, and settling, but not resuspension of sediments.

For estuaries: One-dimensional steady-state model with constant cross-sectional area, constant tidally and sectionally averaged longitudinal dispersion coefficient, and a constant freshwater velocity for simulating dispersion of substances in estuaries. Settling and decay rates, but not growth, are considered.

For oceans: Steady-state Gaussian-type linear diffusion model. Decay rates are considered, but not growth or settling.

For soil: Modified SESOIL model from TOX-SCREEN to account for growth or decay, and filtration.

Limitations. The air model calculates concentrations every hour and may not be applicable to short, aerosol releases near ground level. The surface-water transport model does not consider resuspension of sediments, which is an important mechanism, at least for fecal bacteria.

Computer requirements. MICROBE-SCREEN requires a complex data management system. To accommodate this, it has been implemented in USEPA's Graphical Exposure Modeling System (GEMS). It is available to the public via USEPA's VAX 11/780 (mainframe computer). A PC version of TOX-SCREEN, but not MICROBE-SCREEN, is available.

GEMS, TOX-SCREEN, and MICROBE-SCREEN are user friendly, menu-driven systems with user-interactive prompts. The computer code is in FORTRAN IV.

Data requirements. MICROBE-SCREEN requires a large amount of data, from site-specific climatic values (accessible in the database) to microbial growth and persistence rates, to loading and emission rates. Default values are built into the underlying data set for all parameters.

Model validation. MICROBE-SCREEN is currently undergoing validation by the USEPA's Office of Toxic Substances (OTS). The first model validation results, based on a field study of *Pseudomonas fluorescens* genetically engineered to metabolize lactose (developed by Monsanto Corp. and tested by researchers from Clemson University, Drahos et al. 1988), appear encouraging (Chiu 1988). However, others have pointed out that *P. fluorescens* data are inappropriate to use for the validation, as this microbe is known to be root-associated and the transport is more likely to be by root growth than by soil infiltration, as specified in the model.

MWASTE. MWASTE was developed to evaluate the dispersal of fecal coliform bacteria through soil and water (Moore et al. 1983, 1987). Figure 11.5 provides a schematic diagram of the pathways considered in the model.

MWASTE was developed using existing field and laboratory data. It consid-

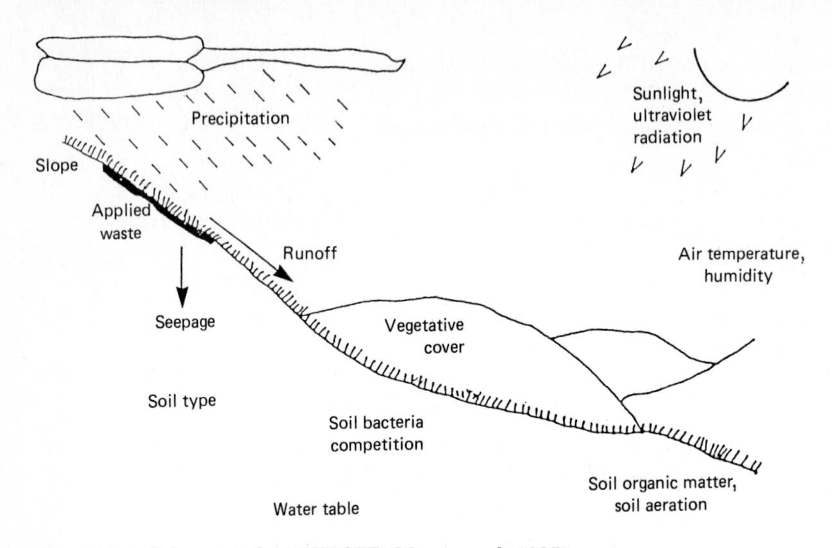

Figure 11.5 Schematic for MWASTE (Moore et al. 1987).

ers source strength, bacterial die-off, topography, precipitation, soil characteristics, and agricultural management practices. It was originally developed for use with dairy cattle in Tillamook Bay, Oregon, and has been successfully extended and applied to other livestock and locations (Moore et al. 1987).

MWASTE assumes passive movement of bacteria and first-order kinetics to estimate partitioning of the coliform bacteria. It uses the CREAMS model (Kneisel 1980, and see below) to generate the runoff component. Growth is not considered since coliform bacteria do not multiply in soil or water. Figure 11.6 provides a flowchart of the MWASTE simulation. The model calculates the number of bacteria expected at a given site. It operates on a daily basis.

Computer requirements. CREAMS is written in FORTRAN, and can be run on an IBM or compatible PC with at least 256K RAM.

Data requirements. There are numerous data requirements such as animal types and number, housing type, and pasture size, as well as detailed meteorological data required by the CREAMS model (see below). Default values have been programmed for bedding values for a variety of livestock.

CREAMS and GLEAMS. Scientists at the USDA developed the CREAMS (Chemicals, Runoff, and Erosion from Agricultural Management Systems) model to evaluate the impact of different agricultural management practices on non-point-source pollution from field-size areas (Knisel 1980, Leonard et al. 1987). The model was originally developed to simulate edge-of-field loadings of sediments and agricultural chemicals. Thus, the model contains a chemistry component as well as extensive water balance calculations to predict the amount of water in runoff, while considering alternate fates such as percolation in the root zone or evapotranspiration. The CREAMS model has been successfully used under a variety of conditions, including as a key component of the MWASTE model to estimate coliform runoff, as described above.

GLEAMS (Groundwater-Loading Effects of Agricultural Management Sys-

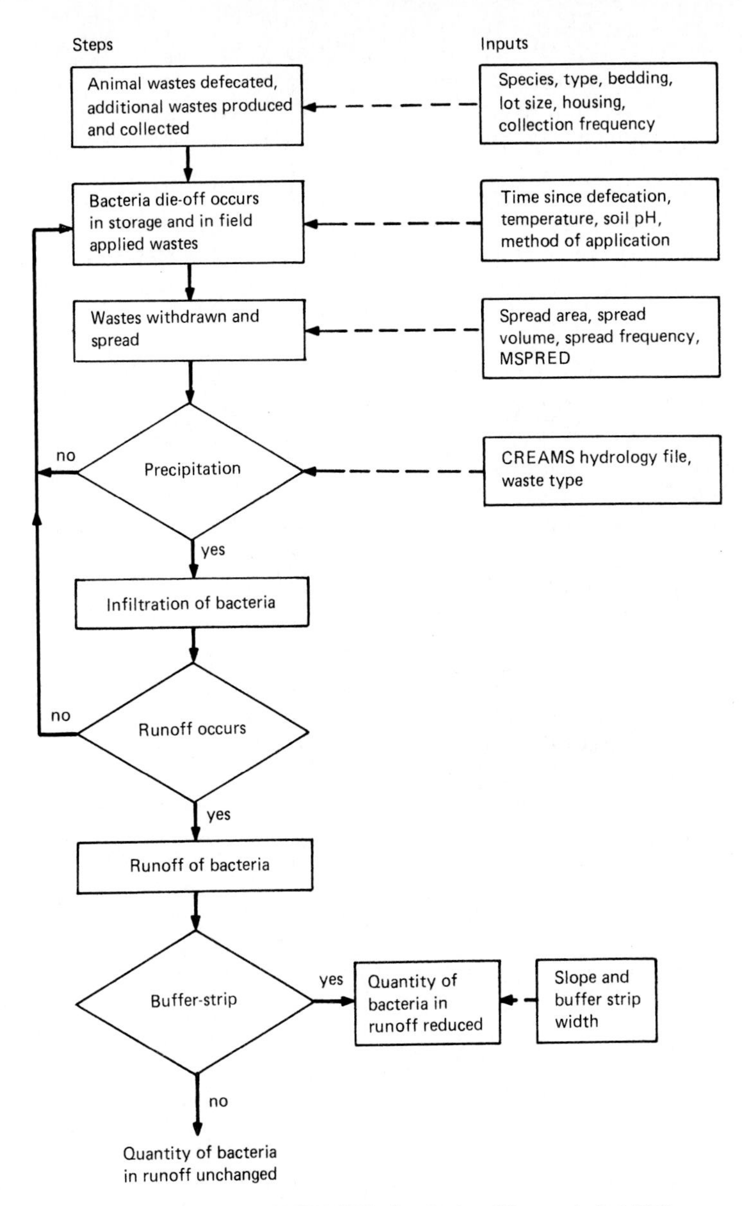

Figure 11.6 Flowchart of MWASTE simulation (Moore et al. 1987).

tems) represents an expansion of the CREAMS model to include an evaluation of the impacts on groundwater in field-size areas of the movement of agricultural pesticides through the plant root zone (Leonard et al. 1987). GLEAMS includes a hydrology component, an erosion component, and several pesticide components. Although many of the pesticide elements are not likely to be necessary for modeling the movement of microorganisms, the combination of the runoff and root-zone-transport components may be useful.

GLEAMS is undergoing continuous updating and expansion by the USDA (Davis et al. 1990). Publicly available components have been validated using field-based pesticide data.

Computer requirements. Both CREAMS and GLEAMS are written in FORTRAN, and can be run on an IBM PC or compatible computer, as well as mini- and mainframe computers. User support manuals are available for both programs.

Fate models

Two fate processes are commonly modeled for microorganisms: growth and decay. Decay is most often modeled as a simple first-order decay rate. Growth has been modeled either as a simple first-order growth rate (simply reversing the sign on the first-order decay rate), or as a Monod growth process, which latter assumes that growth can occur until the available substrate (nutrient) is depleted.

Mathematically, the Monod equation is expressed as

$$\mu = \frac{\mu_{max} S}{K_s + S}$$

where μ is the specific growth rate of the microbial population
S is the concentration of the limiting nutrient
μ_{max} is the maximum specific growth rate
K_s is the saturation constant

As pointed out by Bazin et al. (1976), the Monod equation gives an adequate fit to data from systems in which microbial populations are at steady state. However, transient behavior, such as would be expected of a microbial population newly introduced into soil, or being transported through soil, is not as well described. In addition, the Monod equation does not account for the effects of temperature, pH, and other environmental effects on microbial growth.

Dose-response models

Dose-response models focus on the relationship between the size of the dose of microorganisms entering the target (host) organisms of concern, and the proportion of target organisms (hosts) responding. The models described here were developed for pathogenic microorganisms with known hosts.

Zanetos (1984) described four major classes of dose-response models, based on the underlying mathematical assumptions. The classes are (1) birth-death models, (2) log-normal models, (3) single-hit exponential models, and (4) β-distributed "infectivity probability" models. These models are described briefly below.

"Birth-death models" are stochastic models which assume that the outcome [which was infection for the example developed by Armitage et al. (1965), but it could be colonization of any new niche] is determined by a chain of successive random events. The chain continues until not a single organism remains viable in its host. Thus, in each small time interval dt after inoculation, each organism has a probability λdt of reproducing, a probability μdt of dying, and a probability $1 - (\lambda - \mu)dt$ of doing neither. The model postulates that $\lambda > \mu$, and $p = 1 - (\mu/1)$ where p is the probability of a response (infection). In this model, an increase in the probability of response with an increase in dose is interpreted as reflecting the chance that one or more microbial clones will be "lucky" enough to increase to the critical number (Armitage et al. 1965, Zanetos 1984).

Predicting the probability of infection from exposure to low doses of disease-causing organisms from data on disease incidence at higher doses requires a low-dose-extrapolation model. Haas (1983) has compared the predictions of three such models, the log-normal, single-hit exponential, and β-distributed "infectivity probability" model.

The log-normal model assumes that each host organism has a single-valued minimal infective dose (MID), but that the MID for a population of host organisms is distributed log-normally. Thus, if every member of a population is given the same dose, and the MIDs for each member of the population are log-normally distributed, the fraction of responders in the population is given by

$$P^* = \frac{1}{\sqrt{2\pi}} \int_{-\infty}^{z} \exp\left(-\frac{z^2}{2}\right) dz$$

$$z = \frac{N - \mu}{\sigma}$$

where P^* is the probability that a single individual exposed to a dose of N organisms will become infected, and μ and σ are the logarithmic average and log-standard deviation of the normal distribution, respectively. This model may be fit to experimental data using log-probit paper. Log-probit models have been used to characterize the population variability of responses to toxic chemicals as well (cf., Finney 1971).

The single-hit exponential model (Haas 1983, described by Zanetos 1984) is based on the notion that there are two steps in the disease process: (1) the host is exposed to one or more organisms capable of causing disease and (2) only a fraction of the original dose of disease-causing organisms actually reaches the target tissue in a viable form. The probability of response is the joint probability of these two steps. This is expressed mathematically as follows.

Let $P_1(j)$ be the probability that a single dose will contain j organisms, and $P_2(j,k)$ be the probability that, once having ingested (or inhaled) j organisms, k organisms will survive and reach a location within the host where they are capable of causing infection ($k \leq j$). The probability that k organisms will survive and produce infection following a single exposure is given by

$$P_k^* = \sum_{j=1}^{\infty} P_1(j)\, P_2(j,k)$$

This model explicitly includes host resistance, nonspecific pathogen decay, and the possibility of viable noninfective organisms. If the two probability functions and the minimum infectious dose k_{min} are known, then the probability that a single exposure will result in disease is given by

$$P^* = \sum_{k = k_{min}}^{\infty} \sum_{j = k}^{\infty} P_1(j)\, P_2(j,k)$$

If the disease-causing organisms are distributed randomly in the environmental medium of exposure (e.g., air, water), then the probability that a given number of organisms will be ingested (or inhaled) is given by a Poisson distribution:

$$P_1(j) = \frac{N^j \exp(-N)}{j!}$$

where N is the average number of organisms per dose

If it is assumed that the probability r that a given organism reaches its target organ in a viable state is independent of the presence of other organisms, then $P_2(j,k)$ can be represented by the binomial theorem:

$$P_2(j,k) = \frac{j!(r)^k(1-r)^{j-k}}{(j-k)!k!}$$

Combining the equations to obtain P^*, the probability that an individual exposed to N microorganisms will become infected, and assuming that $k_{min} = 1$, yields the single-hit equation:

$$P^* = 1 - \exp(-rN)$$

Comparison of the predictions of the single-hit equation with experimental data indicates that the actual dose-response curve is flatter than the prediction. It has been proposed that this discrepancy can be accounted for by variation in the virulence of the individual pathogenic organisms, variation in the sensitivity of the host, or both (Haas 1983).

The β-distributed "infectivity probability" model is based on the single-hit model, but allows r to have a predetermined distribution rather than a single value. If $f(r)$ represents the distribution of r, then the single-hit equation can be rewritten as

$$P^* = 1\int [1 - \exp(-rN)]f(r)\, dr$$

If $f(r)$ is taken as a β distribution which is defined only for values in the range of 0 to 1, then this equation can be integrated to yield

$$P^* = 1 - \left(1 + \frac{N}{\beta}\right)^{-\alpha}$$

where α and β are parameters of the β distribution.

Epidemic models

Epidemic models were originally developed to describe the spread of communicable diseases throughout a household or community. Early models assumed that disease transmission occurred only through direct contact, and are thus best suited for use in exposure scenarios where direct contact is important. Later models, such as the FMD model described previously, incorporate air and water transmission routes when appropriate.

There are both deterministic and stochastic variants of the simple and general epidemic models that are widely used by epidemiologists. There are also discrete time models, spatial models, carrier models, host-vector models (e.g., insects as disease-carrying vectors), and models involving epidemics in competition. A wealth of information about these models, including the mathematical details, can be found in the monograph by Norman Bailey entitled "The Mathematical Theory of Infectious Diseases and Its Applications." Many of these models have also been summarized by Zanetos (1984). Epidemic models used to describe plant epidemics are described in Chap. 12.

Summary of models

Each of the models described in this section has its own strengths and weaknesses in regard to predicting microbial dispersal and spread. No model will meet the needs of all users, or be of use in all situations.

MICROBE-SCREEN is intended to be the most comprehensive model for microorganisms. However, as discussed above under "MICROBE-SCREEN," this model is still undergoing development and validation. Thus, the components of importance to a particular release site may or may not be fully incorporated into the model at a particular time.

None of the models described in this chapter adequately considers the complexity of microbial growth and die-off during transport. For example, only the air dispersal model put forward by Lighthart and Mohr explicitly includes the influence of environmental factors on the rate of die-off, although it requires that they remain constant after initial aerosolization. However, as noted earlier, the rate of bacterial die-off is very sensitive to certain environmental factors, such as sunlight (UV), temperature, and humidity.

Two models, the soil component of MICROBE-SCREEN and the model proposed by Corapcioglu and Harida, allow for microbial growth. In both cases, the growth is described by the Monod equation, which assumes that growth can occur until the available substrate (nutrient) is depleted.

Using Models to Integrate Risk Assessment Information

An important component of a risk assessment for the introduction of microorganisms into the environment is the extent of microbial dispersal and subsequent spread. Fate and transport models can play a key role in assessing microbial dispersal. Furthermore, in conjunction with dose-effects and other models, fate and transport models contribute directly to predicting the likelihood (risks) of specified effects. The FMD model described under "Aerial Dispersal Model of Foot and Mouth Disease" is an example of a successful incorporation of a transport model in a prediction of risk from microbial effects. Use of models in this way is the microbial analog of chemical risk assessment, as discussed in Chap. 13.

Epidemic models, which consider the sensitive receptor as well as the released microorganism, are more likely to provide information about microbial spread. Epidemic models have traditionally been used to predict the effects of disease-causing microorganisms, and may be more useful than fate and transport models in some instances. For example, it may be more appropriate to use epidemic models if there is a high degree of adventitious transport (e.g., on surfaces or via insects), or if the receptor population has a high degree of spatial or temporal heterogeneity, is highly variable in its host susceptibility, or is extremely rare. On the other hand, fate and transport models may be more useful than epidemic models in some situations. For example, fate and transport models are more useful in answering the questions "Where are most likely locations for the dispersed microorganism?" and "When is it most likely to get there?" This information can then be used in the design of monitoring or intervention (mitigation) strategies.

Transport models can provide information about spatial and temporal dispersal patterns, while fate models can provide estimates of the number of viable microorganisms that may be present at the new locations. In order to provide the best predictions of microbial dispersal, its dependence on many environmental and microorganism-specific variables must be considered. Fate and transport models can be used to integrate available data to predict microbial dispersal, and to point the way to critical data collection needs. In part, critical data needs can be determined through sensitivity analyses of the models intended to be used. The factors to which the model is most sensitive should be known with the most accuracy to maximize the predictive value of the model.

Fate and transport models alone are not sufficient to assess microbial risk. In principle, however, they can be used in conjunction with models that predict the dose of the microorganisms received by sensitive receptors, and with models that predict the dose dependence of observed effects (dose-effects or dose-response models) to provide risk estimates. Several dose-response models were discussed in this chapter; others are also available. Models that convert exposure to dose depend upon the sensitive receptor being considered and the route by which microorganisms are taken into the receptor (for example, humans commonly take in microorganisms by ingestion or inhalation). These models are discussed further in Chap. 13 on chemical risk assessment.

References

Allen, M. J., and Morrison, S. 1973. Bacterial movement through fractured bedrock *Ground water* 11:6–10.

Ambrose, R. B., Jr., Vandergrift, S. B., and Wool, T. A. 1986. WASP3, A hydrodynamic and water quality model: theory, user's manual, and programmer's guide. EPA 600/3-86/034, U.S. Environmental Protection Agency, Environmental Research Laboratory, Athens, Georgia.

American Public Health Association (APHA). 1970. *Control of Communicable Diseases in Man*, 11th ed. APHA, New York.

Armitage, P., Meynell, G. G., and Williams, T. 1965. Birth-death and other models for microbial infection. *Nature* 207:570–572.

Bailey, N. 1975. *The Mathematical Theory of Infectious Diseases and Its Applications*, 2d ed. Hafner, New York.

Barnthouse, L. W., and Suter, G. W., II. 1986. User's manual for ecological risk assessment. ORNL-6251, Oak Ridge National Laboratory, Oak Ridge, Tennessee.

Bazin, M. J., Saunders, P. T., and Prosser, J. I. 1976. Models of microbial interactions in the soil. *CRC Crit. Rev. Microbiol.* pp. 463–498.

Biskie, H. A., Sherer, B., Moore, J., Buckhouse, J., and Miner, J. R. 1988. Behavior of indicator bacteria in rangeland streams—manure spike experiments. Draft. Oregon Agricultural Experiment Station, Corvallis.

Bovallius, A., Bucht, B., Roffey, R., and Anas, P. 1978. Long range air transmission of bacteria. *Appl. Environ. Microbiol.* 35:1231–1232.

Bovallius, A., Roffey, R., and Henningson, E. 1980. Long range transmission of bacteria. *Ann. N.Y. Acad. Sci.* 353:186–200.

Bruce, N. G., Arnette, S. L., and Dibner, M. 1989. Finding biotech information: Databases and vendors. *Bio/Technology* 7:455–458.

Burns and Cline. 1985. (From Technical Resources, Inc., 1987).

Camann, D. E., Sorber, C. A., Sagik, B. P., Glennon, J. P., and Johnson, D. E. 1978. A model for predicting pathogen concentrations in wastewater aerosols. In *Proceedings of the Conference on Risk Assessment and Health Effects of Land Application of Municipal Wastewater and Sludges*, University of Texas at San Antonio, Dec. 1977, pp. 240–271.

Camann, D. E. 1980. A model for predicting dispersion of microorganisms in wastewater aerosols. In *Wastewater Aerosols and Disease*. H. Pahren and W. Jakubowski (eds). EPA 600/9-80-028, U.S. Environmental Protection Agency, Cincinnati. pp. 46–70.

Chiu, N. 1988. Applicability of microbial fate models in biotechnology monitoring studies. Unpublished report to the modeling section, exposure assessment branch, exposure evaluation division, OTS/OPTS/USEPA, July.

Corapcioglu, M. Y., and Haridas, A. 1984. Transport and fate of microorganisms in porous media: a theoretical investigation. *J. Hydrol.* 72:149–169.

Corapcioglu, M. Y., and Haridas, A. 1985. Microbial transport in soils and groundwater: a numerical model. *Adv. Water Res.* 8:188–200.

Cox, C. S. 1987. *The Aerobiological Pathway of Microorganisms*. Wiley, New York.

Crane, S., and Moore, J. 1984. Bacteria pollution of groundwater: a review. *Water, Air, Soil Pollut.* 22:67–83.

Crane, S., and Moore, J. 1986. Modeling enteric bacterial die-off: a review. *Water, Air, Soil Pollut.* 27:411–439.

Craun, G. F. 1984. Health aspects of groundwater pollution. In *Groundwater Pollution Microbiology*. Bitton, G., and Gerba, C. (eds.). Wiley, New York.

Davis, F. M., Leonard, R. A., and Knisel, W. G. 1990. *GLEAMS User Manual Version 1.8.55.* USDA-ARS Southeast Watershed Research Laboratory, Tifton, Georgia, March 1.

Drahos, D., Barry, G., Hemming, B., Brandt, E. J.,Skipper, H., Kline, E. L., Kluepefel, D., Hughes, T., and Gooden, D. 1988. Prerelease testing procedures: U.S. field test of a lac ZY-engineered soil bacterium. In *The Release of Genetically Engineered Micro-Organisms* M. Sussman, C. H. Collins, F. A. Skinner, and L. Stewart-Tull (eds.) Academic, San Diego, pp. 181–191.

Fattal, B., Davies, M., and Shuval, H. 1986. Health risks associated with wastewater irrigation: an epidemiological study. *Am. J. Public Health* 76:977–979.

Finney, D. J. 1971. *Probit Analysis*, 3d ed. Cambridge Univ. Press, Cambridge, U.K.

General Sciences Corporation. 1987. *User's Guide to Microbe-Screen Execution in GEMS*. Principal contributors, S. Wollman, S. Rheingover, J. Chen. Report no. GSC-TR8753, Nov. 30.

Gloster, J., Blackall, R. M., Sellers, R. F., and Donaldson, A. I. 1981. Forecasting the airborne spread of foot and mouth disease. *Vet. Record* Apr. 25, pp. 370–374.

Haas, C. N. 1983. Estimation of risk due to low dose of microorganisms: a comparison of alternative methodologies. *Am. J. Epidemiol.* 118:573–582.

Hussong, D., Colwell, R. R., O'Brien, M., Weiss, E., Pearson, R., and Weiner, R. 1987. Legionella pneumophila not detectable by culture on agar media. *Bio/Technology* 5: 947–953.

Johnson, L. M. 1989. End-user searching of biotech databases. *Bio/Technology* 7:378–379.

Kelch, W. J., and Lee, J. S. 1978. Modeling techniques for estimating fecal coliforms in estuaries. *J. Water Pollut. Control Fed.* 50:862–868.

Knisel, W. (ed.). 1980. CREAMS: A field-scale model for chemicals, runoff, and erosion from agricultural management systems. U.S. Department of Agriculture, Science and Education Administration, Conservation Research Report no. 26.

Leonard, R., Knisel, W., and Still, D. 1987. GLEAMS: groundwater loading effects of agricultural management systems. *Trans. ASAE* 30:1403–1418.

Lighthart, B., and Frisch, A. S. 1986. Estimation of viable airborne microbes downwind from a point source. *Appl. Environ. Microbiol.* 31:700–704.

Lighthart, B., and Kim, J. 1989. Simulation of airborne microbial droplet transport. *Appl. Environ. Microbiol.* 55:2349–2355.

Lighthart, B., and Mohr, A. J. 1987. Estimating downwind concentrations of viable airborne microorganisms in dynamic atmospheric conditions. *Appl. Environ. Microbiol.* 53:1580–1583.

Lindow, S. E., Knudsen, G. R., Seidler, R. J., Walter, M. V., Lanbou, V. W., Amy, P. S., Schmedding, D., Prince, V., and Hern, S. 1988. Aerial dispersal and epiphytic survival of *Pseudomonas syringae* during a pretest for the release of genetically engineered strains into the environment. *Appl. Environ. Microbiol.* 54:1557–1563.

Mollison, D. 1977. Spatial contact models for ecological and epidemic spread. *J.R. Statist. Soc. B.* 39(3):283–326.

Mollison, D. 1987. Modeling biological invasions: chance, explanation, prediction. In *Quantitative Aspects of the Ecology of Biological Invasions*. H. Kornber and M. H. Wilkinson (eds.). The Royal Society, London.

Moore, B., Camann, B. D., Turk, C., and Sorber, C. 1988. Microbial characterization of municipal wastewater at a spray irrigation site: the Lubbock infection surveillance study. *J. Water Pollut. Control Fed.* 60:1222–1230.

Moore, J. A., Grisner, M., Crane, S., and Miner, J. 1983. Modeling dairy waste management system's influence on coliform concentration in runoff. *Trans. ASAE* 26:1194–1200.

Moore, J. A., Smyth, J., Baker, S., Miner, J. R., Department of Agricultural Engineering, Oregon State University. 1987. Final report: Evaluating coliform concentrations in runoff from various animal waste management systems, Soil Conservation Service, U.S. Department of Agriculture, Corvallis, Oregon.

Quentin, G. H. 1979. General approaches to modeling aerobiology systems. In Redmonds, R. L., *Aerobiology* Chap. 7, pp. 279–284. Dowden, Hutchinson & Ross, Inc. Stroudsburg, Pennsylvania.

Reichenbach, N., Wickramanayake, G., Lordo, R., and Hetrick, D. 1987. Biotechnology model: Microbe-Screen. Draft final report. Battelle Laboratories, Columbus Division—Washington Operations to U.S. Environmental Protection Agency, Nancy Chiu, Task Manager; Elizabeth Margosches, Project Manager, March 16.

Sherer, B., Miner, J. R., Moore, J. A., and Buckhouse, J. C. 1988. Resuspending organisms from a rangeland stream bottom. Trans. ASAE 31:2082.

Sticksel, P. R., Cornaby, B. W., and Reichenbach, N. G. 1984. Evaluation of potential biotechnology models. Unpublished report from Battelle Laboratories. Prepared for the U.S. Environmental Protection Agency under contract no. 68-01-6721.

Strauss, H., Hattis, D., Page, G., Harrison, K., Vogel, S., and Caldart, C. 1985. Direct release of genetically engineered microorganisms: a preliminary framework for risk evaluation under TSCA. Report CTPID 85-3, Center for Technology, Policy, and Industrial Development, Massachusetts Institute of Technology, Cambridge.

Technical Resources, Inc. 1987. Ecosystem model criteria selection. Draft report to the Exposure Assessment Group, Office of Research and Development/U.S. Environmental Protection Agency, Cincinatti, Ohio.

U.S. Environmental Protection Agency (USEPA). 1985. Pathogen risk assessment fea-

sibility study. Unpublished report from Environmental Criteria and Assessment Office, Cincinnati, and Office of Water Regulations and Standards, Washington, D.C.
van Genuchten, M. Th. Sept. 13, 1988. (Personal communication.)
Versar, Inc. 1987*a*.Ambient exposures to recombinant microorganisms intentionally released to municipal and pulp and paper industry wastewaters. Draft final report prepared for Exposure Assessment Branch, Exposure Evaluation Division, Office of Toxic Substances, U.S. Environmental Protection Agency, under EPA Contract no. 68-02-4254, Task no. 45, Sept. 30.
Versar, Inc. 1986*b*.Ambient exposure assessment methods for intentional releases of recombinant microorganisms. Draft final report to the Office of Toxic Substances, U.S. Environmental Protection Agency, under EPA Contract no. 68-02-3968. Task no. 161, Sept. 30.
Wallace, H. R. 1979. Dispersal in time and space: soil pathogens. In *Plant Disease: An Advanced Treatise*, vol. II: *How Disease Develops in Populations*. J. G. Horsfall and E. B. Cowling (eds.). Academic, New York, pp. 181–202.
Yates, M. V., and Yates, S. R. 1988. Modeling microbial fate in the subsurface environment. *CRC Crit. Rev. Environ. Control* 17(4):307–344.
Zanetos, M. A. 1984. Epidemiologic models: applicability to risk assessment for biotechnology products. Unpublished report from Battelle Laboratories, Columbus Division—Washington Operations. EPA Contract no. 68-01-6721.

Chapter

12

Epidemic Models: Lessons from Plant Pathology

Paul S. Teng

Department of Plant Pathology
University of Hawaii
3190 Maile Way
Honolulu, Hawaii 96822

Jonathan E. Yuen

Department of Plant Pathology
University of Hawaii
3190 Maile Way
Honolulu, Hawaii 96822

Introduction

Plant disease epidemics are the result of a series of successful colonizations of a pathogen on host plants, in which the environment (physical, biological) has shown itself conducive for multiplication, dispersal, and spread of the pathogen over a population of hosts. In order to understand the role of modeling in disease epidemiology, it is first necessary to discuss the salient features of disease epidemics which make them complex ecosystems in otherwise simple agricultural systems. Epidemics are commonly characterized as population growth curves called "disease-progress curves" (DPCs), in which either the pathogen or the disease (i.e., symptom caused by the pathogen that is measured) is quantified. Epidemics may simplistically be of the "monocyclic" or "polycyclic" type (Zadoks and Schein 1979). Monocyclic epidemics rely on a fixed amount of inoculum for infection and disease during the crop growth period, whereas in polycyclic epidemics, infection and disease result from both the initial inoculum and additional inoculum produced either by diseased plants or by external sources of inoculum. When viewed as a natural hierarchy, an epidemic results from the integration of pathogen life cycles, in which each life cycle repeats itself from infection to infection. An illustrative life cycle is the infection cycle of *Xanthomonas campestris* pv. *oryzae*, which causes bacterial blight of rice (Fig. 12.1). Here the processes would be the dispersal of inoculum from a source plant, entry into the plant, infection, multiplication, and survival. A fungal pathogen, such as the uredial stage of the barley leaf rust pathosystem, would have processes of sporulation, spore liberation, spore dispersal, spore deposition, spore germination, germ-tube penetration, colonization, and latency (Teng et al. 1980). DPCs are thus symbolic representations of pathosystem dynamics, showing the change in amount of disease with time. Models of this representation of an epidemic are abundant in the literature (Kranz 1974), for example the paralogistic model (Zadoks and Schein 1979).

The ecosystem from which epidemics are observed and measured is the "pathosystem" (*sensu* Robinson 1976), a system characterized by parasitism. The uniqueness of the pathosystem is the close interaction between the pathogen and the host, commonly dictated by the virulence of the pathogen and the resistance of the host; this interaction may result in an infection, but infection does not mean that the disease will always increase in intensity, for example when a weakly virulent fungal race infects a moderately resistant crop cultivar. Thus, with pathosystems, in contrast to other microbial ecosystems, a population increase does not always have significance, as there need not be a corresponding increase in disease and therefore increased impact on the host. Genotype-environment interactions furthermore exert a major influence on the course of an epidemic. This implies that the behavior of a pathosystem is dictated in part by the population genetics of the two key biological entities—the pathogen and the host—and in part by the environment, respectively the system structure and the system environment.

Pathosystem structure furthermore has a phenotypic component in terms of the dispersion characteristics of the pathogen or the host. In crop pathosystems, host dispersion is dictated by the demands of husbandry practices such as plant density, inter- and intra-row spacing, and slope of land. It is also common that crop pathosystems (e.g., a cornfield with common rust disease in the midwestern United States) show genetic homogeneity over a relatively large area, in contrast to natural pathosystems (e.g., a taro plant with phytophthora leaf blight in a mixed garden of a Pacific island). Because of the above features, management strategies which have been proposed for pathosystems have included "gene pyramiding" (placing several specific resistance genes into a single cultivar), use of "multilines" (a group of plants where different plants have different resistance genes but are otherwise identical), "multivars" (cultivar mixtures planted together), and cultivar rotations. The environment in which a pathosystem functions includes the physical (soil, atmosphere, weather) and biological (hyperparasites, competing organisms, plant feeders); each of these is itself a complex, with potential interactions. The environment has a profound effect on the epidemic through its effects on the processes detailed above that make up the epidemic.

Modern plant disease epidemiology may be considered the quantitative

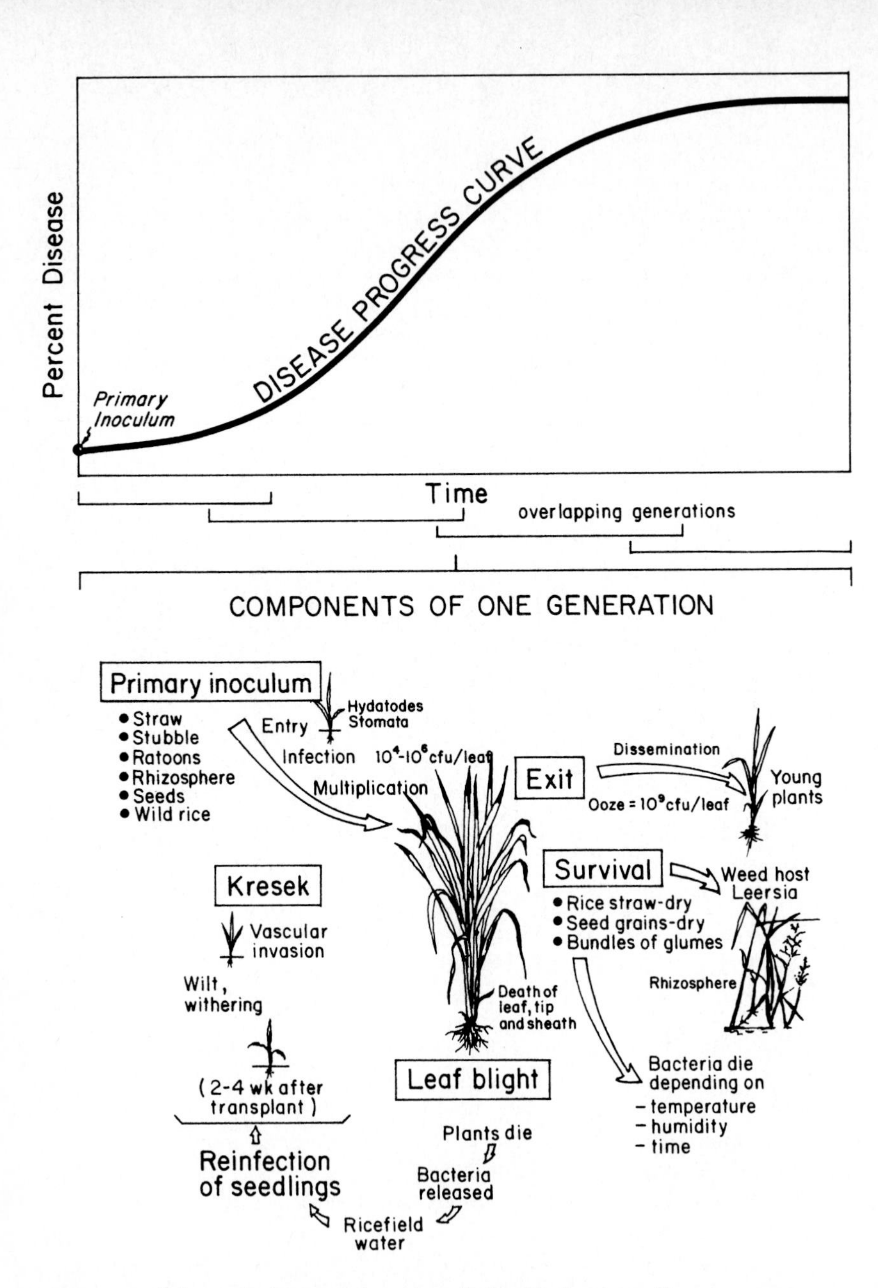

Figure 12.1 Schematic diagram showing relationship between disease-progress curve and its biological components relevant for modeling.

ecology of pathosystems, albeit with a wider scope than traditional ecology. Although disease epidemiology shares many common concepts with ecology (e.g., diversity, aggregation), it has been influenced by the broader field of study of crop pathosystems and currently has a strong economic, management, and sociological component. Because epidemiology grew as a subdiscipline within the larger discipline of plant pathology, it has had to address the issue of its relevance to resolving plant disease problems as they impact crop productivity. Therefore, modern epidemiology includes topics like crop loss assessment, economics of disease control, predictive systems, and disease-management tools. Modern epidemiology is generally considered to date from the pioneering work of J. E. VanderPlank (1963) and J. C. Zadoks (1961), both of whom proposed and demonstrated the use of the exponential and logistic models (described below) for describing DPCs and for understanding the progression dynamics of epidemics. The literature shows much activity for the next two decades aimed at fitting various mathematical models to DPCs (Teng 1985). The limitations inherent in using a single equation to capture the complexity and stochasticity of epidemics lead to the development of epidemic-simulation models, the first of which appears to have been EPIDEM for tomato early blight caused by *Alternaria solani* (Waggoner and Horsfall 1969). Many simulation models have since been developed for diverse pathosystems (Teng 1985), although not many of the workers concerned have explicitly practised a "systems approach" to model development. This aspect of modeling disease epidemics will be discussed in detail in the next section.

The evolution of epidemic modeling from single-equation, heuristic approaches to more holistic, systems-simulation approaches has coincided with the increased use of quantitative techniques and microprocessor-based instruments in epidemiology (Kranz and Hau 1980). The "information explosion" in epidemiology has also fueled the need for methods to synthesize data and to reconcile conflicting hypotheses, both of which are accommodated by system-simulation models. The inherent flexibility of system models makes them appealing as tools for incorporating new information on aspects of the pathosystem that may potentially change epidemic dynamics. A case in point is when an identified gene for resistance is engineered into a host plant, changes the quantitative nature of the pathogen-host interaction at the level of the infection cycle, and then affects multiplication, dispersal, and spread over a large area. Simulation models, which have been described as "If-Then" calculators (Dent and Blackie 1979), are ideal for exploring changes in system structure or system environment. Furthermore, where there is no information on any aspect of the system, assumptions can be tested as stated hypotheses of system function, and recommendations for action taken based on the assumptions. This is particularly important for newer areas of scientific endeavor, such as the topic of this book, in which practical needs for decision making far outstrip the information available.

The issue of risk is one that pathologists have attempted to address in disease management. Epidemic models developed to guide decision making can conveniently be divided into "non-goal-seeking" and "goal-seeking" according to their application. A goal-seeking model attempts to mimic or analyze a particular pathosystem and make recommendations regarding control tactics, whereas a non-goal-seeking model only attempts to mimic the pathosystem. Simulation models are normally non-goal-seeking, since the models predict a response (or responses) according to changes in the stimuli (driving variables). They do not produce a management recommendation as such, but simply reflect whatever changes are made to their operational environment. That this is so has probably limited the use of simulation models for decision making (Teng 1985). To overcome this, various algorithms have been suggested for analyzing the output of simulation models so that an interpretative summary is made (Thornton and Dent 1984*a*). This area of research and application is still relatively new in plant pathology, and in a later section we discuss the available techniques and exemplify the use of one for risk analysis.

This chapter is not intended to provide a comprehensive review of epidemic models, for which the reader is referred to the publications by Teng (1985), Kranz and Hau (1980), Gilligan (1985), and Berger (1989). Rather, we have set out to provide background on the role of epidemic models in plant pathology and to evaluate their potential applicability for assessing the risk of releasing genetically engineered organisms.

Models of Plant Disease Epidemics

There is no perfect way to classify all the models and modeling approaches used in agriculture. Several workers (Mundt 1989, Jeger 1986, Teng 1985) have argued that most epidemic models are either "analytic" or "simulation" models. An "analytical model" is one which is relatively simple in form, often a single equation with few biological parameters, and often can be mathematically solved. "Simulation models" commonly comprise a series of equations which describe the behavior of subsystems and explicitly account for the influence of environment at the subsystem level. These models commonly cannot be solved using analytical (mathematical) techniques and require numerical solution with a computer algorithm. Berger (1989) has observed that some workers (e.g., Teng and Zadoks 1980) have blended the two approaches by starting with an analytic model and then gradually increasing its degree of realism and representativeness of the real world until the model is no longer capable of analytical solution. It is unfortunate that in plant pathology we have seen some modelers take strong positions for or against either of these approaches (Jeger 1986), while in practice, as Mundt (1989) noted, the approach should be influenced by the objective for model building, a view which has repeatedly been emphasized by others (Teng 1985, Kranz and Hau 1980). A detailed discussion of model types and their relationship to different philosophies of scientific endeavor was presented by Teng (1981, 1985), who put it simply when he wrote that there appears to be a division of the "splitters" versus the "clumpers," approximately analogous to those taking a "top-down" versus a "bottom-up" approach towards modeling. To explain why pathologists have used both these approaches, it is first necessary to discuss the object being modeled—the pathosystem.

The plant pathosystem

Most modeling efforts have considered a plant pathosystem (Robinson 1976) to consist of a single pathogen species and a single host species, interacting under the influence of humans and environment. The result of the interaction is a measurable phenomenon—the epidemic—representing the phenotype as pathogen or disease. Kranz (1974) defined epidemiology as the study of "dynamic interactions between the pathogen and the host, under the influence of man and environment." As noted earlier, when the pathogen or disease occurring in a discrete ecosystem is measured over time, the growth in population may be described by a DPC. Because disease is conveniently measured as either incidence (proportion of infected host units) or severity (proportion of infected tissue), these two variables have commonly been used for characterizing the course of a field epidemic. Epidemic curves commonly show a sigmoidal pattern based on symptoms, with undulations in the general trend caused by environmental discontinuities (Fig. 12.1). At any time during the course of the epidemic, pathosystem activities include any one of the following components of the infection cycle—dispersal of bacterial cells, deposition on a host part, multiplication on host part, survival as epiphytic population on host part, entry into host, infection of cells, multiplication within cells, symptoms caused on host part, infectious cells—and the cycle repeats itself. The intensity and duration of each activity are strongly influenced by the environment and by host physiology. For economic impact assessment, the magnitude of the disease symptoms and their relationship to host growth and development must be known. For ecological studies, the pathogen population is of more interest.

Analytic models of disease epidemics

Models of entire epidemics. The classic work using analytic models is by VanderPlank (1963) and Zadoks (1961), who advocated the exponential and logistic models for describing DPCs based on measurements of disease incidence or severity. The exponential model in differential form represented by

$$\frac{dy_t}{dt} = ry_t$$

where y_t is disease amount, r is population relative growth rate, t is time, for population growth under unrestricted conditions appears to apply only during the early stages of an epidemic or in explosive epidemics occurring under optimal conditions. The logistic model, in differential form, represented by

$$\frac{dy_t}{dt} = r\,(1.0 - y_t)y_t$$

describes DPCs of epidemics when restricting conditions operate, as when the amount of host tissue starts to operate as a negative feedback to slow down infection. The factor 1.0 in parentheses is the maximum carrying capacity of the host, i.e., 100% for healthy tissue. VanderPlank (1963) used the logistic in an elegant, heuristic manner to explain the observed dynamics of selected epidemics. One of his expanded logistic models was a model with time delays,

$$\frac{dy_t}{dt} = R_c\,(y_{t-p} - y_{t-p-i})(1.0 - y_t)$$

where p is latent period, i is infectious period. This analytic model has a second-order time delay which requires substantial effort to solve. VanderPlank used the model to explain the roles of latency, infectivity, and tissue death in driving epidemic dynamics. In more recent work, VanderPlank (1975) advocated the "splitting" of these three processes into more detail if necessary, such as when building simulation models. The logistic model has found many practical applications in plant pathology for evaluating tactics such as different degrees of host-plant resistance, different fungicide schedules, and different cultural control practices (Berger 1989). VanderPlank (1963) provided examples for transforming the DPC into a straight line, which could then be described using linear regression,

$$\text{logit}(y) = rt + c$$

where logit(y) represents the transformation $\log[y/(1 - y)]$, r is the apparent infection rate for the epidemic, and c represents the amount of disease at epidemic onset. These "disease-progress lines" (DPLs) have frequently been used by pathologists to compare epidemics occurring under different situations. It is conceivable that changes in a minor gene for resistance may result in a change in disease severity, reflected in the r value (a change in the slope of the DPC in Fig. 12.1), which is measurable. This could then be used to compare two populations of genetically engineered plants with the same pathogen population, or, conversely, two strains of a pathogen on the same host genotype.

Other models have been used to describe population growth as measured using a DPC. These include the Gompertz model, with the generalized form

$$\frac{dN}{dt} = r_g N \ln \frac{K}{N}$$

where K is the maximum population size, N is present population size, and r_g is rate of population growth (Berger 1989). The Richards model is another model which has been used to describe the DPC; it has the form

$$\frac{dN}{dt} = \frac{r_h N \left(\frac{K}{N}\right)^{1-m} - 1}{1 - m}$$

where m is a power parameter and r_h is the Richards rate of population growth. Apart from the logistic, Gompertz, and Richards models, other flexible models such as the Gaussian and the Weibull have also been used to describe the epidemic curves of different pathosystems. Plant pathologists saw a flurry of curve-fitting, empirical modeling research in the 1970s, leading Waggoner (1986) to issue the caution that even if many models can be found to fit the curves of a pathosystem, it does not mean that any model has captured the biological essence of population growth in that system.

Models of epidemic components. Apart from analytic models for describing entire epidemics, models have also been proposed for the component processes of epidemics such as germination, sporulation, or dispersal. In most cases, these are single-stimulus–single-response models, such as the model for germination at elevated temperatures (Waggoner and Parlange 1974):

$$H = H_0 - [t_{45}(0.15 - 0.6t_d)]$$

where H_0 is the normal germination, t_{45} is the number of hours at 45°C, and t_d is the delay in hours before exposure to the elevated temperature. Another component that has received much attention is that of spore dispersal from a point or area source. This component is of great interest, as it is one that applies to models which emphasize temporal or spatial aspects of epidemics. Implicit in considerations of spore dispersal is the question of disease gradients, with early work having been done by Gregory (1968), who proposed an equation of the form

$$y = ax^{-b}$$

where y is number of spores or lesions per unit area, a is number of spores per unit area at one unit of distance from the source, x is the distance from the edge of the source to a receptor of spores, and b is a coefficient describing the slope of the gradient. The model EPIMUL (Kampmeijer and Zadoks 1977) used a Gaussian distribution to determine spore dispersal of stripe rust around an inoculum source:

$$X = \frac{1}{2\pi\sigma^2} \exp\left(\frac{-x^2 + y^2}{2\sigma^2}\right)$$

where X is the concentration of spores at a point in space, and x and y are the distance from the source. Models based on a modified Gaussian distribution have been used extensively to describe air dispersal in many other systems as well (see Chap. 11). The Weibull model has also been used by Shrum (1975) and Teng et al. (1980) to calculate spore numbers dispersed from different sources. Two chapters in a recent book on disease epidemiology cover this topic (Minogue 1986, Fitt and McCartney 1986) although both fail to show how this component is integrated into a more holistic framework for disease spread and increase. In a series of recent papers (van den Bosch et al. 1988 *a,b,c*), models were developed for spatial dispersion of infectious agents. In these studies, the spatial distribution of propagules in two-dimensional space (estimated with a Bessel function) was linked to their infection and subsequent reproduction over time (estimated with a delay and a Gamma function) to produce models that enabled estimation of asymptotic wave speed and slope of the epidemic front as it moved through the field. Spore-dispersal modeling and disease-gradient analysis are activities at best addressing a subsystem of the epidemic—an important one, but one which is subject to highly stochastic influences such as wind frequency and direction. Except in situations where these have quantifiable patterns, dispersal may have to be treated as a random event.

Regression equations have been used to develop simple predictive models of the effect of one or more environmental variables on epidemic components. For example, sporulation of barley leaf rust was described by the equation

$$\ln N = 7.32308 - 2.8477D + 0.11585T + 0.38132D^2 - 0.011108T^2 + 0.19451DT$$

(Teng and Close 1978), where N is number of spores produced per square millimeter of lesion per day, D is age of lesion, and T is ambient temperature in degrees Celsius. The workers concerned attempted to introduce some degree of variability into their model by incorporating a randomly distributed correction to each predicted outcome using the standard error of estimate of the model. Subsystem (component) models are at best indicators of what could potentially be the dynamics of the entire population. Because of their simplicity, some have found application in plant pathology by being the basis for predictive systems to time fungicide sprays (Krause and Massie 1975).

No analytic model can account for the variable effect of environment on epidemics, nor do they explicitly attempt to do so. Rather, analytic models are more heuristic in nature, predicting general behavior and providing insights into the overall dynamics of epidemics.

Simulation models of disease epidemics

To provide an accounting of progression dynamics in space and time of disease epidemics requires that a *systems approach* (see next section for discussion) be adopted for model building. This inevitably leads to the construction of a simulation model of the pathosystem, in which as much as is feasible of current knowledge on the pathosystem is incorporated into the model. A "simulation model," or "system model," is therefore commonly an assembly of linked models or equations, each of which represents some subsystem of the pathosystem. The number and nature of these subsystems has depended on the researchers concerned. The early system-simulation models of plant disease epidemics were developed from a perceived need to synthesize the vast amount of data that had accumulated on the epidemiology of certain diseases such as tomato early blight (Waggoner and Horsfall 1969) and potato late blight (Waggoner 1968). These models did not have a discernible philosophy behind their development. In later work, workers acknowledged the influence of systems thinking on their model building, and the models and their uses reflected this (Teng et al. 1980).

A basic premise for building system-simulation models is that the pathosystem can be dissected into interacting subsystems, each of which has a quantifiable state at any time during an epidemic. This is otherwise known as the "state-variable approach" to modeling. For example, the barley leaf rust model, BARSIM-I (Teng et al. 1980) had the following state variables and associated set of driving variables in its repeating cycle:

State variable	Driving variable(s)
Germinated spores	Temperature, solar radiation, free moisture
Penetrated germ tubes	Temperature, free moisture
Latent infections	Temperature, light
Infectious lesions	Host receptivity
Spores produced	Temperature, light, lesion age, lesion size
Liberated spores	Wind speed, free moisture
Spore mortality	Spore age, solar radiation
Deposited spores	Wind speed, dispersal gradient

The sequence of state changes in a simulation model has been compared to a series of valves regulating water flow in pipes, and is often diagrammatically represented as such (see next section). Thus, at any one time, any or all of the states may exist depending on prevailing conditions. The magnitude of each state variable is dictated by the rate of change of that variable from its quantity at the time period immediately preceding the present. This may be represented in differential equation form as

$$y_t = y_{t-1} + dy_t$$

$$\frac{dy_t}{dt} = r_t \times y_t \times F$$

where y_t is a state variable at time t, y_{t-1} is the same state variable at the preceding time period to t, dy_t is the change in state y between time t and time $t - 1$, r_t is the relative rate governing the change between the two times, and F is a function or parameter governing the particular process. The form of F may be a negative feedback, as with germ-tube penetration on a leaf surface as the number of uninfected sites decreases due to increasing disease, a situation analogous to the $(1.0 - y)$ of the logistic equation. The value of r is commonly between 0.0 and 1.0, and is the outcome of the integrated effects of all the driving variables. Each equation is solved using a numerical integration technique such as Euler integration for each time period. The total of all the solutions then becomes the state of the pathosystem at a particular time.

Although there have been differences between epidemic modelers in how much detail and to what subsystem level to model a pathosystem, most simulation models contain the above breakdown of an infection cycle. A partial chronological listing of epidemic simulation models is given in Table 12.1, which shows that most effort has been with annual crops and foliar fungal pathogens. We are not aware of a fully developed simulation model for a bacterial or a soilborne pathosystem. Most of the simulation models developed in the 1970s treated the host in a superficial manner, commonly using empirical functions for biomass or tissue area changes. A recent development is to couple epidemic models to crop models, so that variables in each will influence the other (Johnson et al. 1987). The potential for this kind of simulation mod-

TABLE 12.1 **Chronological Listing of Selected Epidemic Simulation Models**

Pathosystem	Model name	Computer language	Reference
Potato late blight	EPIDEM	FORTRAN	Waggoner 1968
Tomato early blight	EPIDEM	FORTRAN	Waggoner and Horsfall 1969
Wheat rusts	none	CSMP	Zadoks 1971
Southern corn leaf blight	EPIMAY	FORTRAN	Waggoner et al. 1972
Apple scab	EPIVEN	FORTRAN	Kranz et al. 1973
Wheat stripe rust	EPIDEM	FORTRAN	Shrum 1975
Barley leaf rust	BARSIM-I	FORTRAN	Teng et al. 1980
White pine blister rust	None	FORTRAN	McDonald et al. 1981
Barley powdery mildew	EPIGRAM	FORTRAN	Aust et al. 1983

eling is great since the coupling of one organism to a host would logically lead to more complex models with more organisms, and there is currently much research directed at this (Teng 1988).

There is disagreement among plant pathologists as to whether a "top-down" or a "bottom-up" approach should be used for building epidemic simulation models. VanderPlank (1975) argues that most epidemics can be explained by the logistic equation expanded to account for initial inoculum, a progeny-parent ratio, and a latent period. Teng and Zadoks (1980) illustrated this approach by starting with the logistic and then increasingly building more biological detail until a full simulation model for wheat leaf rust was obtained. It was obvious to the authors that the analytic, single-equation logistic model (Eq. 2), even when modified and numerically solved for latency and infectious period (Eq. 3), could not explain all that was known about wheat leaf rust epidemics, or incorporate a fraction of the available knowledge on environmental influences on observed epidemics. On the other hand, simulation models such as EPIDEMIC and BARSIM-I have been constructed based on an initial definition of the pathosystem and a survey of available literature on the epidemiology of the disease. This "bottom-up" approach tries to reconcile all known information at the subsystem level, and for each subsystem, a module is developed to synthesize available data. Experiments were conducted by the researchers, or preliminary hypotheses formulated, to fill in knowledge gaps for those modules where none existed. Simulation models of epidemics may therefore be viewed as "skeleton models," containing only known logic and unchanging parameters (Dent and Blackie 1979), but would not predict behavior for a specific location unless real world data is fed to the model. In contrast to analytic models, simulation models are useful for understanding why some locations are more conducive for epidemics of a certain pathosystem.

The division of epidemic models into analytic and simulation is admittedly a simplistic one. Another major division is into models which are mainly spatial versus temporal, although all models have aspects of both. The model EPIMUL (Kampmeijer and Zadoks 1977) simulates a spatial arrangement of host units, with the epidemic in each host unit modeled using an expanded logistic equation. The units affect each other through dispersed spores driven by wind direction and speed. The majority of simulation models emphasize temporal dynamics, and account for space by considering average values of state variables sampled from a normal distribution. The ideal, with no constraints of computational power, would be to have simulation models which address population units distributed in space, each unit with its own epidemic system simulation, and the epidemic in the sampling frame made up of all the unit epidemics. To our knowledge this has not been done for any pathosystem.

Techniques and Data Requirements for Modeling Epidemics

Systems analysis

Recent work on simulation modeling of epidemics and disease management has stressed a "systems approach," by which is meant a conscious attempt to determine the main components of a pathosystem, their driving variables, the interactions between state and driving variables, and the relationship of all these to the entire epidemic. The approach proposes a holistic view of epidemics and the need to explain why "the whole is more than the sum of its parts." In practice, this approach requires that a descriptive analysis of the system be done which commonly includes the pathosystem in relation to its wider operational environment (Teng 1986, Kranz and Hau 1980). The steps commonly adopted in using a systems approach are (1) system definition, problem identification, and objective specification; (2) accumulating knowledge about the system; (3) formulation of an initial system model and identification of knowledge gaps; (4) empirical gathering of knowledge to fill gaps; (5) synthesis of knowledge into a detailed system model for computer simulation; (6) model evaluation, including sensitivity analysis, verification, and validation against real world data, and (7) experimentation with the system model for purposes of increased understanding of the system and to design new versions of the system. The term "systems analysis" has been used synonymously with "systems approach" although we consider systems analysis as being limited to steps 1 through 3. The steps are commonly practised sequentially, with feedback. Some of these steps apply to the development of analytic models as well, although there is a fundamental conceptual contradiction. Analytic models appear to have been developed deductively, in which a theoretical form is postulated and empirical data is found to support that form. Simulation models appear to be mainly inductive in their entirety, since subsystem data and theory are assembled without a strong bias for the complete pathosystem. Again, it is unfortunate that strong stands have been taken by plant pathologists in support of either approach (Teng 1985), as this has probably hampered efforts to apply models for decision making.

Collection of data and hypotheses

In plant disease epidemiology, as in studies of other types of ecosystems, much of the empirical data deals with stimulus-response relationships at different levels of biological organization and different trophic levels. Thus the data collected for modeling a particular pathosystem reflect the level of organization and trophic level of the model. A model of the same pathosystem on a different trophic level would have different data requirements. Most of the analytic epidemic models have been concerned about generalized behavior (response); for example, the logistic model is based on population growth with negative feedback exhibited when host tissue becomes limited. The growth curve, plotted using time-series data collected from real world epidemics, is fitted to the model, and a goodness-of-fit test used to determine adequacy of the model. This procedure tends to disregard the "kinks" in the data that may be caused by real variation in response, or by variability in the environmental stimulus (stimuli).

For system-simulation models, data needed to build the model are derived at the subsystem level. Data on the entire epidemic are commonly used for evaluating the model. At the subsystem level, the data are needed to predict response of a process such as germination to a stimulus like temperature. Most of the data are collected in controlled-environment experiments. A dilemma that epidemic modelers face is how to compare and use data that have been collected in different laboratories and often under different sets of conditions. Data collection and model structure become interrelated when model construction becomes dependent on data availability and data collection for validation becomes dependent on model structure. Another problem is that there is little multiple-stimuli–single-response data; such data cannot be developed without performing multifactorial experiments using many controlled-environment chambers. Most of the components of epidemics have multiple stimuli (or driving variables) which are considered important in determining the magnitude of the response. Because much of the literature available is on single-stimulus responses, it is important to know the prevailing conditions under which the data is collected so that corrections may be made to the influence of the first stimulus as affected by the second. An example of a function describing a single-stimulus response is that of temperature T – latent period L for barley leaf rust (Teng et al. 1978), where $L = 19.69 - 0.085T^2 + 0.0025T^3$. Plant pathologists have generally found that, despite reservations about interpolating between sets of different stimuli derived from different experiments, the variability encountered in doing so has been acceptable and has not influenced the outcome of most simulations. In relation to this, building a simulation model in modules also allows testing of each module for its sensitivity to assumptions such as the above. The problem of variability is also present in subsystem data. One way in which pathologists have dealt with this is to use curve-fitting procedures such as least-squares regression to smooth out the stimulus-response relationships and at the same time, to judge the significance of the relationships and their degree of variability.

Model structure

The "model structure" is the entity which ultimately contains all the hypotheses and data about the pathosystem. Model structure is very much determined by the modeler, although, in general, it is true that model structure is often poorly described by modelers. This problem has even led to new models being developed from the same set of data because of personnel changes!

Computer language. Plant pathologists appear to have as many individual preferences on how to program models for simulation as there are models. Apart from which general-purpose language to use—e.g. FORTRAN versus PASCAL versus C—there is disagreement about whether to use this or a simulation language such as CSMP (Continuous System Modeling Program) or DYNAMO. Proponents of the first argue that there is more flexibility and better error detection. Proponents of the second stress the ease of building model structure and providing visually appealing output. We feel that the modeler should use the language type he or she is most familiar with, and that the more important issue is that documentation on the model program be clear so that others can understand it enough to make changes.

Numerical integration and time steps. In a simulation model, the values corresponding to the various states that characterize the pathosystem have to be recalculated at regular intervals. This is done by first determining the rate of change of the state in question, and then calculating the integral of this rate over a small period of time (the integration time step). Most plant-disease-simulation models use Euler, or rectangular, integration. With this method, the amount whereby a state variable changes is considered to be equal to the rate of change multiplied by the length of the time step. This time step is equal to 1 day for the vast majority of plant-disease-simulation models.

For example if the diseased proportion is presently equal to 0.5 and the rate of change is 0.1/day, then with an integration time step of 1 day, the amount of disease after 1 day would equal

$$0.5 + (0.1 \text{ day}^{-1} \times 1 \text{ day}) = 0.6$$

If the integration time step were smaller, say ½ day, then the result would be

$$0.5 + (0.1 \text{ day}^{-1} \times 0.5 \text{ day}) = 0.55 \text{ (after } \tfrac{1}{2} \text{ day)}$$

and

$$0.55 + (0.1 \text{ day}^{-1} \times 0.5 \text{ day}) = 0.60 \text{ (after 1 day)}$$

With a smaller time step, two calculations are necessary to find the amount of disease after 1 day, since each calculation "moves" only ½ day. In this particular case, no additional accuracy is gained with a smaller time step since the rate has remained constant during the integration period. More commonly, rates are a function of the various state variables, and changing the time step also changes the accuracy with which the integrals are estimated.

Some components within a daily model have been integrated using hourly time steps because of the nature of the process (e.g., germination). An important lesson that can be drawn from plant pathology is that often the data available for modeling dictate the time step used. Disease epidemics in the field are seldom monitored daily over the course of a season; a comprehensive database would be measurements done twice weekly. Given the nature of the data available, the increased numerical accuracy from a smaller integration time step would be lost due to the uncertainty of the other measured variables.

Implementations of certain delay processes, such as "boxcar chains" (see below) will often require an integration time step smaller than 1 day. In this case, it is important to distinguish between the time step of the model required for mathematical purposes (integration time step), and the "time step" of the model as determined by the rates of the processes being simulated, usually daily.

Incorporating uncertainty. Uncertainty implies randomness and unpredictability in the occurrence of environmental variables or the response of state variables; it also includes the variability to be expected in a response of the system to the same magnitude of input given at different times. There is no truly stochastic epidemic simulation model of a plant disease epidemic that we know of. Rather, workers have attempted to introduce stochasticity into their models using the variability encountered in the data sets that were used for building the models. For a discussion of variability and errors in a data set due to sampling, see Chap. 10. The empirical stochasticity in a response to various inputs (Teng 1985) is achieved by first multiplying the standard error in the response (state) variable by a random variate drawn from a uniform distribution. The result of this is then either added to or subtracted from the expected value of the state variable. Although imperfect, the procedure introduces an element of variability into otherwise deterministic models. This variability is also essential when models are used as experimental tools, as it is a means to produce replicate model runs, which can then be analyzed further, using techniques discussed below.

Time delays. Delay in response to a driving variable is common in pathosystems. An example of delay is the incubation period or latent period, i.e., the time between when a pathogen infects its host and when symptoms appear (incubation period) and infectious propagules are produced (latent period). The length of this period is commonly variable, being dictated by ambient temperature and host physiology. There is also a mortality factor during latency. Time delays have been programmed by several workers using the "boxcar train" concept (Dent and Blackie 1979), in which the potential maximum length of the delay is treated as the length of a series of boxcars. This length can be shortened in response to changes in driving variables, and the contents of each boxcar modified according to mortality factors. It is generally difficult to program delays with dedicated simulation languages such as CSMP. In some cases special algorithms have been written in general-purpose languages. "Boxcar train" delays can be programmed in FORTRAN and incorporated into a model otherwise written in CSMP.

Validity of models

A central issue in epidemic modeling is the validity of models. Plant pathologists have adopted attitudes ranging from no validation to complete empirical testing of a model's predictive ability (Teng 1981). To a large extent, these attitudes have reflected the different objectives of epidemic models. Teng (1981, 1985), in examining this issue, noted that four basic philosophies existed regarding validation: rationalism (in which models are considered to comprise sets of unquestionable premises, and acceptance of these means acceptance of the model and its validity), positivism (in which models are accepted as valid if they accurately predict, irrespective of what is in the model structure), empiricism (in which models are considered valid only if all aspects of the model have empirical proof), and utilitarianism (in which all of the previous three philosophies are utilized to evaluate a model). There is also the question of terminology regarding model validation. Some workers consider "validation" to be the process of showing that the model predicts real world behavior under similar conditions, while "verification" is the process of ensuring that the model has been programmed in the intended manner (i.e., logical consistency and accuracy). We feel that both verification and validation have to be conducted on any model, and that some testing of a model against actual field data is necessary. Few of the epidemic models used in plant pathology have undergone rigorous testing, and some workers have even argued that their simulation models have been built to improve their understanding of the pathosystem and not for predictive purposes. Statistical and subjective tests have been suggested for testing model validity, including regressing model output against field data for similar inputs, use of the Smirnov test, and use of Delphi procedures (Teng 1981).

Utility of Epidemic Models

Epidemic models can be used to make both strategic and tactical decisions. Strategic decisions are made on a higher level of integration than tactical decisions. For example, a strategic decision might be which disease-control methods (e.g., pesticides, fallow, crop rotation) should be used throughout a single growing season or several growing seasons. Epidemic models are not often used to make these strategic decisions. Tactical decisions are more immediate, and would be whether or not to apply those control measures today—i.e., "Should I spray those pesticides today?"

The strategic versus tactical dichotomy is reflected in the origin of the epidemic models. Simulation models having their origins in modeling coupled processes, and subprocesses fall more naturally into being used as strategic decision tools. If the various subprocesses of the simulation model accurately reflect reality, such models may have utility when run with data that comes from other areas or climates. Thus, they have more validity when run with the diverse weather data sets needed to evaluate different strategies. Analytic models with empirical origins are more useful for immediate, tactical decisions within the production framework from which they were derived. While such empirical models may give reliable predictions for specific areas, there is little guarantee that an empirically derived model will give meaningful results when used with data from other areas or climates.

Applications of tactical decision models typically encompass single growing seasons with user input on a daily or weekly basis. Strategic applications for epidemic models require larger "weather" databases. Several growing seasons of epidemics (replication) are simulated, sometimes with no user intervention during the simulation run.

Models can either be used as learning tools, which generally means implementation as a "game" situation, or implemented as decision tools to aid in answering immediate questions regarding the crop. Learning tools provide an opportunity for the user to learn about the consequences of different farming practices without the economic risks and time commitment of actually growing the crop. Decision tools are generally programs that provide advice to farmers regarding application of disease-control measures such as pesticides. The integration of the strategic versus tactical dichotomy with the learning versus decision difference leads to four different types of model uses, which are best discussed in turn.

Types of model uses

Strategic learning tools. Few epidemic models have been packaged as strategic learning tools. DSSAT (Decision Support System for Agrotechnology Transfer), when run with pest models, will be such a tool. Currently, the only pests modeled with this system are leaf folders and blast added to the CERES RICE crop model (Teng 1988). Within this system, the user decides on certain management strategies, and then long-term simulation runs (25 to 30 years) are made of each strategy, followed by a comparison of results from these strategies (yield, net return, etc.) made with stochastic dominance. A complete strategic learning tool will require

1. Crop-based model with added pests
2. Different strategies for pest control or crop production
3. Large weather database or weather generator
4. Large computer or ample computing power to handle complexity of system

Strategic decision tools. Epidemic models packaged as strategic decision tools are not available. They would function in a similar manner to the strategic learning tools, but would provide answers to higher-level decisions instead of merely allowing the user to learn about the consequences of different cropping strategies. One could envision DSSAT used as a decision tool, but such applications would require better-coupled pest models followed by verification of the coupled models. The models that simulate development of resistance to fungicides in the fungal pathogen population can be used as decision tools to decide whether it is better to alternate different chemicals or to mix both chemicals in a single spray. One problem with strategic decision tools is the actual quantification of risk. Stochastic dominance can identify strategies that are inferior, but as a technique it has more difficulty in identifying those that are clearly superior.

Tactical learning tools. Several epidemic models have been applied as tactical learning tools. APPLESCAB (Arneson et al. 1987) is one example. This program is essentially a game where the user strives to maximize return. The consequences and utility of immediate decisions regarding crop production become evident throughout the simulated growing season. The program prompts the user to supply decisions as to pesticide use, etc., several times during the growing season. The resulting disease increase (or lack thereof) and subsequent loss of quality and quantity of apples, along with the economic consequences, demonstrates the utility of these disease-control measures.

Tactical decision tools. These tools give immediate results regarding crop production during the growing season for a single pathogen or a number of pathogens. EPIPRE, an optimizing system for winter wheat (Zadoks 1981), supplies advice regarding pesticide applications. Within this system, a farmer can receive advice on pesticide applications after supplying information on his or her particular farming practices. The model calculates pest growth, calculates the biological and economic results of this damage, and compares this result to pesticide costs before recommending pesticide application. If pesticide applications are not recommended, the farmer is told when additional information on the field should be supplied.

Use of epidemic models in risk assessment

For an epidemic model to be useful in risk assessment, it must produce information on disease and crop loss given weather, cropping practices, etc. An epidemic model that attempts to optimize production (such as EPIPRE or Blitecast) does not necessarily produce this information. However, coupled-process models that simulate disease development may provide the information needed for risk assessment.

The output from these epidemic models is typically the amount of disease present. During the simulation run the needed information is calculated every time step, but the model may produce summaries at different intervals. Alone, however, this information is of limited utility, since disease alone is not of prime interest to a producer. Since farmers are more concerned with yield, disease information is converted to yield (and/or yield loss), either through empirical yield loss functions or via the coupling of the pathogen model to one of the crop itself. Most farmers do not try to maximize yield levels while ignoring costs, and the expenses involved in crop production and disease control also have to be included so that the net return from a crop can be calculated. If net return was the sole criterion used by a producer to choose between two strategies, then the one with the highest expected monetary value (EMV) would be the one chosen.

Simulation models can help identify the EMV from various cropping strategies. By using historical or simulated weather data, many replicates of the model (along with a specific cropping strategy) can be run, and the resulting output used to produce the probability density function (PDF) of that particular strategy. The disease-simulation model can either be deterministic (in which case the output reflects only variation to environment) or stochastic (in which case the output reflects stochasticity due to both environment and the pathosystem model). This procedure can then be repeated for each strategy of interest. The disease-simulation model transforms the variation in the environment (and variation in the disease model for stochastic models) into variation in disease level, and then to variation in net return. The EMV for each strategy could be then calculated by determining the mean of the simulated return.

The EMV of a strategy, however, is not usually the sole criterion used in choosing a particular strategy. Assume that a farmer can choose from two strategies, called A and B. They could be simple acts such as planting variety A or variety B, or could just as well be "spraying the crop" and "not spraying the crop." If the return of strategy A ranges from $190 to $210 (with an EMV of $200) and the return of strategy B ranges from $0 to $380 (with an EMV of $190), they present very different situations, with almost identical EMV. EMV alone carries insufficient information regarding preferences for these two strategies.

Calculation of the variance of the EMV for a particular strategy is one method of incorporating the additional information on variability. Simply stated, a strategy (A) will be preferred to another strategy (B) if the EMV of A is greater than the EMV of B, but the variance of A is less than the variance of B. In the simple example above, strategy A has a smaller variance and a larger EMV when compared to strategy B. It would be the preferred strategy by the mean-variance criterion.

Analysis of the mean and variance of simulation model output was used by Fohner et al. (1984) in evaluating fungicide timing for late blight control. They used a simulation model for potato late blight and a model for fungicide dispersal within the canopy. Simulation runs were done with historical data. The different strategies compared were different fungicide application rules. An explicit PDF for each strategy was not presented in their data, but they presented summaries in the form of the mean and standard deviation of percent defoliation, assuming that potato growers would attempt to minimize defoliation. A weakness in this approach is that the same percent defoliation may give different amounts of yield loss at different phenological stages in potato (Johnson et al. 1986), and hence greatly change the economics of spraying.

Another technique used to assess risk is stochastic dominance (Hadar and Russell 1969, Meyer 1977). This technique allows prediction of how a decision maker might choose between two different alternatives without knowing the decision maker's utility function (see below). The probability of a given return from each of the alternatives to be compared is essential information in stochastic dominance, and this information can be obtained from a number of replicate runs from a simulation model. If strategies A and B above were to be compared with stochastic dominance, the first step is to convert the PDF into a cumulative distribution function (CDF). These cumulative probability distributions are obtained by integrating the PDFs of A and B respectively and can be abbreviated A_1 and B_1. First-degree stochastic dominance (FSD) can be used to compare the two strategies within a given range, and A is said to dominate B in the first degree if $A_1(x)$ is less than or equal to $B_1(x)$ for x within the range of interest. An additional requirement is that there be at least one inequality so that A_1 and B_1 cannot be equivalent at all points, but A_1 must be less than B_1 at some point. Strategy A would then be preferred to strategy B. The riskiness of prospects can be examined with second-degree stochastic dominance (SSD), whereby the integrals of A_1 and B_1 are compared instead of the CDFs. Strategies that dominate in SSD are said to be preferred by decision makers that are risk-averse. By extension, third-degree stochastic dominance (TSD) and even fourth-degree stochastic dominance can also be calculated, though they are more difficult to interpret. SSD is related to the mean-variance analysis described above. If the PDF is assumed to be normal, a policy that dominates in the case of SSD will also be preferred under the mean-variance rule. For all ordering rules, a dominant distribution must also have a greater mean value and a larger smallest value than the distribution that is dominated. Thus, if A dominates B (in either FSD or SSD), A must have a larger mean value than B and the smallest value of A must be larger than the smallest value of B.

In practice, the PDFs are derived for each strategy using historical or simulated weather data as mentioned before. This is repeated for each strategy to be evaluated. The resulting PDFs are then integrated to convert them to CDFs, and comparisons of these indicate which strategies dominate stochastically in the first degree. Integration of the CDFs produce functions which can be used in SSD comparisons. This technique is used to evaluate cropping strategies in DSSAT, though the technique is easily applied in other situations.

Figure 12.2 shows FSD and SSD comparisons of two strategies for controlling late blight on potato. A weather generator (Bruhn et al. 1980) was used to produce a 30-year weather file, and this was then used to simulate the disease development using a simulation model for late blight (Yuen 1985). This model was based on a simulation model originally written by Bruhn and Fry (1981), with added routines for fungicide distribution (Bruhn and Fry 1982*a*,*b*). The model was run with several different methods of scheduling fungicide applications as the different strategies, though the curves presented here show only no applications and applications with a standard 10-day interval. The disease-free return was assumed to be $200/acre, and the losses due to late blight and the costs of spraying were deducted from this figure to produce a net return for each model run. Figure 12.2*a* shows the FSD curves for no sprays and a 10-day spray schedule. The cumulative probability (ranging from 0 to 1) is the Y axis, whereas the X axis is the net return. Neither strategy dominates in FSD, since the two lines cross. Analysis of these lines using SSD is presented in Fig. 12.2*b*. In this figure the Y axis is the SSD cumulative value (the integral of the curves in Fig. 12.2*a*), and the X axis remains the net return. Strategy 2 (spraying every 10 days) gave returns over a smaller range than strategy 1 (no sprays). Within this range, all points corresponding to strategy 2 lie below points corresponding to strategy 1. Strategy 2 (10-day spray schedule) dominates strategy 1 (no spray) in SSD, thus indicating that it would be preferred by persons averse to risk. Computer programs are available that will compare a number of strategies with stochastic dominance (Anderson et al. 1977).

An additional problem in evaluating risk is that the preference for certain strategies varies from person to person and also varies over time. Economists

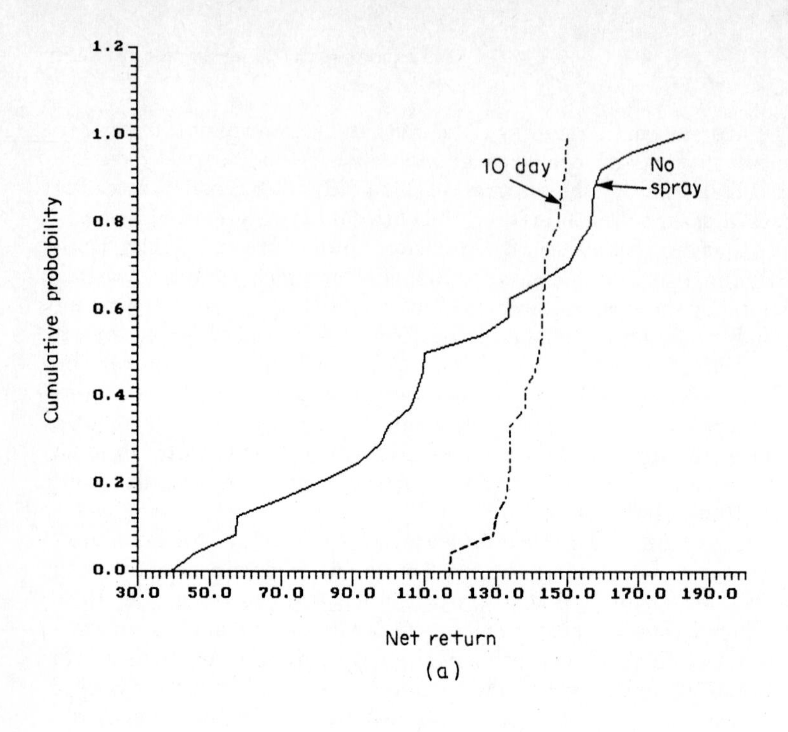

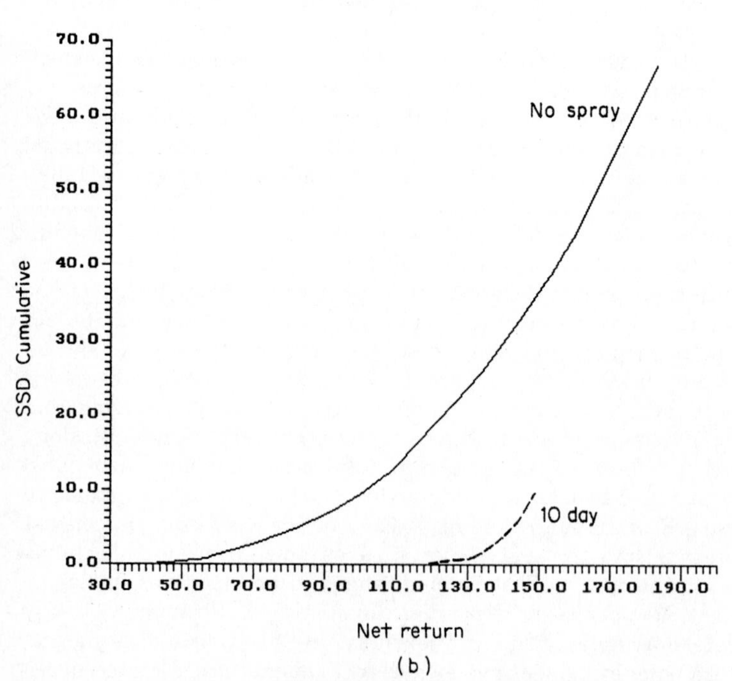

Figure 12.2 Comparison of net return for no spray applications and 10-day spray intervals for control of potato late blight using first-degree stochastic dominance (*a*) and second-degree stochastic dominance (*b*). Net return was assumed to be $200/acre for completely healthy potatoes, with deductions for costs of sprays and losses due to disease.

use the concept of utility to encapsulate information about an individual's preferences for certain acts, given that person's current situation, the EMV for the given acts under different situations, and the perceived probabilities for each of the situations. A lengthy discussion of utility is beyond the scope of this chapter, but it is a dimensionless, real number that reflects the user's preference for risky situations. It should be emphasized that the functions that derive utility from, say, monetary returns, are related to individual preferences, and must be derived for each individual. These functions may also vary over time. Zadoks (1989) presented a series of curves representing utility-loss functions for aphid control in wheat derived from a sample of Dutch farmers. His curves showed that certain farmers were risk-avoiding, while others were risk-neutral. It is easy to imagine that the economic risk one might take would be larger, for instance, if a large amount of money had been saved. With knowledge of the utility function of a particular individual, one can then attempt to optimize utility instead of EMV. For further reference to utility and risk analysis, the reader should consult a reference on risk assessment such as Anderson et al. (1977) or Gold (1989).

Thornton and Dent (1984*a*,*b*) used a simulation model of barley rust (BARSIM I) to produce an information system for the management of *Puccinia hordei*. Within their system, several simulators were used to produce PDFs for yield loss in the presence and absence of spraying at various times throughout the season. They did initial comparisons with first-, second-, and third-degree stochastic dominance, but found that these rules lacked the power to make unequivocal decisions as to whether or not to spray fungicides. Utility functions were then derived for a group of growers, and the strategy or strategies that maximize utility for these individuals could then be identified.

To our knowledge, risk assessment has not been done using ecological criteria such as proportion of survivors after field release of different microorganism phenotypes. The approach is certainly feasible and could provide quantitative comparisons of strategies used to limit risk in field testing of bioengineered organisms. At the time of writing, we have embarked on a 3-year project to test this approach using several strains of *Xanthomonas campestris* pathovar *campestris* in a cabbage ecosystem in Hawaii.

Conclusion

Plant pathologists have had only a relatively short history of quantitative modeling of epidemics, with most accepting that it probably started with VanderPlank's (1963) pioneering book. This book had a profound influence on generations of epidemiologists and fostered much work on the use of analytic, single-equation models characterizing the epidemics occurring in pathosystems. The limitations of these for explaining and predicting epidemics at specific locations were soon realized by workers at Connecticut (Waggoner and Horsfall 1969) and Wageningen (Zadoks 1971), who respectively pioneered computer simulation and use of systems analytic approaches. Formalization of concepts and procedures for using simulation models as strategic tools for managing pathosystems came much later. As with other disciplines, plant pathologists found that the normative nature of simulation models, while presenting flexibility to interpret model output, was tedious to use in guiding tactical or strategic decision making. Although some work has started on using optimization and risk assessment algorithms, no conclusive statements can yet be drawn on their effectiveness. However, from the viewpoint of providing preliminary assessments of environmental risk from bioengineered organisms, simulation modeling would appear to be the ideal tool (Teng 1989). As Andow et al. (1989) stated: "Accurate assessment of risk requires testing in the field, but testing in the field requires prior estimation of risk." Because of their capabilities to incorporate assumptions and testable hypotheses into model structure, simulation enables *ex ante* estimation of potential impact.

References

Anderson, J. R., Dillon, J. L., and Hardaker, J. B. 1977. *Agricultural Decision Analysis.* Iowa State Univ. Press, Ames, 344 pp.

Andow, D. A., Teng, P. S., Johnson, K. B., and Snapp, S. S. 1989. Simulating the effects of bioengineered non-ice nucleating bacteria on potato yields. *Agric. Syst.* 29:81–92.

Aust, H. J., Hau, B., and Kranz, J. 1983. EPIGRAM—A simulator of barley powdery mildew (*Erysiphe graminis*). *Rev. Cresterea Anim.* 90:244–250.

Berger, R. D. 1989. Description and application of some general models for plant disease epidemics. In *Plant Disease Epidemiology*, vol. 2 (ed. K. J. Leonard, W. E. Fry), McGraw-Hill, New York, pp. 125–149.

Blaise, Ph., Arneson, P. A., and Gessler, C. 1987. Applescab: A teaching aid on microcomputers. *Plant Dis.* 71:574–578.

Bruhn, J. A., and Fry, W. E. 1981. Analysis of potato late blight epidemiology by simulation modeling. *Phytopathology* 71:612–616.

Bruhn, J. A., and Fry, W. E. 1982*a*. A statistical model of the spatial and temporal dynamics of chlorothalonil residues on potato foliage. *Phytopathology* 72:1301–1305.

Bruhn, J. A., and Fry, W. E. 1982*b*. A mathematical model of the spatial and temporal dynamics of chlorothalonil residues on potato foliage. *Phytopathology* 72:1306–1312.

Bruhn, J. A., Fry, W. E., and Fick, G. W. 1980. Simulation of daily weather data using theoretical probability distributions. *J. Appl. Meteorol.* 19:415–420.

Dent, J. B., and Blackie, M. J. 1979. *Systems Simulation in Agriculture.* Applied Science, London, 189 pp.

Fitt, B. D. L., and McCartney, H. A. 1986. Spore dispersal in relation to epidemic models. In *Plant Disease Epidemiology*, vol. 1 (ed. K. J. Leonard, W. E. Fry), McGraw-Hill, New York, pp. 311–345.

Fohner, G. R., Fry, W. E., and White, G. B. 1984. Computer simulation raises question about timing protectant fungicide application frequency according to a potato late blight forecast. *Phytopathology* 74:1145–1147.

Gilligan, C. A. (ed.). 1985. *Mathematical Modeling of Crop Disease. Advances in Plant Pathology*, vol. 4. Academic, New York.

Gold, H. J. 1989. Decision analytic modeling for plant disease control. In *Plant Disease Epidemiology*, vol. 2 (ed. K. J. Leonard, W. E. Fry), McGraw-Hill, New York, pp. 84–122.

Gregory, P. H. 1968. Interpreting plant disease gradients. *Annu. Rev. Phytopathol.* 6: 189–212.

Hadar, J., and Russell, W. R. 1969. Rules for ordering uncertain prospects. *Am. Econ. Rev.* 59:25–34.

Jeger, M. 1986. The potential of analytic compared with simulation approaches in plant disease epidemiology. In *Plant Disease Epidemiology*, vol. 1 (ed. K. J. Leonard, W. E. Fry), McGraw-Hill, New York, pp. 255–284.

Johnson, K. B., Johnson, S. B., and Teng, P. S. 1986. Development of a simple potato growth model for use in crop-pest management. *Agric. Syst.* 19:189–209.

Johnson, K. B., Teng, P. S., and Radcliffe, E. B. 1987. Coupling feeding effects of potato leafhopper, *Empoasca fabae*, nymphs to a model of potato growth. *Environ. Entomol.* 16:250–258.

Kampmeijer, P., and Zadoks, J. C. 1977. EPIMUL, a simulator of foci and epidemics in mixtures of resistant and susceptible plants, mosaics and multilines. PUDOC, Wageningen, The Netherlands, 50 pp.

Krause, R. A., and Massie, L. B. 1975. Predictive systems: modern approaches to disease control. *Annu. Rev. Phytopathol.* 13:31–47.

Kranz, J. 1974. Comparison of epidemics. *Annu. Rev. Phytopathol.* 12:253–276.

Kranz, J., and Hau, B. 1980. Systems analysis in epidemiology. *Annu. Rev. Phytopathol.* 18:67–83.

Kranz, J., Mogk, M., and Stumpf, A. 1973. EPIVEN—ein simulator für apfelschorf. *Z. Pflanzenkr. Pflanzenschutzer.* 80:181–187.

McDonald, G. I., Hoff, R. J., and Wykoff, W. 1981. Computer simulation of white pine blister rust epidemics. U.S. Department of Agriculture Forest Service Research Paper, Intermountain For. Range Exp. Stn., Ogden, Utah, 136 pp.

Meyer, J. 1977. Second degree stochastic dominance with respect to a function. *Internal Econ. Rev.* 18:477–487.

Minogue, K. P. 1986. Disease gradients and the spread of disease. In *Plant Disease Epidemiology*, vol. 1 (ed. K. J. Leonard, W. E. Fry), McGraw-Hill, New York, pp. 285–310.

Mundt, C. A. 1989. Modeling disease increase in host mixtures. In *Plant Disease Epidemiology*, vol. 2 (ed. K. J. Leonard, W. E. Fry), McGraw-Hill, New York, pp. 150–184.

Robinson, R. A. 1976. *Plant Pathosystems*, Springer-Verlag, Berlin.

Shrum, R. 1975. Simulation of wheat stripe rust (Puccinia striiformis West.) using EPIDEMIC, a flexible plant disease simulator. Penn. State Univ., Agric. Exp. Stn., University Park, PA. Prog. Rep. no. 347, 68 pp.

Teng, P. S. 1981. Validation of computer models of plant disease epidemics: A review of philosophy and methodology. *Z. Pflanzenkr. Pflanzenschutzer.* 88:49–63.

Teng, P. S. 1985. A comparison of simulation approaches to epidemic modeling. *Annu. Rev. Phytopathol.* 23:351–379.

Teng, P. S. 1986. Integrating crop and pest management: the need for comprehensive management of yield constraints. *J. Plant Protec. Trop.* 2:24–39.

Teng, P. S. 1988. Pest and pest-loss models. *Agrotechnol. Transfer* 4:1–10.

Teng, P. S. 1990. Assessing the impact of planned releases of bioengineered organisms using modeling and non-modeling approaches. *Phytopathology* 80 (in press).

Teng, P. S., and Close, R. C. 1978. The effect of temperature and uredinium density on urediniospore production, latent period and infectious period of Puccinia hordei Otth. *N.Z. J. Agric. Res.* 21:287–296.

Teng, P. S., and Zadoks, J. C. 1980. Computer simulation of plant disease epidemics. In *McGraw-Hill Yearbook of Science and Technology 1980*, McGraw-Hill, New York, pp. 23–31.

Teng, P. S., Blackie, M. J., and Close, R. C. 1980. Simulation of the barley leaf rust epidemic: structure and validation of BARSIM-I. *Agric. Syst.* 5:85–103.

Thornton, P. K., and Dent, J. B. 1984*a*. An information system for the control of Puccinia hordei. I: Design and operation. *Agric. Syst.* 15:209–224.

Thornton, P. K., and Dent, J. B. 1984*b*. An information system for the control of Puccinia hordei. II: Implementation. *Agric. Syst.* 15:225–243.

van den Bosch, F., Zadoks, J. C., and Metz, J. A. 1988*a*. Focus expansion in plant disease. I: The constant rate of focus expansion. *Phytopathology* 78:54–58.

van den Bosch, F., Zadoks, J. C., and Metz, J. A. 1988*b*. Focus expansion in plant disease. II: Realistic parameter-sparse models. *Phytopathology* 78:59–64.

van den Bosch, F., Frinking, H. D., Metz, J. A., and Zadoks, J. C. 1988*c*. Focus expansion in plant disease. III: Two experimental examples. *Phytopathology* 78:919–925.

VanderPlank, J. E. 1963. *Plant Diseases: Epidemics and Control*. Academic, New York.

VanderPlank, J. E. 1975. *Principles of Plant Infection*. Academic, New York.

Waggoner, P. E. 1968. Weather and the rise and fall of fungi. In *Biometeorology* (ed. W. R. Lowry). Oregon State University, Corvallis, pp. 45–66.

Waggoner, P. E. 1986. Progress curves of foliar diseases: their interpretation and use. In

Plant Disease Epidemiology, vol. 2 (ed. K. J. Leonard, W. E. Fry), McGraw-Hill, New York, pp. 3–37.

Waggoner, P. E., and Horsfall, J. G. 1969. EPIDEM, a simulator of plant disease written for computer. Conn. Agric. Exp. Stn., New Haven, Connecticut. Bull. no. 698, 80 pp.

Waggoner, P. E., Horsfall, J. G., and Luken, R. J. 1972. EPIMAY, a simulator of southern corn leaf blight. Conn. Agric. Exp. Stn., New Haven, Connecticut. Bull. no. 729, 84 pp.

Waggoner, P. E., and Parlange, J. Y. 1974. Mathematical model for spore germination at changing temperature. *Phytopathology* 64:605–610.

Yuen, J. E. 1985. A simulation model for potato late blight. *Växtskyddsrapporter, Jordbruk* 42:149–153.

Zadoks, J. C. 1961. Yellow rust on wheat, studies in epidemiology and physiological specializations. *Tidschr. Plantenziekten* 67:69–256.

Zadoks, J. C. 1971. Systems analysis and the dynamics of epidemics. *Phytopathology* 61: 600–610.

Zadoks, J. C. 1981. EPIPRE: A disease and pest management system for winter wheat developed in the Netherlands. *EPPO Bull.* 11:365–369.

Zadoks, J. C. 1989. EPIPRE, A computer-based decision support system for pest and disease control in wheat: its development and implementation in Europe. In *Plant Disease Epidemiology*, vol. 2 (ed. K. J. Leonard, W. E. Fry), McGraw-Hill, New York, pp. 3–29.

Zadoks, J. C. and Schein, R. D. 1979. *Epidemiology and Plant Disease Management*. Oxford Univ. Press, New York.

Chapter

13

Lessons from Chemical Risk Assessment

Harlee S. Strauss

H. Strauss Associates, Inc.
21 Bay State Road
Natick, Massachusetts 01760

Introduction

"Risk assessment" is an analytical tool that facilitates the organization of large amounts of diverse data with the goal of estimating the potential risk posed by a process (or event) of interest. There are many methods, or frameworks, that can be employed to assess risk. For example, risks of failures in complex systems such as the space shuttle or nuclear power plants can be assessed by a methodology known as "fault tree" or "event tree" analysis. In this framework, a series of events, or failures, is hypothesized, the probability of each is estimated, and the final probability of one or more failures is calculated. This type of analysis has been applied to events with a low probability of occurrence, but with important consequences if they do happen.

A different risk assessment framework has been developed to evaluate the health effects associated with toxic chemicals in the environment. In this framework, the risk assessment process is commonly divided into four components: hazard identification, exposure assessment, dose-response assessment, and risk characterization. This framework was endorsed by a well-received National Academy of Sciences (NAS) report (NAS 1983), and further developed at the U.S. Environmental Protection Agency (USEPA), where it has been used in many programs for a variety of purposes. For example, USEPA has used chemical risk assessment methodology to determine site-specific cleanup levels in the Superfund program, to issue discharge permits under the Clean Air Act and Clean Water Act, and to develop proposed regulatory standards for sewage sludge disposal under the Clean Water Act. Chemical risk assessment methodology has also been adopted by many states for use in their waste management programs. Thus, there is a great deal of familiarity among regulators and consultants with this type of risk analysis.

The use of the chemical risk assessment framework is now being expanded to include ecological risks, although alternate frameworks for assessing ecological risks are also available. In this chapter, the applicability and utility of adapting the chemical risk assessment framework to evaluate risks posed by the release of genetically engineered microorganisms (GEMs) into the environment will be examined. The utility of the chemical risk assessment model will also be compared with that of other risk assessment frameworks that have been proposed specifically for genetically engineered organisms.

An underlying assumption of this chapter is that the ultimate goal of a risk assessment is to provide reliable, quantitative predictions of the risks associated with releasing microorganisms into the environment. These quantitative predictions are particularly difficult to make based on our current experience, because of the low probability of the chain of individual events that may actually lead to a hazard. However, because of the large number of microorganisms that are likely to be released each time a product is used and the frequency with which the product will be used, especially following commercialization, even low-probability events may occur at an observable frequency.

Frameworks for Risk Assessment

Throughout the 1980s, various proposals were made for evaluating risks associated with environmental release. Several major reports or papers appeared in the mid-1980s (Alexander 1985, Covello and Fiksel 1985, Gillett et al. 1985, Strauss et al. 1985, Suter 1985, Dean-Ross 1986). Several of these authors (e.g., Covello and Fiksel 1985, Gillett 1985, Dean-Ross 1986) based their recommendations on the framework developed to assess the risk associated with chemicals in the environment, especially on the four-step approach endorsed by the National Academy of Sciences (1983). Other authors, such as Strauss et al. (1985) and Alexander (1985), proposed a framework that did not explicitly incorporate the NAS risk assessment formulation, but could be readily adapted to that model.

In 1989, the NAS (NAS 1989) and the Ecological Society of America (Tiedje et al. 1989) each published new frameworks for risk assessment, specifically directed toward GEMs and plants. These frameworks differ substantially from the chemical risk assessment model, as discussed below.

Alexander—microbial ecology

Martin Alexander, in a widely read and cited paper, presented a six-step framework for examining the risk of releasing GEMs into the environment (Alexander 1985):

- Will the organism be released?
- Will it survive?

- Will it multiply?
- Will it spread to other sites?
- Will it be harmful?
- Will it transfer genes to other, nontarget organisms?

Alexander suggested that if the answer to any of the first five questions is no, then no risks are associated with releasing that organism into the environment. If the organism is capable of transferring genes to other, nontarget organisms, then the first five questions must be reanswered for each new organism which may receive the genes. The Alexander paper was intended to provide a broad overview, with no specific guidelines as to how to answer these questions.

Gillett et al.—the Cornell group

James Gillett led a group of environmental researchers at Cornell University Ecosystems Research Center and the Institute for Comparative and Environmental Toxicology in a study of the potential impacts of environmental release of biotechnology products. In the chapter on risk assessment, Gillett (1985) presented the following criteria as necessary for regulation of biotechnology products:

- Regulatory end points and definitions of unreasonable and unacceptable risk
- Exposure scenarios with critical pathways
- Temporal and spatial distribution of the released organisms with respect to the affected systems, processes, or organisms
- Exposure-response relationships for adverse effects
- Quality control measures to ensure the validity and applicability of data used in the assessment
- Statistical tools for assessing testing and monitoring data
- Evaluation of whether immediate control, further testing, or waiver of testing is most appropriate before environmental use

Gillett (1985) presented schematic views of the components necessary for risk assessment and identification of critical exposure pathways; these are reproduced here as Figs. 13.1 and 13.2, respectively. Gillett (1985) further detailed the data requirements for the exposure assessment. They include:

- Details of how the organism is released, including the locale, rate, and form of release
- Survival characteristics of the introduced organisms, including the range of physiological and nutritional stresses tolerated
- Growth and propagation of the organism and its genetic material, including physiological and nutritional requirements and interaction with other organisms

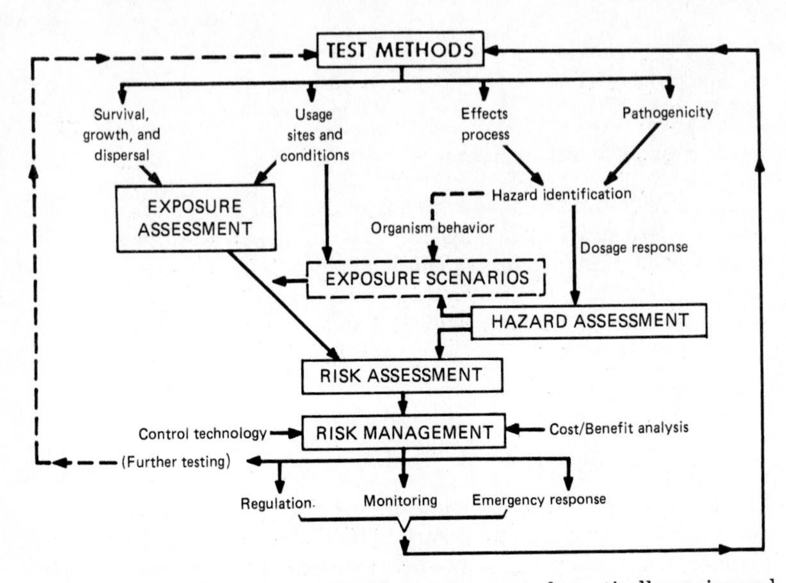

Figure 13.1 Risk assessment and risk management of genetically engineered organisms. In this model of assessment, patterned after that for toxic substances, hazard assessment is combined with exposure assessment in realistic scenarios to yield a risk assessment, which is employed in risk management. These latter activities feed back to alter methods and criteria for decisions as experience is gained in the overall assessment process. (*Gillett 1985*)

- Transfer of genetic information to other organisms
- Dispersal of organisms beyond the site of release
- Interaction of the characteristics and activities of the introduced organism with components of the natural system.

It should be noted that these data will also answer the first four questions, and the last question, posed by Alexander.

Gillett (1985) recommended the development of methods to test the survival and growth of the GEM under a range of environmental conditions. He suggested that primary productivity, growth, respiration, secondary productivity (growth and reproduction of consumers), and nutrient fluxes are the chief candidates for ecological test method development and application in biotechnology risk assessments. Finally, Gillett suggested that microcosm-based tests may be used to assess many of these end points.

Strauss et al.—the Massachusetts Institute of Technology group

A group of scientists, engineers, and environmental lawyers and policy analysts at the Massachusetts Institute of Technology (MIT) Center for Technology, Policy, and Industrial Development, funded by the USEPA's Office of Toxic Substances (OTS), developed and described categories of information that are important to assessing the risk of releasing a GEM into the environ-

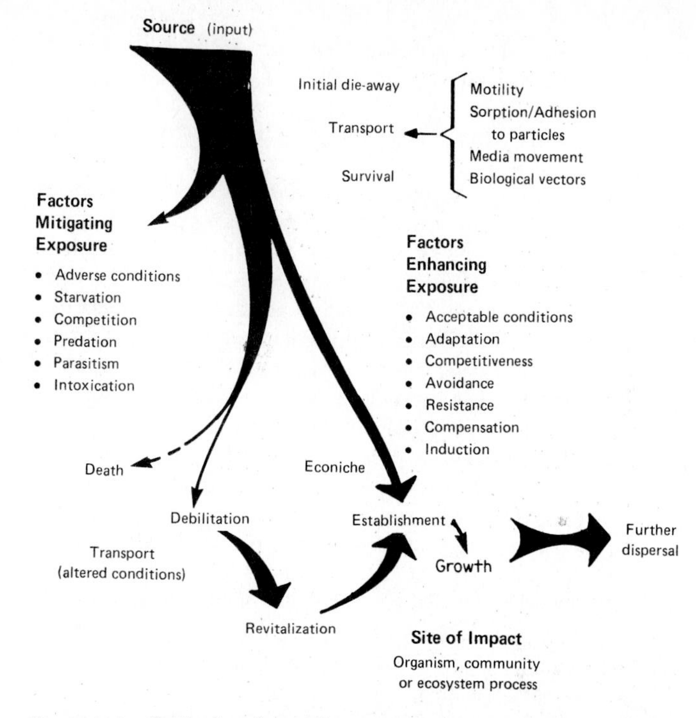

Figure 13.2 Critical pathway for exposure to a GEM. Factors mitigating or enhancing exposure (i.e., relative magnitude of the population of candidate organisms, indicated by the relative width of arrows) affect the potential for adverse effects to be caused by an organism released from a source and following a critical pathway (indicated by broken arrow) of transport and dispersal in the environment. Not shown is the potential interaction of the genetic material of the released organism with species already in the environment, potentially acting as a revitalizing mechanism. (*Gillett 1985*)

ment. They also developed a qualitative, preliminary framework for integrating this information for risk evaluation (Strauss et al. 1985, 1986*a,b*).

Strauss and her colleagues considered five categories of information about the GEM to be important for risk assessment. These categories are similar to those previously discussed:

- Identification, classification, and strain history
- Survival, growth, and genetic transfer
- Transport through the environment
- Intended and unintended impacts and exposures
- Monitoring, containment, and control of the released GEM

Strauss et al. argued that a rigorous identification and a reliable classification system are key elements in the risk assessment of GEMS. They pointed out that the name of an organism provides access to the relevant information from diverse branches of science. They suggested that taxonomic relationships can be used to make predictions about the behavior of relatively unstudied organisms on the basis of what their better-studied taxonomic neighbors do. This is similar to the use of structure-activity relationships when assessing the risks of chemicals.

The Strauss group made similar statements to those of the Gillett group (Gillett et al. 1985) in regard to the importance of developing methods, most likely using microcosms, to predict the survival and growth of GEMs under a range of environmental conditions, including those to which the GEM will be subjected during transport (dispersal). They also suggested that microcosm studies can be used to predict GEM stability and genetic transfer. Additionally, they suggested that a qualitative hazard assessment can be conducted based on a thorough description of the intended use of the GEM and a careful review of hazards that may arise as a result of deviations from the intended behavior. They recommended that these potential hazards be evaluated by suitable test methods. Strauss et al. made a similar argument for the development of the most likely exposure scenarios. Specifically, they suggested that exposure scenarios can be hypothesized based on the planned methods and sites of introduction of the microorganism: "Knowledge of the volume and density of the inoculum, the frequency of the inoculation, the methods of inoculation and the location of inoculation in combination with survival information obtained in microcosm and other tests, can be used to estimate microbial dispersal and hence assess exposure."

In subsequent work, the MIT group (Strauss et al. 1986*c*), funded by Health and Welfare Canada, developed a questionnaire for obtaining information about genetically engineered viruses and bacteria prior to environmental release. The questionnaire requested specific information about the molecular biology, microbiology, and ecology of the microorganism of interest. It particularly focused on ensuring the full and correct identification of the organism, and on specializing questions, depending upon whether the microorganism was a virus or bacteria (fungi and other microbes were not included in the questionnaire).

Covello and Fiksel—the National Science Foundation study

Vincent Covello and Joseph Fiksel edited a study, sponsored by the National Science Foundation (NSF) at the request of the Office of Science and Technology Policy (OSTP), that evaluated the suitability and applicability of scientific methods for risk assessment of environmental applications of biotechnology (Covello and Fiksel 1985, 1986). Senior researchers in a variety of scientific disciplines were members of an expert committee and/or contributed background papers for the report. This study has circulated widely in book form, and is frequently cited.

Covello and Fiksel (1985) used a risk assessment framework similar, but not identical, to that proposed by the NAS (1983) and widely used by the USEPA. Covello and Fiksel considered risk assessment to be a five-stage process, comprising risk identification, risk-source characterization, exposure assessment, dose-response assessment, and risk estimation. The NAS framework combines the first two elements, "risk identification" and "risk-source characterization," into one element called "hazard identification." Table 13.1, which is a reproduction of Table 1 of the Covello and Fiksel report, summarizes the risk assessment methods which may be utilized to measure or estimate values for the various terms in the risk assessment.

The experts at the workshop reached consensus on several points, including the following (Covello and Fiksel 1985):

- The development of a generic approach to risk assessment for environmental applications of biotechnology is both feasible and desirable.
- An important requirement in the risk assessment process is detailed knowledge of the microorganism that is to be modified.
- At present, empirical methods, such as microcosm testing, are indispensable for purposes of risk assessment, but must be supplemented by predictive modeling methods.
- Risk assessment of environmental applications of GEMs should include analyses both of expected impacts and of scientifically plausible low-probability outcomes.
- Several alternative risk assessment approaches are possible, including
  - Deterministic consequence analysis with confidence bounds
  - Qualitative screening
  - Probabilistic risk assessment

The American Society for Microbiology Symposium

In June 1985, the American Society for Microbiology (ASM) convened a cross-disciplinary symposium entitled *Engineered Organisms in the Environment: Scientific Issues*. Symposium speakers included molecular biologists, ecologists, geneticists, and microbiologists. The objective of the symposium was to review the knowledge base regarding engineered organisms in the environment, identify areas of uncertainty, and encourage cross-disciplinary communication in order to elicit a synthesis of thinking on the scientific issues. It included a session on the prospects of a predictive capability for risk analysis of environmental releases.

The ASM symposium proceedings (Halvorson et al. 1985) were published, widely circulated, and have been extensively quoted. In a paper entitled "Application of Environmental Risk Analysis to Engineered Organisms," Suter (1985) compared the available methodologies in environmental risk analysis for chemicals with the methodological requirements for microorganisms. He pointed out that the available chemical models would work in some circumstances, such as prediction of risks from immediate, short-term exposures. However, the two characteristics that distinguish organisms from chemicals, the capability to reproduce and the habitat specificity (the ecology of the organism), require that risk assessment models include the interactions of the ecologies of the introduced organism and the receiving system. These models are not currently available for chemical

TABLE 13.1 Summary of Risk Assessment Methods

Methods directed at:			
Risk-source characterization (measuring the degree of danger associated with the source of risk)	Exposure assessment (estimating the intensity, frequency, duration, etc. of human and other exposures to the risk agent)	Dose-response assessment (characterizing the relationship between the dose of the risk agent received and the health and other consequences to exposed populations)	Risk estimation (developing overall measures of the level of risk)
Monitoring Equipment monitoring Environmental status monitoring Performance testing Accident investigation	Monitoring for exposure assessment Biologic monitoring Food crops, livestock, fish, wild animals, indigenous vegetation, etc. Remote geologic monitoring Aerial photography, multispectral overhead imagery Media contamination (site dose monitoring) Air, surface water, sediment, soil, groundwater Individual dose monitoring, dosimeters, film badges	Short-term tests Molecular structure analysis Tests on humans Animal bioassay	Statistical analysis
Statistical methods for risk-source characterization Statistical sampling Component failure analysis Extreme value theory Codified engineering methods	Calculation of dose Based on exposure time, coexisting or decay substances, material deposition in tissue	Epidemiology Cohort vs. case-control Retrospective vs. prospective Pharmacokinetics	
Modeling methods for risk-source characterization Engineering failure analysis Simulation models Logic trees, event trees, fault trees Analytic models Industrial effluents, biological models for pests, containment models	Exposure modeling Air Analytic models, trajectory models, transformation models Surface water Dissolved oxygen models, etc. Groundwater Absorption models, travel time models Food chain chemical migration models Population-at-risk models Census, sensitive groups, population estimation, trip generation models, etc.	Low-dose-extrapolation models Animal-to-human extrapolation models Ecological effect models	Worst case analysis Sensitivity analysis Confidence bounds Probability distributions Monte Carlo analysis Event tree analysis Probability tree analysis

SOURCE: Covello and Fiksel 1985.

or microbial systems, although Suter (1985) discussed the prospects for their development. Suter also pointed out that since microorganisms can reproduce, the transport of a "colonizing unit" (a number of viable microorganisms sufficient to invade and colonize a new habitat) is important. This means that minor pathways, such as adventitious transport by animals and humans, must be considered in the exposure assessment.

The Ecological Society of America

In mid-1989, the Ecological Society of America (ESA) published in their journal, *Ecology*, a special feature entitled "The Planned Introduction of Genetically Engineered Organisms: Ecological Considerations and Recommendations" (Tiedje et al. 1989). It focused on the ecological and evolutionary aspects of planned introductions of genetically engineered organisms, and included microorganisms, plants, and animals. The report was initially drafted by a workshop committee, and extensively reviewed by the Public Affairs Committee of the ESA.

One accomplishment of the ESA report was to provide a preliminary set of specific criteria for the scaling of regulatory oversight. Tiedje et al. (1989) divided the attributes of organisms and environments into four categories:

- Genetic alteration
- Parent (wild-type) organism
- Phenotypic attributes of engineered organism in comparison with parent organism
- Environment

For each category of attribute, Tiedje et al. (1989) defined specific attributes, and a scale on which to base the level of scientific consideration necessary for a risk assessment. For example, under "environment," the first specific attribute is selection pressure for the engineered trait. The scale for the attribute ranged from absent to present, corresponding to a scale of less to more scientific consideration. However, the scales for each attribute are independent, and not able to be combined. The researchers suggested that if a release was high on the scientific scrutiny scale for one or more attributes, then the appropriate level of regulatory scrutiny should be initiated. Such regulatory scrutiny could include prerelease testing requirements or high levels of biological confinement (see below).

The National Academy of Sciences

In late 1989, the NAS published a report entitled *Field Testing Genetically Modified Organisms, Framework for Decisions* (NAS 1989). The study was requested by the Biotechnology Science Coordinating Committee (BSCC), on behalf of the U.S. regulatory agencies that comprise the BSCC. Its objectives were to identify criteria for defining risk categories and to recommend ways to assess the potential risks associated with introducing modified plants and microorganisms. The study was limited to small-scale and intermediate-scale introductions; it excluded assessment of the risks associated with large-scale commercial uses of genetically engineered organisms. Other limitations of the study include its consideration of only potential environmental effects (and not potential human health effects), and only field conditions in the coterminous United States. The latter point is particularly important for assessing risks associated with genetically engineered plants, and their potential for introducing new traits into weeds (see Chap. 9).

The framework developed by the NAS committee for both microorganisms and plants had three basic criteria (NAS 1989):

- Are we *familiar* with the properties of the organism and the environment into which it may be introduced?
- Can we *confine or control* the organism effectively?
- What are the probable *effects* on the environment should the introduced organism or a genetic trait persist longer than intended or spread to nontarget environments?

According to the NAS, an analyst proceeds through the criteria one at a time, going only so far as necessary to evaluate risk. Thus, the familiarity criterion is the most essential criterion, because it is always evaluated. If the organism, its intended use, and the environment into which it is intended to be introduced pass the familiarity criterion, then no further assessment is necessary. The idea is that if it is familiar in all of these respects, there should be sufficient information available regarding whether or not the risk is negligible to make a decision regarding whether a field trial of the organism should be permitted. If the risk is not necessarily negligible, or if the organism does not pass the familiarity test, then the second criterion, the ability to confine or control the organism, applies. For microorganisms (but not for plants) the second and third criteria are closely linked. That is, because of the difficulty in controlling the dispersal of microorganisms, the level of control necessary and the probable effects of the release are considered together.

This framework differs substantially from the chemical risk assessment (CRA) framework proposed by the earlier committee of the NAS (NAS 1983). One fundamental difference can be characterized as differing judgments as to whether risk assessment and risk management components should be integrated into one scheme. The NAS (1983) report on chemical risk assessment clearly advocated a separation of risk assessment from risk management, in an effort to separate the underlying science of risk assessment from regulatory decisions. However, the first two criteria in the NAS (1989) report on biotechnology risk assessment, familiarity and confinement or control, integrate risk assessment and risk management considerations.

The familiarity criterion is a risk management tool in the sense that it is used to make a decision regarding how much testing and analysis is necessary prior to a field trial. If there is experience with relevant aspects of the microbe, its intended use, and intended location, and no adverse effects have been noted, then no further analysis is necessary. If any of these tests are not met, then further analysis is necessary.

An important difference between the chemical and biotechnology risk assessment frameworks proposed by the NAS is the incorporation of "confine and control" criteria in the biotechnology risk assessment framework. The 1989 biotechnology report used the terms "confine" and "control" in the sense of limiting persistence (time of survival) and spread of the organism, respectively. In the CRA framework, controlling exposure (including dispersal) would be considered a risk management option. However, the fundamental differences between living organisms and chemicals allow limitations to dispersal to be directly incorporated in the product (organism) under consideration, and thus they are appropriate to include in the risk assessment, even under the CRA rules. For example, a microorganism can be biologically "confined" by incorporating a "suicide" component that is activated under certain conditions, by limiting its ecological niche through nutrient requirements or other means, or by otherwise making it "less fit" for survival in the open environment. "Control" options include limiting the possibility of horizontal gene transfer by various genetic engineering techniques, somehow limiting spread to nontarget environments, or using biological confinement strategies to limit persistence.

The third component of the biotechnology risk framework proposed by the NAS in 1989 is an evaluation of the probable effects should the introduced microorganism persist or spread. This evaluation is similar to the hazard identification component of a chemical risk analysis. The NAS suggested that the intended function of the introduced microorganism provide the basis for determining appropriate questions for the hazard analysis.

There is little consideration of quantitative risk assessment or how risks should be presented in the NAS biotechnology framework. Furthermore, it is unclear how one would approach these tasks in the NAS framework, although it could be argued that the available data and methodology are too primitive to even consider quantification of risks. However, these aspects have proven important to the implementation of chemical risk analysis in USEPA's regulatory programs.

A Detailed Look at the Chemical Risk Assessment Framework

Several individual chapters in this book cover, in considerable detail, many of the information needs identified in previous risk assessment frameworks. In addition, Chap. 11 describes and critiques several mathematical models that may be incorporated into microbial risk assessments based on the chemical risk assessment model.

In this section, the components of the risk assessment will be discussed in terms of the overall framework used by the USEPA for chemical risk assessment. Table 13.2 summarizes the information needed for each of the four components of the risk assessment: hazard identification, exposure assessment, dose-response, and risk characterization. Each component will be discussed in

TABLE 13.2 Information Needed for Risk Assessment

HAZARD IDENTIFICATION	EXPOSURE ASSESSMENT

Identification of parent microbe
Strain history
Genetic engineering history
Intended use, likely unintended effects
Testing protocols
  Toxicity of the microorganism
  Toxicity of the intermediary metabolites
  Ecological effects
	Method of introduction
	Amount and frequency of introduction
	Site characteristics
	Transport models in appropriate environmental media
	Monitoring data (statistically valid)
	Fate of the microorganism
	  Survival during transport
	  Reproduction/persistence in new location
	Fate of the introduced DNA (including horizontal genetic transfer)
	Identification, location, and size of susceptible populations
	Method to convert exposure to dose

DOSE-RESPONSE ASSESSMENT	**RISK CHARACTERIZATION**
Quantitative data on dose and effects Models for low-dose extrapolations	Framework for data integration Presentation methods

a separate subsection below. The final subsection briefly discusses the limitations and uncertainties of the currently available information.

Hazard identification

Table 13.2 indicates that several types of information are required to identify the hazards associated with releasing microorganisms into the environment. The information includes identification of the parent microorganism, the strain, and genetic engineering history; the intended use of the microorganism and the most likely unintended effects; and the results of a variety of toxicity-testing protocols. The toxicity-testing protocols to be utilized depend, in part, on the description of the original microorganism and the most likely unintended effects.

The chapters in the first section of this volume provide an overview of available data and methods for assessing the hazard of various end points involving human health and ecological effects. The hazard may result from the introduced organism itself; transfer of the introduced gene into other organisms, where it may cause harm; or potentially toxic intermediary metabolites. Other data requirements have been described in other reports (Strauss et al. 1985, 1986*a*,*b*,*c*, Covello and Fiksel 1985, Gillett et al. 1985) and were reviewed in the first section of this chapter.

Major human health effects may arise from the ability of the introduced microorganism (1) to cause infectious disease; (2) to cause opportunistic infections in immune-compromised populations; (3) to cause delayed hypersensitivity (allergic) reactions; (4) to transfer drug resistance to medically important microorganisms, thus reducing their treatability; or (5) to generate toxic and persistent intermediary metabolites. Major ecological hazards may arise from (1) adverse effects to individual organisms or populations of organisms, (2) disruption of biological communities, or (3) alteration of biogeochemical processes.

Whether human health hazards may exist from a particular microorganism can be determined, to a first approximation, by an examination of the existing scientific and medical literature, the strain history, and from testing protocols. The literature, and perhaps a battery of toxicity tests, can be used for initial screening purposes. More extensive testing protocols can be developed if the initial screen suggests potential problems. For example, the scientific and medical literature may indicate that a particular species/biotype/strain of microorganism has been associated with infectious diseases or opportunistic infections in the past. If it has, then further testing in an appropriate animal model system is indicated. To the extent possible, quantitative dose-response information should be obtained from the toxicity and pathogenicity tests (see Chap. 3).

Plasmid-encoded drug resistance genes have been shown to be transferred across genus barriers (cf., Chap. 8, this volume, and Levy and Marshall 1988). Several DNA transfer mechanisms are known, and sites for the transfer to take place (e.g., soils, ponds, the intestines of animals) may be available. The process is probably driven by selective advantage of coding for the resistance genes, despite the "extra genetic burden" caused by the presence of the plasmids.

There are several potential human health hazards that may be caused by the transfer of genetic information between microorganisms. The transfer of drug resistance to clinically important bacteria and fungi has been repeatedly observed among indigenous microbes. These often multiple drug resistances reduce the choices of antibiotics that are available to treat the infection. In part, this hazard can be identified on the basis of the history of the strain and its engineering. For example, if no drug resistance plasmids are associated with the introduced microorganism, then its ability to transfer the resistance to other microorganisms is negligible.

Another potential hazard associated with horizontal genetic transfer is that introduced genes may function differently in different genetic backgrounds. It has been hypothesized that the unanticipated function may create a selective advantage for the modified microbe, allowing it to increase its population and substantially alter the microbial ecosystem. Again, knowledge of the origins and constructs of the strain can help in the assessment of the likelihood of the transfer (see the section titled "Exposure Assessment" below). Actual testing of the introduced genes of concern in microorganisms known to populate the human gut may be useful prior to the use of a gene of particular concern.

A potential hazard of particular relevance for microorganisms intended to biodegrade toxic chemicals is the buildup of intermediary metabolites that are more toxic, mobile, and/or persistent than precursor chemicals. For example, under anaerobic conditions in microcosms and in the field, tetrachloroethylene and trichloroethylene are biodegraded to vinylidene chloride (1,1-dichloroethylene) and vinyl chloride (Barrio-Lage et al. 1986, Bouwer and McCarty 1983). Vinyl chloride is a known human carcinogen (it causes a rare tumor, liver angiosarcoma), and vinylidene chloride is considered a possible human carcinogen. Other microbes can convert metallic mercury to the more toxic methyl mercury. The generality of the accumulation of toxic intermediates is hard to assess, as most biodegradation studies measure the disappearance of the substrate or the evolution of carbon dioxide, but do not identify metabolic intermediates. However, recent in vitro toxicity studies on soils contaminated by polycyclic aromatic hydrocarbons that are undergoing bioremediation have shown that, at least in some cases, the overall toxicity initially increases, and then decreases slowly as the biodegradation is allowed to continue (Irvin 1989).

Microbial action may make toxic chemicals more mobile in the environment, thus enhancing their exposure potential. For example, microbial action may make metal compounds, such as compounds of mercury, more volatile and bioavailable by converting them from inorganic to methylated compounds. In addition, methyl mercury is more toxic to humans and wildlife (cf., USEPA 1980). Under aerobic conditions, the intermediary metabolites of organic chemicals are more oxidized. The addition of oxygen to organic molecules results in the formation of polar functional groups that increase the water solubility and decrease the soil-water partition coefficient, both of which lead to increased mobility. Strauss and Swallow (1988) suggested that microbial biodegradation products are of key importance in assessing the health risks associated with leachates and gases from municipal landfills.

Introduced microorganisms may pose ecological hazards at the individual, population, community, and/or ecosystem levels of organization. For example, at the community level, nonindigenous organisms can influence both the structure (population size and species diversity) and the function (energy and material dynamics) of the initial community by displacing or destroying the indigenous species (see Chap. 1). The hazard identification process for ecological effects requires background information about the biology and ecology of the parent of the introduced microorganism and the nature of the genetic alteration. Important questions about the genetic alteration include its effect on the distribution of the altered organism with respect to its parent, its effect on the organism's relationships with other members of the ecological community, and its influence on the biochemical pathways for the production or degradation of organic and inorganic materials.

If concerns are identified, then actual tests should be performed on the parent and altered microorganism. Potential hazard end points for affected organisms are pathogenicity or toxicity, as manifested by death, delayed or altered growth and development, altered reproduction, or biochemical effects. Potential hazard end points at the population level include measures of age and stage structure, spatial distribution, predator-prey interactions, and keystone species. Potential hazard end points at the ecosystem level include alterations in primary or secondary productivity; nutrient cycles; decomposition; species diversity, richness, or abundance; and biomass. (For additional descriptions of these end points, see Chaps. 1 and 2, and Gillett 1985.)

Exposure assessment

Table 13.2 summarizes the information needed to conduct an exposure assessment. For cases where exogenous DNA has been introduced into a microorganism, the exposure assessment should consider both the transport and fate of the introduced microorganism and the introduced DNA.

To assess potential exposure to the introduced microorganism, information is needed about:

- The method, frequency, and quantity in which the microbe is introduced into the environment
- The climatic, geographic, and hydrogeological characteristics of the site of the introduction
- Statistically valid monitoring data and/or dispersal models in the appropriate environmental media
- Data and/or models regarding the reproduction and persistence of the microbe in the location of concern
- The identity, size, and geographic location of susceptible populations or ecosystems
- Methods to convert the ambient exposure concentrations to doses taken into the organism, population, or ecosystem of concern

To assess the potential exposure to the introduced DNA, information is needed on:

- The loss of the DNA from the microorganism where it was intended to reside
- The frequency of genetic transfer into other organisms under the environmental conditions likely to be encountered by the introduced microorganism
- The population size and distribution of the new microorganisms carrying the introduced DNA

The method, amount, and frequency of introduction of the microorganism will vary depending upon the reason for introducing the microbe, the engineering design of the process, and the efficacy and persistence characteristics of the introduced microorganism(s). This information is critical to the exposure assessment because it provides the input into the "source" term of the exposure calculations.

The spatial and temporal location(s) of the introduced microbe can be determined by monitoring, by modeling, or by a combination of both techniques. At least in the early stages of environmental releases, a combination of monitoring and modeling is likely to be required.

Monitoring each introduction can be a costly, labor-intensive process. This is especially true if the monitoring includes several environmental media at numerous locations and time points with sufficient replicates to produce statistically valid sampling (see Chap. 10 for a discussion of the statistics of sampling). The question "How much monitoring is enough?" is difficult to answer, but is likely to depend upon the potential risk associated with the release. Qualitative assessments, based on risk-related attributes such as those set forth by the ESA (Tiedje et al. 1989), may be useful in determining the intensity of monitoring required.

Transport and fate models have the potential to reduce the time and costs required to predict the spatial and temporal distribution of the introduced microorganisms. However, as discussed in Chap. 11, the current state of transport (dispersal) modeling and model validation for microorganisms is not sufficiently advanced to provide reliable and accurate predictions. In addition, the coupling of transport and fate models to provide estimates of the number of microorganisms at a particular location is extremely rudimentary. The fate components of most of the current transport models have static, first-order die-off rates that do not account for the environmental complexities. A few models, such as MICROBE-SCREEN, allow substrate-limited growth rates, as described by the Monod equation. However, this is a steady-state formulation that works well for chemostats. It is still untested under the transient conditions likely to be found after a microorganism is introduced into a new environmental setting.

At this stage of development, transport and fate models are best used as guides for the design of monitoring protocols, and perhaps for initial screening studies. However, further development and validation of microbial growth, fate, and transport models under environmental conditions, and the coupling of these models, would be a useful area of future research. Advances in model development and validation will require the results of field studies under a variety of environmental conditions. Several different microorganisms should be tested to determine the range of validity, the extent of calibration required, and the robustness of the models.

An exposure assessment requires the identification of susceptible populations as well as their size and spatial distribution. Susceptible populations may include humans, plants, animals, and ecosystems, as discussed above under "Hazard Assessment." Certain subpopulations, known as "sensitive subpopulations," may be more at risk for infections and other pathogenic effects of the microorganisms. For humans (and animals), the most susceptible populations are those which are immunocompromised to some extent. These may include the very young or old, those undergoing drug therapy for cancer or other diseases, people with nutritional deficiencies, people with diseases such as AIDS which suppress the immune system, and people exposed to chemicals that suppress the immune system. Workers are likely to be the population with the highest exposure potential during most microbial applications. Although workers are usually considered to be a healthy population, this may not be universally true, especially if concomitant exposures to toxic chemicals (from agricultural use, hazardous waste, or elsewhere) are considered.

The biomass or viable counts of a microorganism arriving at a susceptible receptor is the ambient exposure. However, this may or may not be the correct measure of dose, the unit that was used for testing in the hazard assessment protocol (see the section titled "Dose-Response Assessment" below). Analyses of the relationship between ambient exposures and "dose" seem to have been largely neglected in previous frameworks for adapting chemical risk assessments to microbes released into the environment. However, it will be necessary to quantify the relationship(s) between ambient exposure and dose in order to make quantitative risk assessments.

In principle, the fate of the introduced DNA (if exogenous DNA has been added to the introduced microorganism) should be assessed independently of the fate of the microorganism. In practice, this is difficult to monitor experimentally and difficult to predict on the basis of models and data currently available. The introduced DNA may partition as follows: (1) remain in the introduced microorganism, (2) be lost from the introduced microorganism and degraded, and/or (3) be transferred from the introduced microorganism into other microbes by transduction, transformation, or conjugation. Because both the DNA and microorganism can replicate, conservation of mass does not hold.

As discussed by Olson et al. in Chap. 8, it is known that plasmids can be transferred among microorganisms under suitable, but diverse, circumstances. For example, transfer has been observed under laboratory conditions, including chemostats, as well as in the intestines of humans and animals. In addition, there is a growing database on the transfer of plasmids in aquatic and terrestrial microcosms, including some information on transfer frequencies and the factors that affect transfer frequencies. However, there are few data on the frequency of transfer under actual environmental conditions, such as in water or soil. In addition, once the transfer has occurred, the survival of the recipient microbe, or stability of the DNA within that microbe, is usually unknown. There is also uncertainty as to whether or not the transferred DNA increases or decreases the fitness of the recipient microorganism (see Chap. 1).

Dose-response assessment

Microorganisms, like chemicals, produce their effects in a dose-related manner. In general, higher doses result in a higher probability of an effect, a faster response, a more severe effect, or some combination of these. For some microorganisms, and some end points, there may be a threshold below which there is no hazard, although this threshold may differ for receptor populations with different susceptibilities. For other microorganisms, such as some of the highly infectious bacteria and viruses that have caused disease in laboratory workers, a single microorganism may cause an effect.

In order to assess the risk of a particular hazard quantitatively, an understanding of the relationship between dose and effects is necessary. This requires quantitative data, gathered using appropriate test systems, on the response to several doses of the introduced microorganism. By analogy to test methods for chemicals, the assessment of a low-probability event at low doses may require high doses in a test system. If this is the case, then a methodology for low-dose extrapolation is necessary.

Haas (1983*a*) compared the predictions of three low-dose-extrapolation models, the log-normal, the single-hit exponential, and the β-distributed "infectivity probability" model. (These models are described in Chap. 11.) However, the model predictions were not validated against data. Further work is necessary to identify appropriate extrapolation models and the conditions under which they apply.

Risk characterization

Formally, the risk characterization step of a risk assessment combines the quantitative predictions of the exposure assessment with the quantitative dose-response data for each of the hazard end points. This formulation of risk assumes that hazard and exposure can be separated experimentally and mathematically. Unfortunately, this is not likely to be the case for microorganisms.

In a microbial risk assessment, the ambient exposure to a microorganism is likely to be confounded by the effects of the organism. For example, if one of the potentially hazardous effects is microbial reproduction and establishment (invasion) at a new site, the effect will increase the number (or biomass) of the introduced microorganisms and hence the exposure potential (Suter 1985).

The interaction between exposure and hazard suggests that an iterative procedure is necessary for risk characterization rather than simple multiplication of independently derived functions. Moreover, the iterative procedure will not necessarily converge to a fixed value. The consequences of the population dynamics, and the potential inappropriateness of using a steady-state assumption for the number of microorganisms present, is only beginning to be explored in the context of risk assessment.

Despite the interrelationship between exposure and hazard in some instances, it may be possible to combine the results of exposure, dose, and hazard models under other conditions. The predictions may be most usefully made, at the present time, under conditions where microorganisms are not expected to replicate. Haas (1983*b*) described an approach to combining exposure and hazard assessments for estimating the excess risk of enteric diseases from waterborne viruses predicted to result from relaxing effluent wastewater standards. Briefly, Haas (1983*b*) assumed that

$$P^* = P_1 P_2$$

where P^* is the probability that, after a single exposure to one type of virus, an individual will contract a disease

P_1 is the probability that a single exposure will result in the ingestion of organisms of the specified type

P_2 is the probability that, once ingested, an organism will cause a disease.

Calculation of P_1 requires an estimate of (1) the concentration of organisms at the point of exposure, from either monitoring data or fate and transport models, and (2) the volume of water ingested. Estimation of P_2 can be based on dose-response data or low-dose-extrapolation models (Haas 1983*a*).

Once a risk assessment is conducted, the results have to be presented in a way that is useful for regulators and the public, as well as to risk assessment experts. Presentation of the results of a quantitative risk assessment for the environmental release of microorganisms is likely to be more complex than for chemicals. The results of the assessment are unlikely to be in the form of a single number with confidence limits. Benchmark values for comparisons may be difficult to obtain. However, vague and qualitative judgments of high, medium, and low risk are unlikely to be acceptable to the public. Other methods, such as graphical or other representational approaches, could be explored.

Limitations and uncertainties

As with all risk assessments, there are many uncertainties and limitations associated with the framework and information provided here. Many of the limitations of the individual models and the difficulties of predicting hazard end points are discussed in the relevant chapters of this book. Several of the more global limitations are discussed in this section.

One limitation of the fate and transport models discussed in Chap. 11 and used in the CRA formulation is that most were developed for chemicals or inert particles. A few have been developed or specifically adapted for bacteria, although none accounts for the complexity of bacterial survival under changing and variable environmental conditions. None of the models accounts for the ecology of the microorganism—i.e., the limitations on suitable habitats for reproduction and colonization.

Many of the individual models described, especially in Chap. 11 on fate and transport models, either have not been validated for microorganisms, or have been validated in laboratory but not in field situations. Thus, the predictive value of the growth models, the genetic transfer models, and the transport models in actual field situations is generally unknown. Furthermore, several of these fate and transport models provide generalized "default" values for parameters if the more appropriate site-specific and microorganism-specific data are not available. The sensitivity of the model predictions to the parameter values used, especially for microbial survival, has not been tested for many of these models. However, in cases where limited sensitivity testing has been conducted, as in long-range aerial transport (Bovallius et al. 1980), there is a range of survival rates where the prediction of dispersal is critically dependent upon the value of the die-off rate.

Chapter 5 reviews actual field data on persistence and transport of viruses, bacteria, and fungi. A wide variety in the quality of data is observed. There are important gaps, even with well-studied entomopathogens. Fate and transport models based on chemicals or particulates are probably not appropriate, or at least not adequate, for microorganisms in which vector transport is important. The importance of vector transport, especially by aphids, for many plant viruses is described by Tolin in Chap. 6.

The question of whether all hazards have been identified is an uncertainty that is common to risk assessments in complex systems. However, the hazard identification process may be particularly difficult and uncertain for microorganisms for two reasons. First, microorganisms can be extremely species-specific in their effects. For example, a strain of bacteria may be highly pathogenic to one species, such as pigs, but totally harmless for goats, sheep, and humans. This high species specificity makes hazard testing difficult and not generalizable. The "species extrapolation" procedures used for chemicals are not relevant for an infectivity end point of microorganisms. Second, even low-probability individual events may lead to observable effects because of the large number of individual microorganisms likely to be introduced. (The total probability of occurrence will be a product of the probability of an event per chance and the number of chances.) Thus, low-probability events must be included in the hazard identification and evaluation stages of the risk assessment.

In a quantitative risk assessment, models for dispersal, reproduction, and perhaps dose-response have to be tied together. The output of one model is used as the input to another model. The over- or underpredictions of the individual models may either cancel out or be enhanced when the models are tied together. At the present stage of development, it does not seem to be possible to put boundaries on the uncertainty of the risk assessment.

Summary and Conclusions

This chapter has focused on the paradigm for chemical risk assessment proposed by the NAS and used by USEPA, and its applicability to the environmental release of genetically altered microorganisms. One advantage of using this approach is its familiarity to risk assessment professionals and regulators, and the potential availability of models developed for chemicals. However, if starting "from scratch," this may or may not have been the paradigm selected. The use of models developed in disciplines based on the study of (micro)organisms, such as infectious diseases, evolutionary biology, and ecology, might have led to a different risk assessment formulation and a different set of data requirements.

Several risk assessment frameworks are available to use with microorganisms. For example, epidemiological models developed to predict the spread of infectious diseases may be useful if infectious diseases are of concern. Some of the models that have been applied to infectious diseases of crop plants are discussed in Chap. 12 of this volume. However, these models require prior knowledge of the effects of the introduced microorganism and good information on the location, density, and sensitivity of susceptible populations. These data are not likely to be available for many releases.

Other frameworks for microbial risk assessment include the more deterministic fault tree or event tree models that are used for complex engineered systems such as the space shuttle. They also include the familiarity framework suggested by the NAS (NAS 1989), and the risk attribute approach suggested by the ESA (Tiedje et al. 1989).

The various frameworks for risk assessment may not be mutually exclusive. For example, both the NAS familiarity approach and the ESA risk attribute approach include a means to determine a level of concern based on relatively little data. This, in turn, can lead to a regulatory decision regarding the amount of additional data or testing required prior to environmental release and the monitoring requirements for a small-scale field trial. There is no counterpart to this "a priori level of risk" step in a chemical risk assessment approach. Thus, the risk-ranking portions of the NAS approach, or more easily the ESA approach, can be used to determine whether a particular environmental use of a particular microorganism should be evaluated by a more formal and data-intensive procedure such as a microbial analog of chemical risk assessment.

The chemical risk assessment framework described in this chapter is adaptable to a variety of end points, including disease of nontarget organisms (in some cases), and can be used qualitatively when insufficient data are available for quantitative assessments. The exposure assessment can be used to predict the presence of a microorganism outside of a specified area. If desired, this can be a regulatory end point in the absence of data regarding specific hazards. However, there are some cases where the fate and transport models used in chemical risk assessment are inappropriate. Examples of these cases include viruses for which reservoirs are important and cases in which vector transmission or adventitious transport are important parameters. In these cases, alternative formulations to the CRA approach, such as the epidemiological models used for infectious diseases, are likely to be more predictive of potential risk.

References

Alexander, M. 1985. Ecological consequences: reducing the uncertainties. *Issues Sci. Technol.* 1(3):57–68.

Barrio-Lage, G., Parsons, F. Z., Nassar, R. S., and Lorenzo, P. A. 1986. Sequential dehalogenation of chlorinated ethenes. *Environ. Sci. Technol.* 20:96–99.

Bouwer, E. J., and McCarty, P. L. 1983. Transformation of 1- and 2-carbon halogenated compounds under methanogenic conditions. *Appl. Environ. Microbiol.* 45:1286–1294.

Bovallius, A., Roffey, R., and Henningson, E. 1980. Long range transmission of bacteria. *Ann. N.Y. Acad. Sci.* 353:186–200.

Covello, V. T., and Fiksel, J. R. (eds.). 1985. The suitability and applicability of risk assessment methods for environmental applications of biotechnology. Final report to the Office of Science and Technology Policy, Executive Office of the President. Report no. NSF/PRA 8502286, National Science Foundation, Washington, D.C.

Dean-Ross, D. 1986. Applicability of chemical risk assessment methodologies to risk assessment for genetically engineered microorganisms. *Rec. DNA Tech. Bull.* 9(1):16–28.

Fiksel, J. R., and Covello, V. T. (eds.) 1986. *Biotechnology Risk Assessment: Issues and Methods for Environmental Introductions.* Pergamon, New York.

Gillett, J. W. 1985. Risk assessment methodologies for biotechnology impact assessment, in *Potential Impacts of Environmental Release of Biotechnology Products: Assessment, Regulation, and Research Needs*, J. W. Gillett, A. Stern, S. Levin, M. Harwell, M. Alexander, and D. Andow (eds.). Report no. ERC-075, Ecosystems Research Center, Cornell University, Ithaca, New York.

Gillett, J. W., Stern, A., Levin, S., Harwell, M., Alexander, M., and Andow, D. 1985. *Potential Impacts of Environmental Release of Biotechnology Products: Assessment, Regulation, and Research Needs.* Report no. ERC-075, Ecosystems Research Center, Cornell University, Ithaca, New York; also published in full in *Environ. Mgmnt.* 10(4), July 1986.

Haas, C. N. 1983*a*. Estimation of risk due to low dose of microorganisms: a comparison of alternative methodologies. *Am. J. Epidemiol.* 118:573–582.

Haas, C. N. 1983*b*. Effect of effluent disinfection on risks of viral disease transmission via recreational water exposure. *J. Water Pollut. Control Fed.* 55:1111–1116.

Halvorson, H., Pramer, D., and Rogul, M. (eds.). 1985. *Engineered Organisms in the Environment: Scientific Issues.* American Society for Microbiology, Washington, D.C.

Irvin, R. 1989. Presentation at Hazmat International, Atlantic City, New Jersey, May 1989.

Levy, S. B., and Marshall, B. M. 1988. Genetic transfer in the natural environment, in *The Release of Genetically-Engineered Micro-Organisms*, M. Sussman, C. H. Collins, F. A. Skinner, and D. E. Stewart-Tull (eds.), Academic, San Diego, California.

National Academy of Sciences (NAS), National Research Council. 1983. *Risk Assessment in the Federal Government: Managing the Process.* National Academy Press, Washington, D.C.

National Academy of Sciences (NAS), National Research Council. 1989. *Field Testing Genetically Modified Organisms: Framework for Decisions.* National Academy Press, Washington, D.C.

Office of Technology Assessment (OTA), U.S. Congress. 1988. New developments in biotechnology—field-testing engineered organisms: genetic and ecological issues. OTA-BA-350, U.S. Government Printing Office, Washington, D.C.

Strauss, H. S., Hattis, D., Page, G., Harrison, K., Vogel, S., and Caldart, C. 1985. Direct release of genetically-engineered microorganisms: a preliminary framework for risk evaluation under TSCA. Report no. CTPID 85-3, Center for Technology, Policy and Industrial Development, Massachusetts Institute of Technology, Cambridge.

Strauss, H. S., Hattis, D., Page, G., Harrison, K., Vogel, S., and Caldart, C. 1986*a*. Genetically-engineered microorganisms. II: Survival, multiplication, and genetic transfer. *Rec. DNA Tech. Bull.* 9(2):67–88.

Strauss, H. S., Hattis, D., Page, G., Harrison, K., Vogel, S., and Caldart, C. 1986*b*. Genetically-engineered microorganisms. I: Identification, classification and strain history. *Rec. DNA Tech. Bull.* 9(1):1–15.

Strauss, H. S., Ingram, C., and Hattis, D. 1986*c*. A draft questionnaire for gathering data to assess the risks of releasing microorganisms into the environment. Report no. CTPID 86-8, available from the Center for Technology, Policy and Industrial Development, Massachusetts Institute of Technology, Cambridge.

Strauss, H. S., and Swallow, K. C. 1988. Landfills: environmental and public health issues, to be published in *Proceedings of The Conference on Municipal Solid Waste Disposal: Public Health, Environmental, Economic, and Technological Aspects*. Lewis Publishers, Chelsea, Michigan.

Suter, G. W., II. 1985. Application of environmental risk analysis to engineered organisms, in *Engineered Organisms in the Environment: Scientific Issues*, H. Halvorson, D. Pramer, and M. Rogul (eds.). American Society for Microbiology, Washington, D.C., pp. 211–219.

Tiedje, J. M., Colwell, R. K., Grossman, Y. L., Hodson, R. E., Lenski, R. E., Mack, R. N., and Regal, P. J. 1989. The planned introduction of genetically engineered organisms: ecological considerations and recommendations. *Ecology* 70(2):298–315.

U.S. Environmental Protection Agency (USEPA). 1980. *Ambient Water Quality Criteria for Mercury*. PB81-117699, National Technical Information Service, Springfield, Virginia.

Chapter

14

Using Expert Panels to Assess Risks of Environmental Biotechnology Applications: A Case Study of the 1986 Frostban® Risk Assessments

Jonathan S. Naimon[1]

Investor Responsibility Research Center, Inc.
1755 Massachusetts Avenue
Washington, D.C. 20036

Introduction

This chapter analyzes the advantages and disadvantages of using expert panels (EPs) to integrate information on potential hazards and exposure to genetically engineered microorganisms (GEMs) used in environmental applications. Risk analysis case histories are used to illustrate how EPs were used to analyze potential risks of the first deliberate environmental release of GEMs in the United States. These case histories are based on a review of public documents and interviews with key scientists and regulators in the U.S. Environmental Protection Agency (USEPA), California Department of Food and Agriculture (CDFA), and Monterey County (Naimon 1987).

Scientific background

Advanced Genetic Sciences (AGS) of Oakland, California, planned to market Frostban® as a new method for reducing frost damage to crops at temperatures

[1]The author gratefully acknowledges the support of the National Science Foundation and the University of North Carolina, Institute for Environmental Studies. The views expressed in this chapter are those of the author alone.

just below 0°C. Frostban® is a mixture of two naturally occurring microorganisms, *Pseudomonas syringae* and *P. fluorescens*, that have been genetically engineered by deleting the genes that code for a protein that facilitates ice nucleation at near freezing temperatures. Bacteria that express the ice-nucleation protein are termed ice-nucleation-positive (Ina^+); bacteria that do not are termed ice-nucleation-negative (Ina^-). Ina^+ bacteria facilitate the formation of ice crystals on leaf surfaces and have also been isolated from hail, raindrops, and atmospheric aerosols. Several authors have suggested that low concentrations of Ina^+ bacteria may be important in the formation of rainfall (Vali 1971).

The field test involved spray application of Ina^- GEMs onto a quarter acre of emergent strawberry blossoms. AGS maintains that the Ina^- GEMs would prevent naturally occurring Ina^+ populations from colonizing the sprayed plants, and thus reduce frost damage.

Organization

This chapter has four components. The first three components review the risk analyses undertaken by the USEPA, the CDFA, and Monterey County. Each section contains:

- A review of the role the special review panels played
- A discussion of problems that were adeptly addressed and of problems with which expert panels had difficulty

An attempt is made to distinguish between true expert judgments and opinions of experts—"instant knowledge." The importance of recognizing the limitations of experts' knowledge and the role that their personal preferences can play in decision making is described in *Acceptable Risk* (Fishhoff 1981).

The last section summarizes key issues that arose in the case studies and provides some recommendations for increasing the usefulness of risk assessment EPs that are called upon to review future environmental applications of biotechnology.

U.S. Environmental Protection Agency Risk Analysis

Role of expert panels

Policy background USEPA has legal authority to regulate GEMs under the Toxic Substances Control Act of 1976 (TSCA) and the Federal Insecticide, Rodenticide, and Fungicide Act of 1946 (FIFRA) as amended. Under FIFRA, there was an exemption from the experimental use permit (EUP) requirement for small-scale field tests. USEPA changed its exemption policy for field tests of GEMs as a result of three factors: a letter from the FOET[2] suggesting that

[2]The Foundation on Economic Trends (FOET) is a public interest organization that successfully sued the Department of Health and Human Services to block a field test of an Ina^- GEM by the University of California in 1983.

allowing a small-scale field test of GEMs without requiring an EUP would constitute a violation of the National Environmental Policy Act (NEPA), discussions with the National Institutes of Health (NIH) about a proposed (1983) Ina$^-$ GEM field test, and the opinion of USEPA's legal counsel (Betz 1990). USEPA published an interim policy under which it could require additional information from EUP applicants based on the unique potential risks (if any) posed by specific GEMs (USEPA 1984).

USEPA's analysis of the field-test proposal submitted by AGS had two distinct components. In a 3-month first phase, USEPA determined that an EUP would be required for AGS' Frostban® field-test proposal. In the second stage, a special EP was convened, and USEPA undertook an elaborate, 10-month analysis before approving the EUP. Figure 14.1 diagrams the process USEPA used to evaluate the AGS application for an EUP.

Phase one: USEPA's preliminary analysis Under FIFRA, USEPA had 90 days to determine whether an EUP should be required. USEPA's preliminary review was undertaken by staff with backgrounds in environmental engineering, chemical engineering, chemistry, and ecology. The preliminary review focused on a basic exposure question and a broad hazard question, which were, respectively:

- Could the GEMs survive in the environment outside the proposed test site, and potentially expose other organisms?
- Did the GEMs have the potential to cause adverse health or environmental effects?

The staff analyzed the AGS proposal using information in the AGS proposal, the 1983 review by NIH's Recombinant DNA Advisory Committee (NIH RAC) of a similar experiment proposed by the University of California, and both telephone calls and face-to-face meetings with AGS staff.

USEPA staff divided the first question (GEM survival and possible nontarget exposure) into two components:

- Would the Ina$^-$ GEMs be dispersed by physical, biological, or other factors?
- Would they survive outside the test plot?

The staff concluded that the physical and biological mechanisms by which microbes could be dispersed from the site(s) made off-site exposure to the GEM plausible, and that GEM survival in off-site environments was also plausible because the commercial product was designed to survive in outdoor, agricultural environments that are similar to the test sites.

USEPA's preliminary review of the potential for effects on precipitation used articles and argumentation in the FOET suit and NIH's analysis. USEPA staff did not reach a tentative conclusion on this complex, contentious issue during this preliminary analysis.

Based on the assumption that the traits of GEMs would be similar to those of the parental strains and considering the 1983 NIH review of a similar experiment, USEPA reached a tentative conclusion that the genetically altered

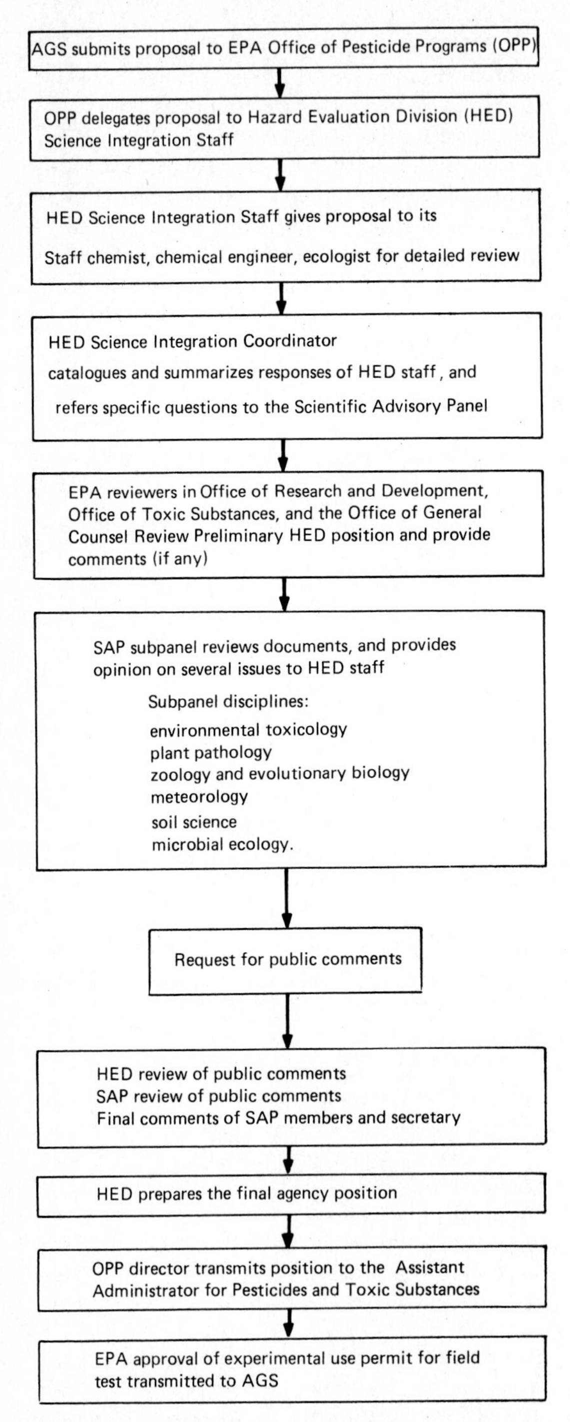

Figure 14.1 Flow diagram for USEPA's review of the AGS proposal to field-test Frostban®.

bacteria posed similar risks to plants and animals as *P. syringae* and *P. fluorescens*.

USEPA's risk analysis framework USEPA staff devised adverse-effects risk scenarios as a framework for its analysis of the AGS proposal based on (1) their procedure for evaluating conventional microbial pesticides, and (2) NIH's review of the University of California's Ina$^-$ proposal. In fact, USEPA's risk analysis framework incorporated three adverse-effect scenarios: infection of mammals, pathogenesis of plants, and disruption of precipitation. In the scenarios USEPA evaluated, three exposure steps are required for a potential hazard to be projected. These exposure steps were:

1. GEM dispersal from test site to adjacent areas
2. Colonization of adjacent areas
3. Displacement of indigenous Ina$^+$ microorganisms by Ina$^-$ GEMs in Frostban®

Following the potential displacement of indigenous Ina$^+$ microflora in various habitats, USEPA considered three types of adverse-effect scenarios, one of which (3) was unique to this proposed application:

1. Infection of mammals and/or humans in proximity to the test
2. Infection of nontarget agricultural plant species
3. Alteration of normal precipitation patterns

Table 14.1 summarizes the parameters considered within the USEPA risk analysis framework, and the information USEPA required the applicant to supply to support evaluation of each.

Role of USEPA's FIFRA SAP subpanel USEPA has formed several panels of outside experts (scientific advisory panels, or SAPs) to provide objective interpretations or advice on issues that arise in implementing the agency's broad charter. The FIFRA SAP is a panel chartered to provide advice to the agency on risks and benefits of agricultural products. The FIFRA SAP subpanel whose activities are discussed in this chapter was convened to provide the agency with an additional scientific review capability for the Frostban® risk analysis. USEPA believed that an additional panel was required because the regular FIFRA SAP did not have experience in areas that were evaluated by NIH for the University of California proposal: molecular biology, atmospheric science, and evolutionary biology (Naimon 1987). It also may have been required because of the intense scrutiny associated with the first environmental application of genetic engineering reviewed by USEPA. The SAP subpanel was chaired by Wendell Killgore, a member of the regular FIFRA SAP and an environmental toxicologist. The specializations of the subpanel members (all academic scientists) are shown in Fig. 14.1. The subpanel reported through the regular FIFRA SAP to the Director of the Office of Pesticide Programs (OPP).

TABLE 14.1 USEPA Risk Analysis Parameters for Frostban®

Parameter	Information required of applicant
Organism characteristics	Genetic engineering methods used to produce Frostban® Genetic and physiological comparisons of the GEMs and the parental organisms Habitat of Ina^- GEMs relative to Ina^+ parental strains
Possible off-site transport and establishment:	
Off-site transport by physical or biological vectors	Monitoring plans for off site dissemination via insects Precise location of field-trial sites and description of nearby crops
Establishment of off-site colonies	Methods of detection of Ina^- GEMs Growth rates and survival rates for Ina^- GEMs Plans for detecting possible off-site dissemination Colonizing "ability" of Frostban® and other Ina^- strains under different experimental application conditions
Atmosphere transport	Upward flux from plants to air Monitoring plans for horizontal drift Vertical flux from thermals not monitored
Adverse-effects scenarios:	
Effects on nontarget species	
Mammalian toxicity	Rabbit challenge experiment
Plant pathogenicity	Parental strain pathogenicity to crops and native plants in the area of the test sites Host specificity of parental strains
Effects on precipitation	
Displacement of atmosphere Ina^+ bacteria by terrestrial Ina^- GEMS	None
Reduction in atmospheric ice nucleation at −5–0°C.	None
Reduction in terrestrial precipitation in areas where most Ina^+ nuclei originate from agricultural flora	None

SOURCE: Information requirements listed in USEPA letter to Steven Cull, Feb. 1, 1985, p. 4., contained in USEPA (1986*a*).

The FIFRA SAP subpanel's scope was limited to helping USEPA analyze a single product. USEPA has subsequently formed a Biotechnology Science Advisory Committee (BSAC) that reports directly to the USEPA Administrator. The BSAC is responsible for addressing cross-cutting issues such as emerging research needs, cross-program and case review consistency, and interactions with other agencies on scientific definitions and problems.

USEPA's procedure for analyzing EUP risks. USEPA adopted an iterative procedure (Fig. 14.1) to prepare a comprehensive risk assessment that could stand up to a potential court challenge by FOET or other opponents of the field test. The procedure gave USEPA multiple opportunities to obtain additional information from AGS to fill information gaps identified by either USEPA staff or the SAP subpanel.

Shortly after determining that an EUP would be required, USEPA formally requested that AGS provide data addressing a host of issues (Table 14.1). These data were required because analogous requirements existed to support EUPs for new chemicals, in order to assess individual end points in the adverse-effect scenarios described above, and to provide the USEPA with adequate background information on the microorganisms.

Data submitted by AGS to support the EUP. In April 1986, AGS submitted approximately 300 pages of data and discussion of the competition and precipitation issues. The bulk of the AGS data was made up of:

- Experimental data and a review of the literature on the ability of the parental strains to cause infections
- Pathogenicity experiments on agricultural plants
- Discussion of risk to atmospheric precipitation processes

Since the capacity of the parental strain *P. fluorescens* to contribute to opportunistic infections in immunocompromised humans is well known, the mammalian infectivity tests performed with the Ina$^-$ GEM were important to the USEPA. Agency staff reviewed the experimental data and decided that the potential for the product to cause infections in mammals was minimal (USEPA 1986*a*).

Because several groups of commercially important plants, such as tomatoes, apricots, and sorghum, are susceptible to infection by *P. syringae*, USEPA was concerned about the plant pathogenesis adverse-effect scenario. Although the literature revealed that there was some plant pathogenesis risk from indigenous, naturally occurring *P. syringae*, the plant pathogenicity tests submitted by AGS indicated that the pathogenic range of the genetically engineered *P. syringae* was similar to the parent strain's. The agency concluded that the release would not present any significant incremental risk.

The discussion of the competition and precipitation issues was not sufficient to convince the USEPA reviewers that the company's position (that the organisms would remain on site and there would be no effects on precipitation) was adequately supported. Agency staff asked the SAP subpanel to review the AGS data and the staff's analysis of the issues.

The SAP subpanel members asked the USEPA staff to obtain additional information from AGS on three experimental questions:

- Competition between Ina^+ and Ina^- bacteria
- Pathogenicity of the GEM toward certain crop families susceptible to naturally occurring *P. syringae*
- Test design and conditions under which greenhouse colonization experiments were performed[3]

These data were provided in June 1986.

In meetings with the SAP subpanel, USEPA staff asked several critical risk assessment questions to help interpret the information supplied by AGS. These questions ranged from analysis of experimental data submitted by AGS to evaluation of current ecological theory vis à vis the competition questions. USEPA staff stated that the subpanel's key contributions to the EUP risk analysis were their interpretations of:

1. The survival and competition data supplied by AGS
2. The precipitation discussions provided by AGS and FOET

The SAP subpanel's consensus on these questions was that:

1. The GEMs would survive off-site, but would not outcompete (displace) the naturally occurring Ina^+ organisms
2. The evidence for involvement of Ina^+ organisms in atmospheric precipitation is circumstantial; therefore, the probability of a small-scale field test affecting precipitation patterns is virtually nil.

Although the agency's staff received a strong statement from the SAP subpanel's sole atmospheric scientist, Randolph Borys, that there was no evidence to prove that Ina^+ bacteria were involved in rainfall, staff contacted other atmospheric scientists by telephone to corroborate Borys' interpretation. One of these scientists, Russell Schnell, has written that Ina^+ bacteria may be critical to atmospheric precipitation processes at temperatures approaching 0°C, and are also likely to originate from terrestrial plant sources (Schnell and Vali 1986).

Although the SAP subpanel had consensus on the issues critical to its risk analysis, members of the subpanel had differences of opinion on several less critical issues. Most, but not all, members believed that previous experiments with Ina^- mutants obtained by chemical mutagenesis were good models for genetically engineered Ina^- mutants because of the phenotypic similarity and open-environment testing. One member, zoologist and evolutionary biologist Robert Colwell, emphasized that the genetic differences between the GEMs

[3]AGS did not report that some of the plant pathogenicity tests supporting this EUP were conducted on plants located on an unenclosed rooftop rather than in indoor greenhouses. The USEPA temporarily suspended AGS' EUP and issued a civil complaint. AGS agreed to pay a fine of $13,000 in 1987 for procedural violations. (USEPA Office of Compliance Enforcement 1986*b*)

and the parental bacterial strains from which they were derived could affect off-site survival and the resulting impacts on nontarget species. Several other members disagreed with Colwell's analysis of the importance of genetic differences to ecological impacts.

Two subpanel members requested that additional experiments be conducted to assess pathogenicity of the GEM toward several additional plant species. AGS performed several of the tests on the commercial plant varieties, but did not perform the pathogenicity tests on the wild plant (native to Monterey County, site of the proposed field test) that is the ancestor of the modern strawberry. Most other members believed that such tests were unnecessary because the GEMs were likely to be very similar to the parental bacterial strains (USEPA 1986*a*).

Exposure information assessment USEPA's integration of each of the adverse-effect scenarios implicitly combined a hazard evaluation with an evaluation of the potential for exposure to the GEMs. USEPA did not perform a quantitative evaluation of the potential for off-site receptors to be exposed to the GEMs, perhaps because no quantitative models of bacterial transport were available. As a result, the agency's exposure analysis hinged on two factors:

- The small land area (0.20 acre) proposed for the test, and the commonsense notion that small tests should result in small off-site exposures
- The SAP subpanel's interpretation of the AGS "cocolonization" experiments that genetically engineered Ina^- bacteria would not "outcompete" naturally occurring Ina^+ bacteria.

The evidence AGS submitted on competition between Ina^- and Ina^+ bacteria in greenhouses and outdoor agricultural settings (with Ina^- bacteria obtained by conventional techniques) enabled USEPA to conclude that the potential for Ina^- GEMs to "outcompete" the Ina^+ bacteria was insignificant.

Subsequent tests by AGS and the University of California in Contra Costa County, California, and Tulelake, California, have thus far borne out these exposure analysis assumptions for Ina^- GEMs applied in agricultural settings (Lindow et al. 1988).

USEPA's final scientific position on the AGS experimental use permit While there were some differences of opinion about what specifically could be concluded from interpreting the data provided by AGS, the panel members all agreed that the test should be undertaken and would not pose an unacceptable risk. Based on this input, agency staff prepared a formal scientific position on the issues assessed in the risk analysis; this position is summarized in a memorandum recommending that USEPA grant a permit to AGS to field test the Frostban® product in the open environment, subject to some siting and notification qualifications. Risk assessment issues were summarized as follows:

> The evidence submitted to support the application showed that, at the "low level" proposed in the AGS EUP application, the genetically engineered organisms were

unlikely to overrun the naturally occurring epiphytic flora present outside the test site, but would, rather, coexist with them. At these low levels, the INA-products would not have any significant impact on the indigenous populations of bacteria that some have suggested may have an effect on precipitation patterns. (USEPA 1986)

USEPA sent a letter to AGS informing them of the EUP approval and describing several permit conditions including the following:

- AGS was required to obtain permission from the State of California for the test.
- AGS was required to clear a 15-m buffer zone between the test site and adjacent properties (to attempt to prevent GEMs from migrating off site).
- AGS was required to conduct the test in a "remote" area.

Advantages of using USEPA's expert panel procedure

Management issues Use of an EP permitted USEPA staff to reach a decision on a precedent-setting GEM permit application that involved scientific issues with which the agency's regulatory staff had little experience. The EP extended the capability of the staff to address issues associated with an emerging technology. Although many of the risk analysis issues assessed in this case are quite similar to questions addressed in conventional microbial pesticide risk analyses, the use of GEMs has brought out some additional concerns. Some examples of issues that received more attention than they would in a conventional analysis are:

- The identity of the organisms
- Influence on geographic ranges of beneficial insects
- How the GEMs will fare in new environments in comparison with the existing microflora
- Whether genetic differences that are not expressed can be ecologically or evolutionarily important
- The involvement of bacteria in precipitation processes
- The effect of altering genotype on phenotypic expression

Although the precipitation issue is not a typical microbial pesticide issue, USEPA had already registered a copper compound used to kill Ina^+ bacteria as a frost protection product. Nonetheless, the Frostban® risk analysis brought out a number of facets of these issues that are uniquely associated with the use of genetic engineering.

The SAP subpanel also helped USEPA to separate critical issues in the biotechnology risk analysis process from noncritical issues. Use of the EP provided key HED staff with an opportunity to learn from face-to-face meetings with leading academic scientists in several relevant disciplines that are not widely represented in USEPA's regulatory arms.

From a broader perspective, use of the EP yielded a rational and defensible basis for regulatory decision making in an area of applied ecology where there are conflicting theoretical models, sparse directly relevant data, and (at the time of the test) no established protocols for even determining the most important factors in integration of exposure and hazard.[4] The panel gave USEPA's ultimate decision the imprimatur of scientific rigor that preempted subsequent risk analyses of most of the scientific questions. However, the expert panel did not confer an imprimatur of regulatory rigor sufficient to preclude an independent and conflicting regulatory review by Monterey County. From a management perspective, this regulatory problem cannot be changed by changing experts, but only by changing either the panel charters or the responsibilities of the staff.

Scientific issues

Pathogenicity The EP provided critical advice to USEPA in interpreting pathogenicity and competition experiments conducted in support of the EUP. The SAP subpanel's evaluation of the incremental risks to plants required an understanding of the existing threats from various strains of *P. syringae*, competition between this species and other epiphytic bacteria, and other factors. The subpanel was able to interpret these tests and thus contributed substantially to reducing the USEPA staff's uncertainty about an important adverse-effect scenario.

Competition The hazard(s) resulting from each adverse-effect scenario evaluated hinged on the "competition or cooperative interactions" between Ina^- GEMs and indigenous microflora. A variety of conflicting ecological models have been postulated to assess the competition between native and exotic microflora. The data presented by AGS did not show a clear competitive advantage or disadvantage for Ina^- or Ina^+ bacteria, and it did not suggest any advantage or disadvantage based on the use of genetic engineering to obtain Ina^- bacteria. The EP provided a critical interpretation of the AGS cocolonization experiments that enabled USEPA to conclude that the probability of significant off-site colonization was very low.

Disadvantages of USEPA's expert panel procedure

Management issues

Scope of analysis An important aspect of early (in a technology's life) risk analyses is their value as precedent. Future risk analyses of environmental

---

[4]There have been some notable, recent developments in the area of identifying factors to consider in the analysis of risks from environmental applications of biotechnology. See *The Planned Introduction of Genetically Engineered Organisms: Ecological Considerations and Recommendations* (Tiedje et al. 1989).

biotechnology applications may be expected to be based on those conducted in the United States and Europe in the late 1980s and early 1990s. This EP did not assess risks to potentially sensitive terrestrial receptors (such as birds, bees, wild plants) from widespread, commercial use of products similar to Frostban® that may be anticipated in the future. If and when this technology is commercialized, USEPA's risk analysis is likely to serve as a baseline analysis.

An alternative approach could have used the agency's BSAC rather than the FIFRA expert subpanel to examine commercial-scale exposures in order to more robustly assess potential risks from the biotechnology application. Considering commercial-scale exposures would result in analysis of the potential for off-site colonization as a result of greater exposure. Such a procedure would reduce the need to convene other EPs in the future, at greater expense, to assess risks from widespread usage of a GEM.

Scenario generation based on administrative precedent as well as scientific plausibility. The USEPA analysis was generally limited to adverse effects in the risk analysis scenario depicted in Fig. 14.1. While a great deal of time and effort was spent evaluating each adverse-effect scenario devised by USEPA, somewhat less effort was apparently spent in generating scenarios. As a result, atmospheric ecological effects of a terrestrial biotechnology application were assessed, but *terrestrial* effects from potential alteration of local frost cycles in regions that do not always have temperatures below −5°C were not. Two terrestrial impacts that seem to have received short shrift in USEPA's risk analysis were:

- Risks to agriculturally beneficial insects that could result from replacement of some Ina^+ bacteria by some Ina^- bacteria in terrestrial environments adjoining test sites
- Risks to wild plant species or the fauna that depend on them

The relatively low level of attention paid these problems in the risk analysis may be a function of many factors. The agency initially identified impacts on ranges of insects and other fauna as an item it wanted AGS to present data on. However, the initial interest displayed in the request to AGS was not followed up when data were not submitted on the subject.

One reason for this may be the dominant influence of administrative precedent in USEPA's formulation of the scientific risk issues for consideration in this review. The USEPA's adverse-effect scenario for effects on precipitation is virtually identical to the adverse-effect scenario described in the FOET's successful suit against NIH over the proposed University of California experiment with Ina^- *P. syringae*. The more plausible terrestrial ecological effects on local frost cycles and beneficial insects were not aggressively pursued because FIFRA permits USEPA to waive requirements on the effects on beneficial insects and ecological processes. Impacts on wild flora and fauna were identified by panelist Robert Colwell but were not pursued. The very tests used to assess pathogenic effects on agricultural plants could have been performed for wild plants. Colwell was the only reviewer with this particular concern and he could not convince the panel to support it. Like other EPs, this EP's scope was restricted by the consensus of its members.

As noted above, the precedent-setting value of small-scale risk analyses provides a reason for assessing potential ecological effects that may not occur until large-scale use is begun and the GEM is exposed to a variety of different environments.

Scientific issues

Atmospheric precipitation USEPA's SAP subpanel did not give any credence to the pre-1986 evidence cited in an Office of Technology Assessment (OTA) report that Ina^+ bacteria are involved in normal precipitation processes, and in this respect its judgment appears to have suffered from selective reading of the literature. This could be attributed to the fact that the subpanel was charged with reviewing a small-scale project and OTA reviewed large-scale, commercial use of Ina^- bacteria (Upper et al. 1986). Alternatively, it could be due to the fact that this subpanel had a preponderance of biologists and perhaps an insufficient internal capability to discuss and referee atmospheric sciences issues. This issue illustrates the differences that may exist between acknowledged experts on areas of scientific controversy, and underlines the importance of assuring that EPs are fairly balanced on such issues. If panels are not well balanced, there is a possibility that the scientific review will be less robust for the issues for which there is reduced disciplinary representation. On such issues, judgments may be based on unverifiable intuitions of individual reviewers, rather than the whole fund of useful knowledge that experts are expected to bring to such issues. If risk analysis panels are split, regulators will be forced to make decisions based on strict interpretation of the relevant regulations.

Probabilistic risk assessment model for integrating exposure and hazard USEPA's analysis relied on the conceptual framework of chemical risk analysis. In this framework, the risk of a hazard is evaluated as nil if exposure values are low enough. There is no evidence that this model is valid for environmental risks of GEMs. Most models of infectious disease spread are not structured like the chemical risk assessment model implicitly used in this analysis. Whether potential adverse environmental effects are related to the initial exposure, or "dose," when the exposure is to organisms that can reproduce, is still an outstanding question.

Conclusions Regarding the USEPA's Expert Panel

The panel provided scientific support for the agency's risk analysis and its EUP approval. The EP was required to use its expert judgment to interpret the microbial competition issue because there are relatively few data and theory on the subject is conflicting. The EP was also required to use its judgment to integrate the exposure and hazard information, because no models or procedures were available for integrating exposure and hazard information into a GEM risk analysis. Because of the absence of critical data, information on ecological relationships, and theory relating genetic changes to ecological changes, it seems unlikely that either computer "expert systems" (see Chap. 15) or routine administrative processes can integrate GEM hazard and exposure data until there is agreement about how ecosystems function and what ecological indicators constitute evidence of disruption.

This case study shows that substantial knowledge and a substantial regulatory commitment are required to effectively utilize EPs. It seems likely that the integration of exposure and hazard information for environmental tests of GEMs will continue to require expert intuitions, as opposed to quantitative, expert-certified, peer-reviewed knowledge, for the foreseeable future. Reducing the subjective component of the work of EPs can only be accomplished by improving the state of the scientific knowledge base on which expert panelists can judge risks of individual applications.

California Department of Food and Agriculture Risk Analysis

Role of expert panel in risk analysis

Background In the state of California, the Department of Food and Agriculture (CDFA) has responsibility for regulating agricultural applications of genetic engineering involving bacteria, fungi, plants, and animals. The CDFA had been following the Ina$^-$ GEM controversy for several years prior to undertaking its review of the AGS Frostban® EUP. CDFA anticipated that they would review and issue a research use permit for a field test of Ina$^-$ GEMs to the University of California or AGS. The university had already provided CDFA with detailed information on the Ina$^-$ GEM.

CDFA had conducted an internal assessment of their ability to respond to the risk assessment issues raised by the permit request. CDFA determined that there were two areas in which outside expertise might be required: molecular biology and microbial ecology. Since USEPA's expert panel had experts with national reputations in microbial ecology, CDFA decided not to replicate that agency's lengthy review process for the AGS field-trial proposal.

CDFA began its formal analyses of the Frostban® proposal upon receiving notification of USEPA's approval of the AGS EUP. CDFA required AGS to secure two permits: a California research use permit under the pesticide registration framework, and a California transportation permit under the exotic species importation framework (Fig. 14.2). Both permit analyses were principally conducted by senior staff members in the Division of Plant Industry Pesticide Registration. CDFA asked several of its sister agencies in the California state government to review the AGS proposal in addition to the review performed by its staff. The individuals who provided this review may be construed as an "internal" expert panel. CDFA delegated the risk analysis to two senior plant pathologists in its Pesticide Registration Branch (PRB). The PRB conducted two parallel reviews: one to register the product, and another to assess the risks (if any) of transporting the GEMs through the state.

Transportation permit review The PRB reviewed the potential risks to Californian agricultural plant species posed by a transportation accident en route

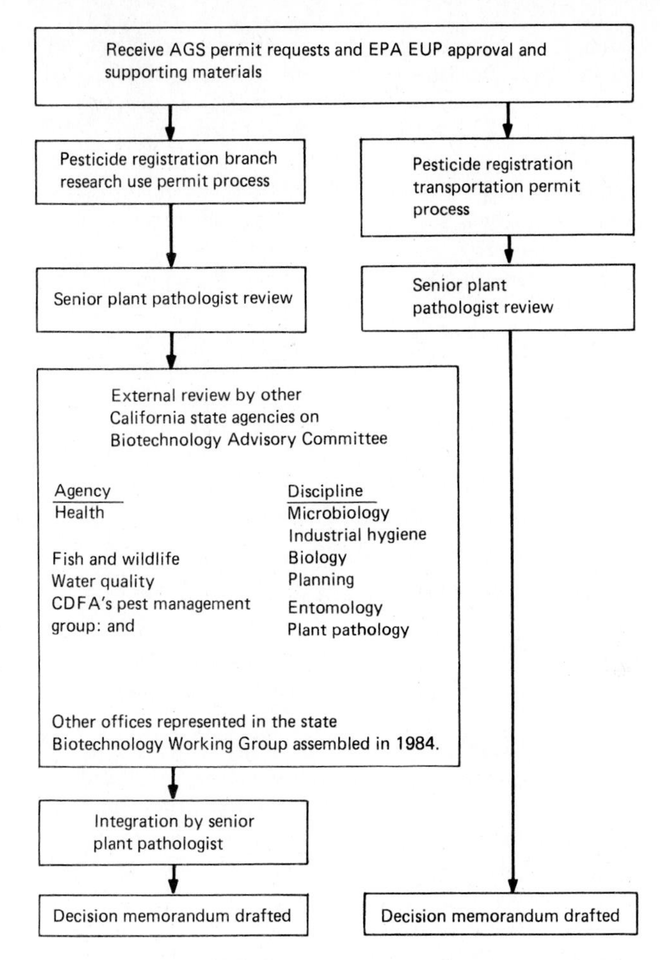

Figure 14.2 Flow diagram for CDFA's review of the AGS proposal to field-test Frostban®.

to the test site. Like the risk analysis conducted by USEPA, this analysis made use of a worst case adverse-effects scenario. Under this scenario, significant damage could result from either of two events:

- An extension of the pathogenic range of the GEM
- An increase in virulence

The PRB's worst case was substantially different from USEPA's, in that it was based on the explicit assumption that the GEMs could be disseminated widely and survive in the environment surrounding the transportation pathway or the test site.

To assess the potential phytopathogenic effects of the GEMs, CDFA undertook a comprehensive literature search. CDFA also reviewed the plant pathogenicity data AGS provided on both the GEMs and the parental strains.

CDFA analyzed the literature on interactions of *P. syringae* with over 100 plant species in most of the plant families represented in California's agriculture, because the genus contains several variants with widely divergent properties.

As a result of the review and reevaluation of the USEPA's EUP data, CDFA concluded that the proposed transport of Ina$^-$ *P. syringae* did not present a novel risk to the state's agricultural system. CDFA was reassured that the GEM would behave like the parental organisms because a variety of physiological indicators unrelated to Ina function were identical in the GEM and the parental strains.

Research use permit review Figure 14.2 shows that the process CDFA employed to generate a position on the issues raised in the research use permit was more complex than the transportation review. In addition to CDFA review, the application was reviewed by experts in other state agencies with biotechnology risk management responsibilities. Unlike the USEPA's earlier review and Monterey County's subsequent one, the CDFA review process did not provide for public comment on the issues before the review was completed.

CDFA's pesticide registration review coordinator, a senior plant physiologist, did not feel that it was necessary to obtain outside expert opinions on the mammalian infectiousness and plant pathogenicity issues raised in the USEPA review. However, he did contact an outside expert to discuss potential effects on precipitation processes.

CDFA's review coordinator asked the representatives of each state agency on the Biotechnology State Working Group to review the proposal and provided an explicit standard by which to evaluate both risk assessment and management issues: "whether there will be environmental or health hazards presented by the proposed small-plot testing on strawberries, the significance of any such hazards, and whether they can be mitigated" (Naimon 1987).

After reviewing the literature on *P. fluorescens*, the state health department microbiologist and environmental hazards specialist suggested that the applicator(s) wear protective coveralls, goggles, and respirators to prevent accidental inhalation of a potentially infectious dose of the genetically engineered Ina$^-$ *P. fluorescens*, even though they saw no significant differences between the GEMs and the Ina$^+$ parental strains.

A CDFA entomologist noted the possibility of adverse effects arising from interactions between the Ina$^-$ GEMs and agriculturally useful insects. He suggested that experiments could be conducted to assess the effects of Ina$^-$ GEMs on selected beneficial insects. The review coordinator responded that the small scale of the field test obviated the need to assess this issue.

Two toxicologists suggested that quality control tests be performed to assure that the inoculum used in the test would be identical to that identified in the proposal. An environmental hazard specialist in the medical toxicology branch of CDFA suggested that a contingency plan be developed for early termination of the experiment in the event that the organisms behaved as pests. CDFA ultimately required that AGS have two chemical bactericides available for use at the site in case termination of the experiment were desired, although the problems associated with field use of bactericides are recognized (Naimon 1988).

The outcome CDFA granted a research use permit to AGS approximately 1 month after it received the USEPA EUP and supporting information. CDFA's approval had three stipulations:

- That the applicators wear full-body protective gear to prevent inhalation of the GEMs
- That a contingency plan (employing chemical bactericides) be devised
- That personnel from CDFA be present at the test

In comparison with the USEPA's SAP subpanel, CDFA's panel had broader responsibilities, more public health and environmental protection specialists, fewer academic biologists, and no atmospheric scientists. This may have been reflected in the risk management recommendations made by the panelists and adopted by CDFA. In contrast with the USEPA process, there was no opportunity for public comment on the analysis or the decision. CDFA placed no additional restrictions on the location of the site.

Advantages of CDFA's risk analysis process

Management issues CDFA's risk analysis was anchored in their staff's interpretation of the literature on the parental strains. This was possible because many members of CDFA's staff were plant pathologists and personally familiar with this type of research. CDFA staff were more familiar with the properties of the bacteria in Frostban® than were USEPA staff when their review began.

Risk management Perhaps because many members of CDFA's expert review panel had administrative authority, several reviewers recommended risk management actions to accompany approval of the test proposal. The CDFA reviewers' recommendations reflected the diverse disciplines from which the members were drawn. Although there may be some technical problems with the recommendations suggested by these panelists, the use of regulatory experts as panelists may have been critical to the generation of the concrete risk management recommendations.

Scientific issues

Plant pathogenicity The conclusion that the risk was marginal depended on the analogy between the model plant species tested and the other members of the plant family and on the similarity of the GEMs to the parental strains. The first analogy has served as a rule of thumb for analyses of potential plant pathogens for several decades. The second is based on CDFA's assumption that the phenotype of the organisms characterized them fully, even though there were genotypic differences. This conclusion is similar to the consensus of the USEPA's SAP subpanel.

For CDFA, the primary issue was the host range of the genetically engineered Ina^- *P. syringae*. CDFA was also concerned that the Ina^- GEM could also infect commercially important crops if it were accidentally released in a transportation accident en route to the test site, or as a result of dispersal from the site by environmental vectors.

Exposure assumptions CDFA's exposure scenario, a release of the entire experimental load in a transportation accident, seems like a reasonable "worst case scenario," because transportation accidents are common. In light of the absence (at the time of the test) of reliable data or models to extrapolate dispersal and survival of Ina^- GEMs, CDFA's use of a conservative exposure scenario seems rational.

Scenario generation The CDFA staff developed a terrestrial effects scenario that USEPA did not: alteration of the range of agriculturally beneficial insects. Since frost sensitivity is characteristic of many groups of insects, this potential impact seems plausible. CDFA did not independently address this issue, but integrated the potential risks from this hazard with a low estimate of potential off-site exposure to yield an overall characterization of this risk as very low.

Integration of exposure and hazard information CDFA handled the beneficial insect question raised in the research use permit analysis by accepting the idea that transport of the GEM would not result in wide-scale colonization. This interpretation is consistent with USEPA's integration of exposure and hazard information. Like USEPA, CDFA staff believed that potential hazards from larger-scale tests need not be evaluated until such tests were proposed. CDFA staff indicated that this assumption is not generic for all GEMs and may be revised depending on the colonization ability of GEMs that are proposed (Koehler 1989).

In the transportation permit analysis, CDFA explicitly assumed that there could be off-site dispersal and colonization. In evaluating the evidence compiled by AGS, CDFA scientists seemed to consider the pathogenic virulence and range questions independently of the scale-of-test and competition questions. Even though USEPA and CDFA reached the same conclusion about the safety of the test, CDFA's analytic assumptions seem to be preferable because each risk factor should be evaluated independently rather than using information about one factor to condition evaluation of an independent factor.

Atmospheric precipitation After examining the information that had been compiled and consulting with the same atmospheric scientist USEPA consulted, CDFA concluded normal atmospheric precipitation patterns were unlikely to be disturbed by the proposed small-scale field test. The colonization experiments were particularly important for CDFA's conclusions because they were consistent with the preemptive-exclusion model of microbial competition that the University of California and AGS propounded. Although its analysis of the small-scale tests was virtually identical to the national agency's, CDFA's willingness to consider potential effects on atmospheric processes from large-scale usage reflects a broader interpretation of the atmospheric science literature.

Disadvantages of CDFA's expert panel use

Management issues

Risk scenario follow-up Although CDFA's EP review process identified an additional plausible terrestrial ecological adverse effect and a method for empirically evaluating it, there was no follow-up. One interpretation of this may be that the constituency for beneficial insects at CDFA is recognized but is not as important as the constituency for commercially important plant families that were tested. A second interpretation may be that CDFA was obligated to assure that agricultural microbial products will not directly harm commercially important plants; however, there was no requirement to assure that such products do not harm beneficial insects. If the scope were broadened to include this issue, a precedent for future risk analyses would be established.

Identification of risk management needs CDFA had no research capability to independently review the adequacy of these measures. Review of USEPA's recommendations by a high-level research committee similar to the BSAC could have alleviated the problem of recommending techniques which could have been ineffective (chemical bactericides) and alarming the public unnecessarily (the use of full-body suits and respirators).

Public involvement CDFA did not provide the public an opportunity to comment on the risks and benefits or provide advance notice of the proposed test. CDFA has since amended its process to provide for public comment and public disclosure of field-test sites (Naimon 1987).

Scientific issues

Integration of exposure and hazard information Like USEPA, CDFA's regulators and review panelists integrated hazard and exposure assessments into overall risk assessments without the benefit of a standardized procedure. Certain ecological adverse-effects scenarios were effectually absorbed or dismissed as a result of the CDFA integrator's assumption that they would not occur with a small-scale field test. CDFA did not require additional testing in this area, although they could have.

Identity of the GEM CDFA staff reasoned that if the underlying physiology of the GEMs was virtually identical to the parental organisms', the pathogenic range and properties of the GEM were also likely to be virtually identical. However, just as genotypic variation confers enhanced survival in some settings, genotypic variation can be expected to confer enhanced survival and increased ecological risks in other settings. Thus, CDFA's tacit assumption that physiological similarities reduce the need to assess risks of environmental effects is worrisome.

Monterey County's Risk Analysis

Role of expert panel in risk analysis

Background Monterey County asserted its right to regulate environmental applications of genetic engineering as a new land use. Federal and state regulators questioned its authority, based on the notion of federal and state preemption of local law on national safety issues. As of this writing, the courts have not overturned Monterey County's actions.

In contrast to the relatively private, consensus-oriented administrative processes used by CDFA and USEPA, Monterey County held hearings in which individuals publicly presented opposing views on several important assumptions contained in the USEPA and CDFA risk analyses. The adversarial public hearings resembled a courtroom, in which experts were pitted against each other, and their statements were open to comments from the public. In contrast to the procedures used by both CDFA and USEPA, the decision makers in Monterey County were not senior staff but were elected officials. Because of the high level of public interest, the county's decision on the AGS field-test application required widespread citizen assent.

Milestones After the CDFA approval, rumor suggested that the test was slated to be conducted in the northern Monterey County property of an AGS field engineer. A politically active resident of the county sent a petition to the county Board of Supervisors, requesting that the board examine the location of the test site and the desirability of having the test in Monterey County. The petition highlighted the secrecy of the proposed test, and the lack of information available to residents of Monterey County on the reasons for, and potential effects of, the test. The board formed a committee to look into questions associated with the proposed test (Fig. 14.3). The county's Environmental Health Director, Walter Wong, was the de facto chairman of this committee, which initially identified agricultural and human health impacts, financial liability, and intergovernmental relations as the most important issues.

Monterey County formally requested the AGS EUP package from USEPA and CDFA in January 1986. Because both USEPA and CDFA believed that the county did not have a right to address risk issues that had been preemptively addressed by federal and state authorities, both initially refused to cooperate. As a result, Monterey County's committee initially obtained information on the proposed test from scientific journals, private citizens, the FOET, and the press.

The information obtained through these channels did not directly address all the risk issues. However, maps and aerial photos did corroborate the rumor that a property owned by an AGS employee, Steven Cull, in the northern, more populous segment of the county was the intended test site.

On the basis of this information, the committee decided that the proposed

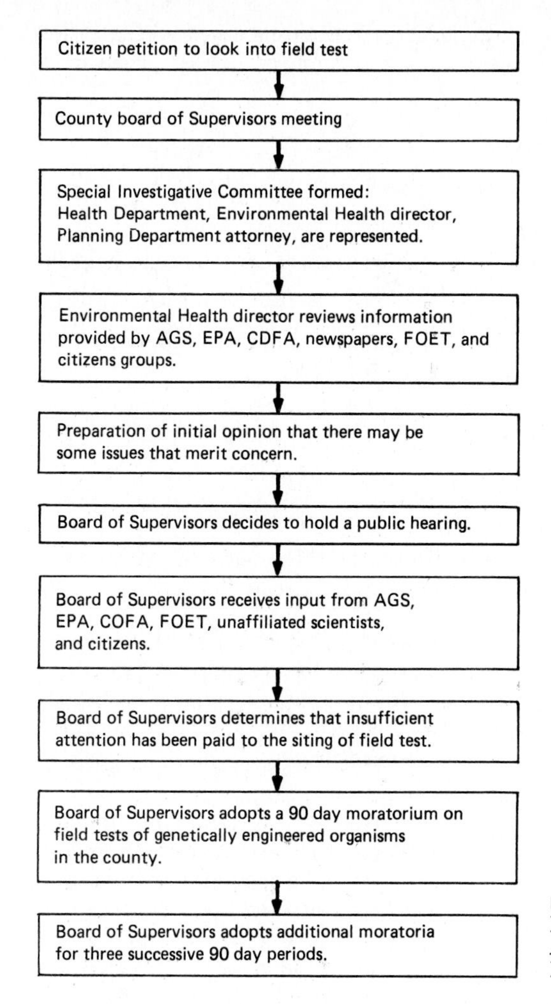

Figure 14.3 Flow diagram for Monterey County's review of the AGS proposal to field-test Frostban®.

application was not certain to bring additional benefits to Monterey County and could potentially cause adverse effects for the county. The committee provided two preliminary recommendations to the County Board of Supervisors:

- That AGS be asked to find a site farther away from residential areas
- That a public hearing be conducted with testimony from AGS, USEPA, CDFA, and residents

The board decided it needed more information before it would request AGS to take any actions and decided to accept only the second recommendation. They planned to use the information obtained in the hearing to help decide whether or under what conditions to permit the field test.

Hazard assessment issues Monterey County was concerned with a different set of scientific and management questions than the USEPA and CDFA, including:

- The identity of the Ina^- GEMs
- The availability of assays for field monitoring
- The likelihood of off-site spread
- The existence (or lack) of any directly relevant local experience to help the county review and manage the proposal

Several secondary issues emerged in the analysis, after Monterey County staff reviewed the USEPA file. These issues included:

- Could antibiotic resistance factors or other deleterious genes be transferred to or from the GEMs?
- Could the *P. fluorescens* sensitize or aggravate existing health conditions among sensitive human populations living near the proposed test site?
- Could the GEM act as a plant pathogen to any of the agriculturally important crops in Monterey County?

Although many Monterey County residents publicly indicated their concern about the potential for the Frostban® organisms to affect precipitation processes, the question of whether Ina^- GEMs could alter normal precipitation patterns was not a major question for Monterey County decision makers. The major question was the problem of human exposure.

Exposure assessment issues The potential exposure of humans, animals, and off-site plants was more important to Monterey County decision makers than any of the specific hazard assessment issues. On the basis of converging evidence, Monterey County concluded that the test would be held less than 200 yd from a neighbor's property, and less than 500 yd from a family residence. Moreover, the test was planned for early spring, which investigators have shown is the optimum period for survival and colonization of the parental strain (Gross 1983).

Public hearing as an evaluation forum Representatives of AGS, USEPA, CDFA, and FOET, as well as interested scientists and interested citizens, testified in a public hearing on January 27, 1986. AGS's Project Director suggested that because of phenotypic similarity with wild-type parental strains, the GEMs should not be considered to be novel organisms. As a result, she contended that analogies to exotic-species introductions such as the kudzu vine and Dutch elm disease are not appropriate. [For a contrasting view, see Sharples (1982) or Pimentel (1989).] An AGS spokesman closed by painting a picture of the test opposition as being duped by the FOET to blindly oppose a new technology (Naimon 1987).

Representatives from CDFA and USEPA described and defended their risk analyses. CDFA emphasized that the parental strains are not hazardous to either humans or commercial crops. USEPA described its review process. When questioned, USEPA's representative admitted that site visits were not required as part of its process.

Although several of the analyses performed by the USEPA and CDFA were well explained, several of the citizens and a few scientists who spoke were critical of

- The lack of adequate follow-through on the part of USEPA on potential site location issues
- The limits of the scope of each analysis
- The site location
- The potential for terrestrial ecological disruption resulting from displacement of Ina^+ bacteria by the Ina^- GEM.

Two scientists not affiliated with AGS testified on the risks of the proposed test. Robert Colwell, a member of USEPA's SAP subpanel, indicated his support for the proposed small-scale field test, although he expressed reservations about widespread use of the GEMs. Liebe Cavalieri, the only molecular geneticist at the hearing, pointedly disagreed with the AGS scientists, saying that the test organisms were genetically different from the naturally occurring Ina^- bacteria. He contended that, as a result, their behavior could not be identical to the behavior of the parental strains. He further suggested that the proposed small-scale field test could have "potentially catastrophic consequences." Dr. Cavalieri's views are a minority view among molecular biologists (Naimon 1987).

The FOET representative outlined a set of scientific uncertainties (particularly in the area of atmospheric ice nucleation) that suggested shortcomings in the USEPA process, and concluded that the proposed test was truly precedent-setting from a legal perspective.

In response to these concerns, AGS offered to postpone the test for 30 days, provide all the information contained in the EUP issued to it by USEPA to Monterey County's government, and relocate the trial to another location within Monterey County. However, AGS refused to state the location of the field test. The continued refusal of USEPA, CDFA, and AGS to publicly state where the proposed test would be conducted reinforced the point that both USEPA and CDFA had not viewed local approval of test sites as a requirement.

Perhaps the most important testimony was the County Attorney's opinion that the county could regulate tests of GEMs as a novel land use, through local zoning ordinances. On the basis of this opinion, and the general consensus among the Board of Supervisors and audience that there were significant holes in the AGS, USEPA, and CDFA claim that all risk issues had been responsibly addressed, the Board of Supervisors decided to adopt a 60-day moratorium on the testing of GEMs in the county.

Examination of the hearing records suggests that the moratorium was based on several assumptions, including the following:

- The GEM could act differently than Ina^+ parental strains.

- The bacteria were likely to spread off-site as a result of wind, rain, and insects, particularly in rainy months.
- The greatest potential for *P. syringae* to infect new plants occurs under damp, cool conditions.
- There was no directly relevant scientific and regulatory experience that Monterey County could use to resolve its scientific and regulatory questions.

Advantages

Management USEPA now visits sites prior to issuing EUPs. For public relations reasons, applicants like AGS no longer try to hold the identity of test plots secret. CDFA has revised its process to add public participation. Each of these changes reflects a tightening of the GEM review process that resulted, at least in part, from the actions of Monterey County.

Science

Integration of exposure and hazard information The potential for exposure to GEMs may have been more important to Monterey County than the identified hazards of those GEMs. The risk of exposure was assessed primarily on the basis of the location of the site, and supported by published reports on the spread of the parental organisms under moist conditions.

Monterey County's risk analysis emphasized exposure analysis in comparison to hazard analysis. By contrast, USEPA's and CDFA's analyses weighted the hazard assessment more than the exposure assessment. The Monterey County moratorium was declared primarily on the basis of exposure. Because there are no formal standards or protocols for integrating exposure and hazard information on environmental releases of GEMs, it is difficult to comment on why one organization's method of integrating seems preferable to another.

Disadvantages

Management

The role of political pressure It seems clear that although the scientific facts of the case were the same, the political pressures were substantially different. In the case of USEPA and CDFA, there was substantial pressure from the regulated community to approve the field-test proposal. In contrast, Monterey County had a substantial and vocal constituency that reacted against both the perceived localized risks of the test and the process by which the USEPA and CDFA reviewed the case but failed to consider their political interests, or even communicate with their political representatives.

Perhaps the simplest explanation for the refusal of the county government to accept the results of prior risk analyses is the notion that the people were whipped into an hysterical frenzy against the test by FOET, and the Board of Supervisors responded by acting against their better judgment to preserve their positions. Interviews with decision makers show that the board and senior staff were skeptical of the hazard claims of FOET. Further, Monterey

County leaders felt slighted by a cavalier federal and state attitude toward the county's interest in evaluating the proposed field test. As a result, Monterey County passed an ordinance regulating the siting of GEM field trials in 1987 (Krimsky 1989). This action may reflect doubt that the risks were adequately managed or involve considerations outside the scope of the CDFA and USEPA reviews.

Doubts about the risk analysis may have been reinforced by the public hearing process, in which there was disagreement among the "experts" over (1) what the primary issues were, (2) what the conclusion that no adverse effects were likely to occur meant in terms of larger-scale application, and (3) the novelty and essential identity of the test organisms. This collision of viewpoints clearly showed that the decisions were "judgment calls." While all regulatory decisions may be judgment calls, in this case, the public was explicitly aware of this.

As a result, Monterey County's political leaders questioned the wisdom of field-testing GEMs before there were legislative, financial, and emergency response infrastructures to deal with potentially unique risks. The importance attached to these policy considerations may have been enhanced as a result of the reduction in public trust that resulted from the public hearing.

The broader communication achieved in the Monterey County public hearing may have been responsible for the emergence of a conclusion that federal and state risk assessment and risk management for Ina$^-$ GEMs were not adequately linked. This does not suggest that the individual risk assessments were flawed, but rather that the regulatory follow-through was not commensurate with the risk analyses.

Communication with the public until the time of the Monterey County review had been perfunctory and not subject to intense scrutiny. The closest thing to a "science court" was the deliberations of USEPA's SAP subpanel on the public comments of FOET. However, the SAP subpanel's meetings were closed to the public when information deemed confidential by the applicant, such as the methods used to construct the GEM, was discussed.

Epilog

In 1987, AGS told Monterey County that the company would drop all plans for its first test site and would concentrate on two possible sites located approximately 20 mi southeast of Hollister, California. Two similar trials were approved in 1987, in San Benito County and Contra Costa County. These experiments were preceded by an expanded risk communication process with extensive and expensive efforts to inform the public as to how the risks of the proposed tests were assessed (Krimsky and Plough 1989).

These experiments have largely borne out the USEPA and CDFA prediction that Frostban® GEMs would not aggressively colonize locations where they were not applied in high concentrations. Tests conducted by the University of California have shown that off-site receptors would be exposed to a low level of *P. syringae* from aerial dispersal (Lindow 1988). Thus, USEPA was borne out in its interpretation of dispersal from small-scale field tests, and CDFA and Monterey County appear to have been conservative in their exposure assessments. The efficacy of the product in the field has not yet been conclusively proved.

CDFA and USEPA have altered their procedures for processing applications, providing public input, and defining the types of information that companies can classify as confidential (e.g., the location of test sites). Monterey County's position that field-test sites should be open to public inspection has become standard operating procedure. The fact that the level of controversy surrounding the field tests of Ina$^-$ has lessened following these field-test experiments is also consistent with the view that the controversy preceding the 1986 proposal was due to the novelty of the test as much as to legitimate scientific controversy and uncertainty.

Conclusions and Recommendations

Conclusions

Although there were differences in the basic assumptions made by each EP, the conclusions of all three were similar. There was substantial discretion in designing methods, applying tacit rules, or using intuition for combining data. All three organizations added stipulations to the test proposal and thus performed some risk management functions. Although the risk analyses were conceptually based on extant models of chemical or microbial pesticide risk assessment, each panel provided a different risk analysis perspective. A possible reason for this divergence may be the paucity of existing theory relating genetic changes to ecological changes.

Differences between GEMs and microbial pesticides: generic GEM questions

Level of scrutiny The Frostban® risk analysis carried the intellectual baggage of all the risk assessment problems for genetic engineering because it was for the first environmental release of a GEM and could set precedent for all subsequent releases. Basic questions about irreversible genetic change, phenotypic versus genotypic similarity, and impacts on evolution contributed to USEPA's perceived regulatory need to identify and address all scenarios, and Frostban® received greater scrutiny than microbial pesticides generally receive. This scrutiny magnified both the scientific and management issues. As the image of a smooth fabric is revealed to be a series of ridges and holes under a microscope, the smooth fabric of microbial pesticide registrations appeared to dissolve into a conglomeration of scientific assumptions in this case.

Although virtually every assumption made by the review panels had a reasonable basis, many alternative scientific assumptions and interpretations could be made from the same facts. In this case, conclusions arose from the ways in which scientific information was compiled. Cases could be made for a number of different conclusions based on different standards for evaluating the same data. In addition, the integrity of data submitted by AGS was called into question. The special attention given this EUP proposal may have been of value, since future risk analyses are likely to begin with an examination of this case.

Novelty of product effects The risks posed by interrupting the frost cycle on plant surfaces with Ina$^-$ GEMs are somewhat different than risks typically evaluated in microbial pesticide registrations. The prevention of frost formation on plant surfaces is not a classical pest management function. Few ecologists recognize frost as a pest, although many farmers view it as an annoyance. USEPA's interpretation that Ina$^+$ bacteria were "pests" and Ina$^-$ GEMs were pesticides strained the credulity of Monterey County citizens and this observer. USEPA's analysis of the ability of the GEM to cause infections in nontarget mammals and selected plant groups was similar to their evaluations for traditional microbial pesticides. However, the potential for a pesticide application to alter seasonal frost and atmospheric processes are novel issues.

Findings

Integration of hazard and exposure information It is clear from the Frostban® case that the way hazard and exposure information are integrated depends on the intuitions and values of the expert panels. Two of three panels came to similar conclusions, suggesting a shared set of values or intuitions. Each organization used different tacit rules to combine information on potential hazards and exposure to Frostban®. It would be useful to explicate these so that future EPs can use them as an explicit example. Without explicit procedures, there is the potential for real or perceived manipulation of the integration phase to benefit a particular position.

Exposure analysis The exposure issue was addressed by the regulatory organizations in diverse ways. USEPA assumed minimal risk of exposure on the basis of a competition model in the literature and AGS data. CDFA assumed that agricultural environments would be exposed through transportation accidents. Monterey County assumed that humans could be exposed as a result of their proximity to the proposed test site.

USEPA used the competition model as a basis for its decision to require a remote test site. CDFA required a contingency plan because of a lack of absolute confidence in the model. Monterey County's insistence on a relatively unpopulated test site assumed off-site exposure. Monterey County viewed CDFA's requirement that applicators wear respirators as a signal that similar protection should be afforded to residents near the proposed field-test site.

The fact that these organizations used three different ways of handling the exposure question suggests a need to develop evaluative methods and criteria for GEMs. Basic and applied scientific questions need to be addressed in order to develop standardized test systems for exposure assessment and standardized criteria for evaluating proposals. As USEPA's BSAC noted in 1989, most of the species that exist in soil ecosystems are not even cultured or named at the present time (USEPA BSAC 1989). Thus, it seems unlikely that generalizations regarding the competitive behavior of a well-characterized bacterium in a sea of uncharacterized bacteria can be made.

Case-by-case approach The case-by-case approach does not require the development of specific criteria for evaluating GEM applications prior to their receipt.

In all three cases, risk assessment review panelists developed their own methods for integrating information into a risk assessment recommendation.

Protocol requirements It is essential to devise experimental protocols to assess potential effects prior to field testing, since reliable data is the anchor of risk analysis. As shown by the use of assumptions in the Ina$^-$ case, it is important that regulatory agencies or EPs spell out the standards for evaluating data prior to conducting assessments.

AGS obtained experimental data to help assess potential pathogenicity, infectivity, and colonization. The data from these tests were directly relevant to the risk analyses. Because there were no clearly specified criteria for evaluation, the data could be (and was) interpreted in divergent ways. Monterey County's request for a repeat of an infectivity test seems reasonable because USEPA did not specify evaluation criteria prior to the performance of the test.

Although experimental protocols for evaluating the potential for adversely affecting precipitation processes could be developed, USEPA determined that expert opinion was sufficient to resolve the issue. Similarly, the potential for ecological disruption of terrestrial environments was not tested. This pattern suggests that development and use of test systems depends on the importance the regulatory framework places on the specific environmental protection goals.

Risk management recommendations The Frostban® case indicates that most expert panels will provide risk management recommendations. Two out of the three risk analysis panels in this case did provide such recommendations. USEPA's SAP subpanel, composed entirely of academic scientists, did not. The USEPA panel was specifically asked to deal only with risk analysis issues. This charge, combined with the academic orientation of the panelists, led to "science only" recommendations. The regulatory background of the CDFA and Monterey County panelists may have contributed to their provision of risk management recommendations. It may be advantageous for agencies to select reviewers with regulatory experience.

Effect of disciplinary base of expert panel members The disciplinary base of panel members greatly affects risk management recommendations. Industrial hygienists recommended personal protective gear similar to that specified by OSHA. Bactericides similar to those used in pathology laboratories were recommended for control purposes by medical personnel, and other specialists recommended the tools of their trade. There seemed to be a correlation between risk management tools selected and the disciplinary background of the panelists.

Since members of risk analysis panels intimately examine the experiments, they may be in an ideal position to recommend appropriate risk management options. Risk analysis panelists should therefore review the emerging GEM safety literature in order to improve their ability to intelligently analyze risk management issues (c.f., Molin 1987, Naimon 1988).

Site selection Consideration of receptor environments (human settlement, agricultural flora, and wild flora and fauna) should be a part of analysis, as should consideration of organism(s) to be introduced. Expression of traits such as pathogenicity and disease susceptibility depend on the "fit" of receptors and agents. Inadequate handling of siting issues may have precluded conduct of tests in Monterey County as much as differences in risk analysis or regulatory philosophy.

Data evaluation USEPA's risk analysis procedure assumed that data supplied by applicants would be reliable. Subsequent investigation by the agency showed inadequate reporting of selected experimental conditions. Similarly, inspection of the test site by USEPA, CDFA, and Monterey County found that the site selected was not the "remote site" agreed upon. Expression of traits such as pathogenicity or disease susceptibility depends on the "fit" of receptors and agents. This suggests that regulators must verify the accuracy of the data submitted by permit applicants prior to convening their EP.

Determining the potential effects of GEM applications AGS presented data to USEPA and CDFA showing that some enzymatic activities and some physiological indicators used in microbial classification and expression of selected proteins are virtually identical to those of the parental strain. This data convinced CDFA that Frostban® would behave the same as the parental organisms in the field. Because little scientific information was available on physiological and ecological functions that are mediated by the *ina* genes, this analysis bypasses question(s) of what differences (if any) would result when the *ina* genes are permanently deleted.

An additional issue not examined in depth by any panel is the interaction between the GEM and indigenous microflora. The exchange of genetic material between microbes is well known and may result in modification of some properties of either GEMs or indigenous microbes. Several authors have suggested that the properties of bacteria may not be well predicted by their formal classification, but rather the properties may be dependent on transient environmental conditions that select the bacteria. Prediction is further clouded by the documented ability of bacteria such as *Pseudomonas* to adapt to selective conditions using genetic material from other organisms (Sonea 1980).

Role of the public The public influences risk analysis and risk management by selecting hazards for review and reviewing the final risk analysis product. The role of the nonexpert public in decision making may have been as critical as the role of experts in defining the issues for the USEPA and Monterey County. CDFA did not solicit public input. USEPA solicited public input in writing only. Monterey County solicited oral public input. It is possible that the increased involvement of nonexperts in the Monterey County hearing brought out the disagreements among experts. This would be consistent with the notion that experts try to maintain respect for one another, but elements of the U.S. public have limited respect for experts (Krimsky and Plough 1989). Public review reinforces the analytic need to define risk analysis issues before risk analyses are initiated.

Scope The USEPA and CDFA risk analyses were limited to small field tests. The Monterey County public insisted that scope and management issues be examined before field tests were conducted. Risk issues arising from widespread use of the Ina$^-$ GEM product were not considered by any organization. Monterey County's actions could also be viewed as a public judgment that the EP's refusal to consider issues outside the scope of their charter was inappropriate, raising the issue of how and when to establish the EP charter.

Limitations of expert panels for GEMs The AGS case showed that there are limitations to the use of EPs in evaluating risks from GEM tests, including:

- EPs cannot resolve certain scientific questions unambiguously without obtaining experimental data (obtaining the opinions of scientists is not a substitute for scientific experimentation).
- EPs cannot resolve questions for which there are few data, such as *significance* of a particular issue, without resorting to their personal values.

Underlying issues Underlying many of the scientific issues discussed in the AGS review are several "metascience" questions. Metascience issues have been defined as scientific issues whose resolution is outside the scope of a single discipline. These issues include questions such as how much information is required to establish causation of an environmental effect. Key metascience questions were:

- Was there sufficient data to evaluate the proposed application?
- Were existing protocols for risk analysis of chemicals or microbial pesticides adequate for GEMs?
- What does the paucity of theory relating genetics and ecology mean for GEM risk analysis?

Each of these problems, and the relevance of these problems to future risk assessment efforts, is discussed below.

Scarcity of data The paucity of scientific data on the environmental properties of GEMs was one of the reasons for the intensive scrutiny by the regulators and the media. Risk assessors dealt with the issue by requesting additional data and developing a series of assumptions relating the data to adverse-effects scenarios. In their risk analyses, USEPA and CDFA discounted reports in unrefereed journals, newspapers, and the marketing literature of the applicant. Like a court, Monterey County allowed such information at its hearings.

While the Frostban® field tests provided relevant data on dispersal and competition, the data are not necessarily relevant to evaluations of GEMs based on other organisms or alternate field-deployment methods. Given refereed journals' review time and technology developers' desire to maintain confidential business information, it is not likely that peer-reviewed information on new GEM applications will be widely available in the future. The absence of sufficient and reliable data may characterize GEM risk analyses unless there is standardization of test organisms or deployment methods.

Lack of protocols The term "criteria" refers to the identification of important effects, and the term "protocols" refers to procedures that can be used to determine whether an effect is or could be caused by application of a given product. Protocols that include specific evaluative criteria should be developed to enable organizations to accomplish risk analyses for environmental applications of GEMs without relying heavily on EPs. The variation in assumptions between two organizations reaching essentially the same conclusion (CDFA and USEPA) provides ample evidence that methods for integrating diverse types of hazard and exposure information are not self-evident even to experts.

All three organizations devised scenario-based procedures. The lack of well-established protocols forced risk assessors to develop their own methods. Unlike fault tree analysis, in which the probability of each branch point occurring is quantified, the Frostban® analyses developed qualitative branch points, using the scientific literature, data from recent experiments, and implicit assumptions about each adverse-effect scenario.

Two general problems occurred with these worst case analyses. Neither staff nor experts followed up adverse-effect scenarios for which few data were available. For example, two risk analysis panelists had concerns about biotic dispersal and terrestrial ecological effects that were not pursued. Second, none of the risk analyses contained explicit criteria for interpreting information. As a result, the evaluations of different branches of the risk scenarios were subject to different standards of proof.

Future analyses of environmental applications of genetic engineering may be able to use several elements of the scenario analysis method(s) used in the AGS case. These include the use of multiple scenarios that address exposure through different routes as well as different types of adverse effects that can result from the same exposure pathway. Within the general framework provided by scenario analysis, there is a need to develop, test, revise, and validate protocols for evaluating the probability and severity of each scenario.

Lack of theory Several competing theories have been propounded to address the general relationship between GEM introduction and environmental changes. The absence of widely accepted theories concerning the relationship between genetic changes and ecological changes (and the lack of agreement on what constitutes integrity of an ecosystem) forced experts to turn to personal assessments of the resiliency of potentially affected ecosystems in order to address the questions associated with the probability and severity of adverse-effect occurrence.

Recommendations

EPs are used by federal, state, and local agencies for many purposes. The usefulness of the products provided by EPs depends on a range of factors including their charge, the qualifications of the panelists, and the policy framework provided. Most of these factors are not unique to risk analysis for GEMs; nonetheless, each of these can have an effect on the quality of the risk analysis process and its acceptance by decision makers and the public they represent.

Structural, operational, and output recommendations for improving the usefulness of EPs in handling GEM risk analyses are outlined here:

- I. Structural recommendations
  - *A.* Charge and specific duties
    1. Regulators should define the scope and regulatory goals clearly
    2. Regulators should define procedure for integrating hazard and exposure
  - *B.* Selection of experts
    1. Credentials must be commensurate with importance of case
    2. Panelists must represent variety of disciplines
    3. Panel membership should be balanced
    4. Panels should be customized for each application
- II. Operational recommendations
  - *A.* Case-by-case scenario development by panelists should be required
  - *B.* Protocol requirements
    1. Regulators should specify experimental protocols
    2. EPs may be used to devise new protocols
    3. Regulators should fund research to validate protocols
  - *C.* Public participation helps ensure full review
  - *D.* Data requirements
    1. Regulators should establish basis for data requirements
    2. Regulators should check data integrity
- III. Output recommendations
  - *A.* Risk assessment
    1. Data and assumptions should be separated
    2. Basis for decisions should be described
  - *B.* Risk management
    1. "Success" should be defined
    2. Needs should be determined
    3. "Appropriateness" should be analyzed
    4. An optimal approach should be identified
  - *C.* Conceptual frameworks
    1. Panels should develop conceptual frameworks for GEMs
    2. Panels should educate decision makers

By virtue of their research experience, experience with other regulatory programs, and their relative independence from the short-term political constraints of regulatory agencies, EPs have many capabilities that regular staff don't have. EPs can and should be employed to

- Educate regulators on the evolving scientific basis for GEM risk analysis
- Develop improved conceptual frameworks for evaluating environmental biotechnology risks
- Develop evaluative criteria and experimental protocols to help agency staff evaluate data
- Identify adverse-effect scenarios for evaluation
- Provide regulatory agencies with assessments of long-term environmental effects from full-scale application of GEMs proposed for field tests
- Recommend optimal risk management techniques to regulators

Because of the flexibility of GEMs, there are a wide set of potential adverse effects. Because there is not yet an adequate scientific basis for pruning scenario lists for individual environmental applications of GEMs (Gillett et al.

1986), risk assessors should be responsible for weaving together a wide set of adverse-effect scenarios to serve as the framework for GEM risk analyses.

Until criteria are established, EPs should provide their own criteria defining the significant environmental effects of a GEM. Since there may be as many definitions of "significance" as there are experts, explicit environmental protection goals are needed to ensure consistent risk management.

Policies for GEM risk analyses should address the types of data required, the way that hazard and exposure data should be integrated, and the type and degree of "conservatism" of the risk assessment that is desired by society. Within this type of framework, regulators can develop consistent procedures for combining hazard and exposure information for the many proposed biotechnology applications.

In setting up EPs, regulators should ensure that the disciplinary base of a panel's membership is sufficiently broad and unbiased to permit the panel to assess all the potential adverse effects identified for a particular GEM application. At the present time, the personal intuitions of experts are critical to the resolution of complex issues. Risk assessors should try to distinguish between judgments that are based on knowledge that has been peer-reviewed by other scientists and opinions based on panel members' best professional judgment. This type of characterization will help policy makers evaluate the input provided by the EP.

Although EPs are critical today, the use of such panels for future GEM risk analyses could be expensive and impractical. Improving the scientific basis of GEM risk analysis would facilitate standardization of the process and potential use of other integrative options such as expert systems. A wide-ranging research effort is required to begin the process of standardization of risk analysis. In particular, advances in "predictive," or "genetic," ecology (defined as the study of the relationship between changes in the genetic endowment of organisms and the ecology of environments into which GEMs or other organisms are introduced) are necessary. Development of this field may also stimulate development of new product concepts as well as more sophisticated, more reliable risk management techniques.

References

Betz, F. 1990. U.S. Environmental Protection Agency Office of Pesticides and Toxic Substances, Health Evaluation Division, personal communication, March 9.

Brill, W. 1985. Safety concerns and genetic engineering, *Science*, 227:381–384.

Fiksel, J., and Covello, V. (eds.). 1985. The applicability of risk assessment methods to environmental applications of biotechnology. NSF-PRA Report no. 8502286, Washington, D.C.

Fishhoff, B., et al. 1981. *Acceptable Risk*, Cambridge Univ. Press, Cambridge, U.K., pp. 9–79.

Foundation on Economic Trends v. Heckler. 1984. 14 *Environmental Law Reporter* 20467, May 16.
Gillett, J. W., et al. 1986. Genetic stability of engineered microorganisms, *Environ. Mgmnt.*, 10(4).
Gross, D., et al. 1983. Distribution, population dynamics, and characteristics of ice nucleation-active bacteria in deciduous fruit tree orchards, *Appl. Environ. Microbiol.*, 46:1370–1379.
Koehler, D., California Department of Food and Agriculture, Division of Plant Industry. 1989. Personal communication, Sept. 19, Sacramento, California.
Krimsky, S., and Plough, A. 1989. *Environmental Hazards: Communicating Risks as a Social Process*, Auburn House, Dover, Massachusetts, pp. 75–121.
Lindow, S. E., et al. 1982. Distribution of ice nucleation-active bacteria on plants in nature, *Appl. Environ. Microbiol.*, 36(6):831–838.
Lindow, S. E. 1985*b*. Ecology of *Pseudomonas syringae* relevant to the field use of ice-deletion mutants constructed in vitro for plant frost control. In *Engineered Organisms in the Environment: Scientific Issues*, Halvorson, H. O., Pramer, and Rogul, M. (eds.), American Society for Microbiology, Washington, D.C., pp. 23–25.
Lindow, S. E., Knudsen, G. R., Seidler, R. J., Walter, M. V., Lambou, V. W., Ammy, P. S., Schmedding, D., Prince, V., and Hern, S. 1988. Aerial dispersal and epiphytic survival of Pseudomonas syringae during a pretest for the release of genetically engineered strains into the environment, *Appl. Environ. Microbiol.*, 54:1557–1563.
Naimon, J. S. 1987. A case study of the first environmental application of biotechnology in the U.S.A., University of North Carolina Department of Environmental Science and Engineering Tech. Rep., Chapel Hill, North Carolina.
Naimon, J. S., Canady, R., and Bridgen, P. 1988. Biological and chemical controls for large scale introductions of engineered microorganisms, presented at the First International Conference on the Release of Genetically Engineered Organisms in the Environment, poster session, Cardiff, *U.K.*, April 1988.
Office of Science and Technology Policy (OSTP). 1985. Coordinated framework for regulation of biotechnology: part 1, *Fed. Reg.*, 51(123):23302.
Pimentel, D., et al. 1989. Benefits and risks of genetic engineering in agriculture, *Bioscience*, 39(9):606–614.
Schatzow, S. Application of advanced genetic sciences for experimental use permits—decision memorandum, U.S. Environmental Protection Agency, Washington, D.C.
Schnell, R., and Vali, G. 1986. Biogenic ice nuclei: terrestrial and marine sources, *J. Atmospher. Sci.*, 33:1554–1564.
Sharples, F. 1982. The spread of organisms with novel genotypes: thoughts from an ecological perspective, Oak Ridge National Laboratory, Environmental Sciences Division, Publ. no. 2040, ORNL/TM-8473, Oak Ridge, Tennessee.
Sonea, S., and Panisset, M. 1980. *A New Bacteriology*, Jones and Bartlett, Boston.
Tiedje, J. M., et al. 1989. The planned introduction of genetically engineered organisms: ecological considerations and recommendations, *Ecology*, 70(2):298–315.
Upper, C. D., Hirano, S., and Vali, G. 1986. An assessment of the impact of large-scale applications of Ice-minus bacteria and other procedures designed to decrease population sizes of ice-nucleation-active bacteria on crops, report prepared for the Office of Technology Assessment Biological Applications Program, July 16, Washington, D.C.
U.S. Environmental Protection Agency (USEPA). 1984. 40 *C.F.R.*, Parts 172 and 158, and *Fed. Reg.*, 49:42856, Oct. 24, 1984.
U.S. Environmental Protection Agency (USEPA), Office of Public Information. 1985. Questions and answers for AGS experimental use permits, Washington, D.C.
U.S. Environmental Protection Agency (USEPA). 1986*a*. Regulatory docket for Advanced Genetic Sciences experimental use permit, Washington, D.C.
U.S. Environmental Protection Agency (USEPA), Office of Compliance Enforcement. 1986*b*. Press release regarding AGS investigation, Washington, D.C.
U.S. Environmental Protection Agency (USEPA), Biotechnology Science Advisory Committee (BSAC). 1987. *Report of the Biotechnology Science Advisory Committee of the U.S. Environmental Protection Agency*, vol. 1, no. 1, p. 1.

U.S. Environmental Protection Agency (USEPA), Biotechnology Science Advisory Committee (BSAC). 1989. *Report of the Biotechnology Science Advisory Committee of the U.S. Environmental Protection Agency*, vol. 3.

Vali, G. 1971. Freezing nucleus content of hail and rain in Alberta, *J. Appl. Meteorol.*, 10:73–78.

Chapter

15

Applications of Knowledge Systems for Biotechnology Risk Assessment and Management

Joseph Fiksel

Cimflex Teknowledge Corporation
P.O. Box 10119
Palo Alto, California 94303

Introduction

Artificial intelligence (AI) technology is a branch of computer science that originated about 30 years ago, and over the last decade has come into widespread commercial usage (Feigenbaum et al. 1988). In essence, AI consists of various methodologies for developing computer programs (often called "knowledge-based systems" or "expert systems") that are capable of logical reasoning for the purpose of interpreting complex information and seeking solutions to problems. Although most current applications of AI are in the area of industrial and financial decision making, some of the earliest AI research was focused on medical diagnosis applications, and this continues to be a thriving area (Buchanan and Shortliffe 1984). Of more relevance to the biotechnology industry, AI-based programs by Intellegenetics, Inc., are routinely used to assist in molecular design.

Recently, a number of investigators have introduced AI methods into the fields of health, safety, and environmental risk analysis (Fiksel 1987*a*) and emergency response (Hushon 1986). The U.S. Environmental Protection Agency (USEPA) and private companies such as E. I. Du Pont de Nemours are investing heavily in expert system applications to health and safety issues such as hazardous waste management (Rossman 1986). Based on these prior experiences, there is a clear potential for AI technology to be useful in supporting the assessment and management of risks associated with the modern practice of genetic engineering (Fiksel 1988). However, no full-scale AI applications of this type have yet been attempted.

The purpose of this chapter is to provide an introductory overview of AI technology in the context of biological exposure and risk assessment, and to identify specific opportunities for applying this technology to support the planning, design, evaluation, and control of environmental releases of genetically engineered organisms.

Limitations of Quantitative Risk Assessment

The traditional framework for risk assessment and management involves a methodical progression through a rigorous sequence of analytical steps, including hazard identification, exposure assessment, risk estimation, and risk-cost-benefit assessment (Merkhofer and Covello 1986). However, the microbiological and ecological phenomena related to environmental releases of genetically engineered organisms do not submit neatly to this quantitative approach, due to both the complexity of the phenomena and the scarcity of relevant data.

Previous risk assessment methods have addressed either equipment failures, which could be mathematically simulated, or biochemical agents, for which statistical estimates of dose-response relationships could be developed. In contrast, when organisms are released into an ecosystem, the various pathways of outcomes and consequences are too numerous for detailed empirical investigation. Instead scientists must rely on judgmental analysis, reasoning by analogy, and, where possible, contained-release experiments. Perhaps the best analog is the introduction of a new vaccine, which requires successively more refined tests to discover any possible adverse side effects.

In a 1986 study sponsored by the White House Office of Science and Technology Policy (OSTP) and the National Science Foundation (NSF) (Fiksel and Covello 1986), available risk assessment methods were evaluated for their applicability to environmental introductions of genetically engineered microorganisms (GEMs). This study concluded that

> existing scientific knowledge and methods are adequate to perform qualitative screening of specific environmental applications using modified microorganisms developed from organisms with well-defined characteristics. Further knowledge is necessary to advance risk assessment methods to the point where quantitative, predictive analysis can be performed for a range of modified microorganisms developed from microorganisms with poorly-defined characteristics. (p. 31)

Since that time, considerable effort has been devoted to improving the state of the art of ecological risk assessment, but it is still far from providing a standardized methodology. Those responsible for evaluating a proposed environmental release must invoke a considerable degree of scientific knowledge and qualitative judgment in order to anticipate its potential harmful consequences and to balance it against the available alternatives. This is precisely the type of application for which AI technology is well suited. Knowledge systems can provide access to a voluminous scientific and regulatory knowledge base regarding past environmental releases, transcending the limitations of specialized individual expertise. They can also assist in the tedious process of interpreting that knowledge and systematically applying it to a specific proposed release.

Qualitative Methods in Risk Assessment and Management

The National Academy of Sciences (NAS) has made an effort to distinguish between "risk assessment" and "risk management," suggesting that risk assessment is largely objective and scientifically based, whereas risk management inevitably incorporates policies and value judgments based on economic, political, and social factors (NAS 1983). The NAS defined "risk assessment" as the use of scientific methods, models, and data to develop information about specified risks, and "risk management" as the subsequent balancing of risks against other criteria in order to make decisions about risk mitigation or control. A subsequent NSF study suggested a more sophisticated three-stage model involving risk assessment, risk evaluation, and risk management, with the middle stage focusing on analysis of risk-benefit trade-offs prior to decision making (Merkhofer and Covello 1986). However, it has become clear that, in practice, the lack of scientific certainty requires the introduction of judgmental factors into the risk assessment process (Russell and Gruber 1987).

Depending upon the availability of data, risk assessment methods can range from "empirical" methods based on scientific evidence and real world experience (e.g., epidemiological studies of human disease patterns) to "model-based" methods, which rely upon predictive models (e.g., toxicological models for dose-response extrapolation), to "qualitative" methods, which use judgmental reasoning to draw approximate conclusions. Generally, qualitative risk assessment is used when there are neither sufficient empirical data nor reliable predictive models available, or when available resources are limited. Instead, qualitative methods rely upon good scientific judgment and common sense to assess the combinations of factors that might contribute to a risk. This method of risk assessment can be just as useful as the other two, even though it does not strive for mathematical precision.

For many innovative types of environmental introductions, empirical risk assessment methods will be impractical due to the lack of an adequate database of organism characteristics and interactions with the target ecosystem. Similarly, model-based assessment will be difficult or impossible due to the current lack of adequate predictive models of environmental outcomes for specific ecosystems. Therefore, for the foreseeable future, it is likely that biotechnology risk assessment will rely heavily upon qualitative screening methods such as those currently practiced by the U.S. National Institutes of Health Recombinant DNA Advisory Committee (NIH RAC), the U.K. Advisory Committee on Genetic Manipulation, and other regulatory bodies.

However, as practiced today, this qualitative approach has some inherent limitations. Assessments are typically carried out by expert panels, based on documents that set forth certain "points to consider" or guidelines. In general, no clear logical scheme or methodology has been articulated for how to interpret, evaluate, or critique the information that is requested in these documents. Thus, each review is dependent upon the judgment of the expert review panel, and there have been few systematic efforts to codify the underlying review principles. In this author's view, it would be helpful to capture the rationales and inferences that have been used in past reviews so that they may be refined and reapplied to future cases (Fiksel 1988). As suggested below, knowledge system technology can provide this type of capability.

Knowledge System Technology

Knowledge system technology involves the computer-aided representation of human knowledge and reasoning in symbolic form. The design and development of knowledge systems has traditionally required the specialized expertise of "knowledge engineers" (Hayes-Roth et al. 1983). However, recent advances in user interface software are making it possible for business or scientific professionals with little or no background in computer programming to develop simple knowledge systems. While knowledge system applications vary widely, they are generally used for providing expert advice, solving problems, or performing other "intelligent" tasks. Just as spreadsheet programs enable professionals to create calculational models for their numerical data, an emerging generation of knowledge system software enables them to create logical, deductive models for their qualitative knowledge.

At the present time, the most common risk-related applications of knowledge system technology involve the use of "expert systems," which provide specialized technical advice based on knowledge extracted from human experts. For example, a number of expert systems have been developed that advise emergency response teams about how to deal with industrial accidents such as chemical spills (Hushon 1986). In the regulatory arena, the USEPA is exploring the use of expert systems to assist in permitting of hazardous waste sites, in water quality modeling, and in a number of other environmental engineering applications (Rossman 1986). Many large manufacturing companies have established internal AI groups that are building expert systems to advise on diagnosis and repair of equipment failures, product safety testing, and a host of similar tasks (Feigenbaum 1988). This rapid expansion of expert system applications has been made possible by the commercial availability of low-priced general-purpose development tools, called "shells," which have shortened the typical system development time by an order of magnitude, from years to months.

Recent innovations have made it possible for knowledge system technology to provide a decision support environment, in which a broad range of knowledge is captured and applied to a variety of ongoing risk management activities (Fiksel and Hayes-Roth 1989). System users can simultaneously access qualitative knowledge bases, quantitative databases, conventional analytic models, and expert advisory systems. This new technology takes us beyond the expert system paradigm, and instead offers knowledge systems that can be implemented rapidly and maintained easily by anyone with a modicum of personal computing experience. These "knowledge assistant" systems (KAs)

TABLE 15.1 Comparison of Expert System and Knowledge Assistant Technology

Expert system	Knowledge assistant
Consultation based	Transaction based
System initiative	User initiative
Self-contained	Fully coupled
Goal directed	Data driven
Narrow scope	Broad scope
Static knowledge	Cumulative knowledge

are a natural extension of the current generation of decision support systems, such as database and spreadsheet programs. However, they complement these conventional systems by focusing on knowledge rather than data. Table 15.1 highlights several important distinctions between KA technology and expert system technology, which are discussed below.

Expert systems tend to follow a "consultation-based" model, in which the system guides the user through a structured dialogue. KAs follow a "transaction-based" model, in which the user retains the initiative.

Expert systems provide a self-contained problem-solving capability, whereas KAs are fully coupled with the user's application data and tools.

The goal-directed nature of expert systems makes them inherently narrow in scope, while KAs can interpret a broad range of data on a variety of subjects.

Finally, the knowledge encoded into expert systems is generally static and difficult to update, while KAs are deliberately designed to accumulate and maintain fragments of knowledge.

In the near future, it is plausible that risk analysts and risk managers will be routinely assisted in their day-to-day activities by knowledge assistants running on powerful workstations. These systems will simplify much of the effort involved in searching for information, running analytic models, and interpreting the results. The following section illustrates a number of specific areas in which these types of intelligent systems can support risk assessment and management for environmental releases of genetically engineered organisms.

Illustration of Knowledge-Based Biotechnology Risk Assessment

This section illustrates how knowledge system technology might be used to support risk assessment and management for deliberate releases (Fiksel 1988). The capabilities described are hypothetical, in that no systems of this type have been constructed to date. However, this type of system is well within the scope of modern knowledge system technology.

There are several types of human knowledge that may contribute to risk analysis in biotechnology applications:

- Empirical knowledge about organisms and their environment
- Situational knowledge about test conditions and objectives
- Judgmental knowledge about human beliefs and priorities
- Theoretical knowledge about ecological relationships
- Normative knowledge about policies and acceptance criteria

A knowledge system that logically introduces these types of knowledge can be useful throughout the various stages of risk assessment and risk management, which are discussed briefly below.

"Risk identification" is the first step of the process, and involves applying screening criteria and logical deduction to the proposed introduction. For example one possible risk identification criterion might be: "The organism was modified with genetic material from a pathogen."

If no possible adverse outcomes are detected in this risk identification step, then there may be no need for further screening. If certain classes of potential risk are identified, then the system can invoke more detailed screening criteria that are relevant to each class. For example, a judgmental screening rule relevant to pathogenic risks might be: "If pathogenic traits are present in the donor or recipient organism and intergeneric transfer of genetic material is plausible, then a high level of concern should be assigned to this proposal."

At this point, specific risks (i.e., end points) might be identified and assigned a sufficient level of concern to warrant further assessment. The knowledge system might suggest that certain scientific models be invoked, such as event tree analysis models, that can produce theoretical estimates of the degree of potential hazard. For example, a model might compute the expected likelihood of an adverse impact on indigenous microorganisms using the following simple formula:

$$\text{Probability of impact} = P_1 \cdot P_2 \cdot P_3 \cdot P_4$$

where P_1 = probability of survival
P_2 = probability of dissemination beyond test plot
P_3 = probability of nontarget plant colonization
P_4 = probability of competitive advantage

Naturally, such estimates are based on numerous assumptions and may have large associated uncertainties. Biotechnology risk assessment methods are relatively immature, and in many cases there will be insufficient data to support quantitative analysis. The system can instead use qualitative models to establish the nature, if not the degree, of possible risks. For example, rule-based reasoning might be used to derive predictions as follows: "If the microorganism survives, and disseminates beyond the test plot, and colonizes nontarget plants, and has a competitive advantage, then there is a moderate likelihood that it may displace indigenous species."

Additional rules can be developed to investigate each of the premises in the above rule, thus providing a chain of inferences. This is typically how rule-based knowledge systems are structured.

Once potential risks have been assessed, it is possible to introduce value judgments regarding the degree of concern about a specific hypothesized end point. The system can assist in this evaluation by providing comparisons with other similar applications.

The final step in the process requires normative judgments about appropriate methods for controlling or reducing risks. Choice of these methods may be based on logical rules, or may require more formal risk-benefit balancing methods. While knowledge system support is certainly appropriate for these latter stages of risk management, it is beyond the scope of this paper.

Suggested Biotechnology Risk Assessment Applications

In the NSF study of biotechnology risk assessment cited earlier (Fiksel and Covello 1986), a number of categories of assessment methods were described which provide a useful framework for discussing the potential applications of AI technology. These categories are:

1. Evaluation of organism properties
2. Human exposure and effects analysis
3. Environmental fate and transport analysis
4. Ecosystem structural and functional analysis
5. Controlled testing and monitoring

It is also useful to distinguish three different categories of generalized AI techniques, each of which has evolved through consideration of numerous application domains. The techniques are the following:

- *Data interpretation:* These techniques involve the screening of data to detect patterns, to identify potential problems or opportunities, or to discover similarities between current and past situations. Example applications are in loan underwriting and geological sample analysis.
- *Problem diagnosis:* These techniques involve the investigation of known problems to recognize characteristic symptoms, to develop and confirm hypotheses about possible causes, and to suggest strategies for repair or recovery. Example applications are in medical diagnosis and treatment, and equipment troubleshooting.
- *Decision support:* These techniques involve the evaluation of alternative choices to explore their possible consequences, to compare their relative costs and benefits, and to recommend appropriate action plans. Example applications include emergency response and investment portfolio selection.

Based on past experience, it appears that the most useful techniques for biotechnology applications will be data interpretation for risk assessment and decision support for risk management. Problem diagnosis is performed after a problem has been manifested, and generally requires a substantial amount of prior knowledge about problem types and characteristics. Biotechnology risk applications will more likely be focused on anticipating and preventing problems before they occur.

Below, each of the risk assessment method categories is discussed in terms of the AI techniques that are relevant and how they might be usefully applied. Note that in many cases these techniques can be usefully combined; for example data interpretation may be used to identify problems, which are then diagnosed in depth, after which a decision is made between several possible solutions. The discussion below focuses principally on microorganism releases, although the methods can also be applied to releases of higher organisms.

1. *Evaluation of organism properties:* When considering organisms for genetic modification and environmental introduction, scientists rely on existing information about the organisms that indicates their potential interactions with the target ecosystem. For example, information about the taxonomy, physiology, pathogenicity, genetics, and ecology of a microorganism can provide insights into its potential fate and persistence in the environment. Because of data scarcity, many of these insights are qualitative in nature, and could be supported by a set of "inference rules" representing reasonable heuristic judgments based on scientific knowledge. While a knowledge system could not be expected to substitute for a competent scientist, it could support scientific review by systematically applying a set of screening criteria, and pointing out potential problems for a given organism-ecosystem combination. This type of data interpretation approach has proved successful in a variety of fields from creditworthiness assessment to medical device regulation.

2. *Human exposure and effects analysis:* In assessing proposals for deliberate releases of microorganisms, an important consideration is the potential for colonization of humans and resulting adverse effects. As with chemical agents, risk assessment for human impacts requires consideration of both exposure and effects. Exposure analysis will evaluate the organism's potential transport pathways, survival traits, invasiveness, and resistance to antibiotics. Effects analysis will evaluate the pathogenicity of the organism, and may need to address a number of subtle mechanisms: for example, viral infections may produce a variety of host responses. Genetic transfer via transmission of naked viral nucleic acids may also need to be considered. Since data on the modified organism will be limited, these types of assessments will need to extrapolate from available knowledge about the unmodified organism. Again, these analyses are heavily dependent on scientific judgment, but data interpretation systems can help to highlight key organism characteristics and suggest important risk pathways and mechanisms.

3. *Environmental fate and transport analysis:* Prediction of environmental fate for microorganism releases must consider the size of the source population, rates of dispersion along various pathways (e.g., airborne, waterborne, animal vectors), and survival during transport. Survival in turn depends on complex interactions of the organism with both climatic conditions and biological phenomena. Due to this complexity, it is unrealistic to expect reliable model-based predictions of fate and transport. However, qualitative assessment using data interpretation techniques can point out more likely fate mechanisms and help to focus the risk analysis effort. Moreover, these types of assessments can influence the design of a proposed release, and decision support techniques can then be applied to

select the best approaches for engineering, introducing, detecting, and monitoring the released organisms.

4. *Ecosystem structural and functional analysis:* Ecosystem structural analysis can provide an understanding of the relationships and matter-energy flows between various compartments (e.g., air, soil, biota), and functional analysis can provide more detailed insight into the fluctuations that may result from introducing a new organism. These techniques require the use of qualitative models that represent the ecosystem as a dynamic network. Empirical support for this type of analysis is still scarce, and its usefulness for risk analysis has not yet been demonstrated. However, a knowledge system could be a useful device for exploring the hypothetical implications of the network model.

5. *Controlled testing and monitoring:* While the above method categories are oriented toward risk assessment, controlled testing and monitoring methods combine risk assessment with risk management. Controlled releases, involving microcosms or small-scale field trials, provide an empirical approach to assessing the fate and effects of engineered organisms. This approach complements the predictive approaches, and may actually reveal consequences that were overlooked or incorrectly analyzed. A combination of data interpretation and decision support techniques can be used to develop knowledge systems that support real-time monitoring of the environmental indicators, and evaluate alternative ways of responding to unanticipated results.

Applications to Biotechnology Safety Assurance

A recent workshop sponsored by the North Atlantic Treaty Organization (NATO) put forth the concept of "safety assurance" as an alternative to the paradigm of "risk analysis" (Fiksel 1988). Safety assurance does not depend upon the identification or quantification of hypothetical adverse events; it merely develops guidelines and criteria for assuring adequate confidence about the safety of a proposed release. The workshop sought to evaluate the available safety assurance and risk assessment methods for environmental introductions of genetically engineered organisms and to establish principles for improving these methods. As a result, a number of opportunities for knowledge system applications were identified. The following suggestions are relevant to government scrutiny of proposed releases, and reflect regulatory approaches already in use in both Europe and the United States:

Capture of scientific knowledge base: Knowledge systems permit the electronic capture, maintenance, and distribution of a large body of scientific knowledge regarding genetic modification and environmental introduction of plants, animals, and microorganisms, and the possible outcomes associated with the introduction of organisms into new ecosystems. Establishment of a single conceptual framework for accumulation of knowledge from multiple sources can help to clarify the nature of scientific controversies and disagreements about facts and assumptions, and can provide a basis for continual improvement of the knowledge base in the light of new information.

Risk identification criteria: Knowledge systems can assist in screening proposed environmental introductions for characteristics that might conceivably lead to adverse outcomes, based on the nature of the organism and of the environment into which it is introduced. They can apply both quantitative and qualitative screening criteria, and can recommend follow-up actions or priorities depending upon the results. Thus, use of knowledge systems as a preliminary review mechanism can assure consistent and systematic application of relevant criteria.

Use of generic guidelines: Generic guidelines can make case-by-case review of proposed releases more efficient, more consistent, and less burdensome, without compromising the safety of approved releases. Knowledge systems can help to implement such generic guidelines by systematically and exhaustively reviewing all proposed introductions against each relevant guideline. More generally, knowledge systems can expedite the often burdensome process of determining the applicable health, safety, and environmental regulations, and ensuring that all requirements have been followed.

Categorization of proposed introductions: Knowledge systems are well suited to the definition and application of qualitative classification schemes based upon the nature of biological function(s) affected or introduced, the environment from which the host organism was taken, the ecological characteristics of the genetically engineered organism, the characteristics of the target environment, and the scale and frequency of the proposed introduction. Once a set of categories has been established, a knowledge system can easily recognize the appropriate classification for a proposed introduction and initiate whatever review procedures are required. For example, a specific category might trigger particular risk identification criteria or generic guidelines, as discussed above.

Analytical logic schemes: Knowledge systems are capable of representing and executing complex logical schemes involving a specific domain of investigation. Analogous to road maps, such schemes would suggest in what order information should be considered, what questions should be asked at each step of the review, and what criteria should be used to initiate additional investigation. Thus logic schemes can subsume both guidelines and categories, as discussed above. A knowledge system can seek the necessary information to pursue a logic scheme from the current knowledge base, and if it is not available, it can request additional knowledge from the system user. Knowledge systems are also capable of working with incomplete or missing information, and can interact effectively with the user in selecting alternative strategies.

Example of a System for Safety Assurance

General description of LIBRA

In order to demonstrate the type of analytical logic scheme described above, we have used Teknowledge's LIBRA® technology to construct a simple prototype system that assists with qualitative and quantitative decision making in the safety assurance process. LIBRA® consists of a family of intelligent decision support systems that help to organize and apply human knowledge. The risk management application discussed below was implemented in M.1, Teknowledge's patented expert system development package; therefore it can run on an IBM PC AT or compatible personal computer. Versions of LIBRA® are also available for UNIX workstations and for advanced PCs with Microsoft Windows.

LIBRA® is designed as a day-to-day decision support tool for professional analysts and decision makers in a wide variety of fields. It allows users to maintain and explore knowledge about their sphere of interest, including possible risk scenarios and decision criteria. Then, through systematic logical reasoning, it can screen, evaluate, or compare alternative scenarios and recommend action priorities.

A central notion in LIBRA® is the use of "scenarios" to describe existing or hypothetical situations. Scenarios are composed of "concepts" which represent aspects of the situation. Concepts, in turn, have "attributes," or properties. "Criteria" can be defined in terms of required attributes, and then scenarios can be tested to see whether they satisfy these criteria.

Example A scenario about environmental introduction of microorganisms might include the concept "soil ecosystem," which might have the attribute "high diversity of microflora." A screening criterion might include "high diversity" as a required attribute, in which case the scenario would satisfy this criterion.

To use LIBRA®, one normally loads a specific knowledge base that has been previously created. For purposes of illustration, a demonstration knowledge base called BIO-Logic was developed for addressing the possible risks of introducing genetically modified organisms into the environment. The knowledge was extracted from various scientific reports and symposia that have focused on this important subject over the past few years.

Knowledge representation

The BIO-Logic knowledge base is organized as a hierarchy of classes, moving from the general to the specific. The user can easily browse through the current knowledge base by selecting the classes they wish to focus on. There are four major types of knowledge represented:

Concepts: The classes of concepts defined in BIO-Logic include organisms, ecosystems, modifications, proliferation stages, and outcomes. Each class is subdivided into more specific subclasses; for example, higher organisms are divided into insects, plants, animals, and humans.

Attributes: Attributes are properties that apply to concepts. For example, attributes that apply to bacteria might include host range, genetic stability, donor organism type, modification, pathogenicity, infectivity, etc. Attributes can be either qualitative or quantitative.

Scenarios: Scenarios are descriptions of particular situations to be evaluated. For example, in BIO-Logic scenarios represent proposed applications of genetically modified organisms.

A scenario might have the following concepts defined as components:

Bacterium: *Rhizobium*
Ecosystem: soil ecosystem
Beneficial outcome: plant growth enhancement
Possible risk: displacement of indigenous species

Criteria: Criteria are sets of conditions that are relevant to decision making or priority setting, including regulatory constraints, review policies, or risk indicators. Criteria are defined by stipulating the attribute conditions that must be present (or absent) in the scenario of concern. For example, a criterion might require that the microorganism be a nonpathogen.

The user can also specify logical inference rules that will deduce attribute values. For example, a sample rule in BIO-Logic is: "If the donor organism of X is A, and A is known to be pathogenic, then X may exhibit pathogenic traits."

While browsing, the user can modify or delete anything in the knowledge base. Moreover, a unique aspect of LIBRA®, in contrast to previous software tools, is that it allows the user to incrementally add new knowledge to the knowledge base. Users simply indicate what type of knowledge they wish to define, and LIBRA® guides them in expressing it.

Screening and evaluating scenarios

LIBRA® can screen any selected scenarios against any selected criteria. LIBRA® tests each scenario to see if it matches the required or excluded attributes, provides ongoing explanation of the findings, assesses the degree of certainty that the criteria are satisfied, and displays an appropriate message. If there are gaps in the knowledge base, then LIBRA® will try relevant inference rules or query the user to determine values of unknown attributes. Moreover, for any set of criteria, LIBRA® can rank several scenarios based on criteria weights assigned by the user, and determine the highest-ranked scenario.

Finally, LIBRA® can help in making qualitative comparisons of various scenarios. For purposes of understanding the potential risks associated with a given scenario, it is often helpful to use comparisons with previously encountered scenarios. For example, if the scenario being evaluated involves an application of modified "*Rhizobium*" in a "soil ecosystem," LIBRA® can seek all scenarios in BIO-Logic that also include either of the concepts "soil ecosystem" or "*Rhizobium*."

LIBRA® can then examine each of these comparable scenarios to see if its attributes are similar to the scenario of interest. For example, the "host range" attribute can be tested for each scenario that includes "soil ecosystem," to see whether the organism in question has a host range similar to the modified *Rhizobium*. Scenarios that are discovered in this way by LIBRA® can then be screened or ranked against the scenario of interest to assist in decisions about acceptability or appropriate safety assurance measures.

Conclusion

This paper has suggested that knowledge system technology provides a number of potentially useful capabilities to support risk assessment and management for environmental introductions of genetically engineered organisms. Because of the limitations of quantitative methods, risk assessment often requires judgmental reasoning, and knowledge systems can provide consistent support to humans in evaluating the various risk factors relevant to proposed introductions. Specifically, knowledge systems can:

- Help to preserve and disseminate specialized knowledge regarding environmental introductions
- Help to implement generic guidelines by providing a systematic road map for review procedures
- Support the practice of risk assessment methodology, including risk identification and trade-off analysis
- Interpret available data about organism-ecosystem combinations to suggest risk considerations
- Support decisions regarding controlling and monitoring of organisms subsequent to release

The LIBRA® demonstration system illustrates the use of an "intelligent" system to provide qualitative screening assistance. It represents only a glimpse into the rich set of possibilities enabled by knowledge system technology.

References

Buchanan, B. G., and Shortliffe, E. H. 1984. *Rule-Based Expert Systems; The MYCIN Experiments of the Stanford Heuristic Programming Project*, Addison-Wesley, Reading, Massachusetts.

Feigenbaum, E., McCorduck, P., and Nii, H. P. 1988. *The Rise of the Expert Company*, Times Books, New York.

Fiksel, J. 1987*a*. Artificial intelligence: Software reasons to analyze risks, Safety and Health, National Safety Council, March, pp. 64–66.

Fiksel, J. 1987*b*. The impact of artificial intelligence on the risk analysis profession, *Risk Anal.* 7(3):277–280.

Fiksel, J. 1988. Potential applications of knowledge system technology to biotechnology safety assurance, in J. Fiksel and V. Covello (eds.), *Safety Assurance for Environmental Introductions of Genetically-Engineered Organisms*, Springer-Verlag, Berlin.

Fiksel, J., and Covello, V. 1986. The suitability and applicability of risk assessment methods for environmental applications of biotechnology, in J. Fiksel and V. Covello (eds.), *Biotechnology Risk Assessment*, Pergamon, New York, pp. 1–34.

Fiksel, J., and Hayes-Roth, F. 1989. Knowledge systems for planning support, *IEEE Expert*, Fall 1989, pp. 16–23.

Hayes-Roth, F., Waterman, D. A., and Lenat, D. B. 1983. *Building Expert Systems*, Addison-Wesley, Reading, Massachusetts.

Hushon, J. 1986. Response to chemical emergencies, *Environ. Sci. Technol.*, 20(2):118–121.

Merkhofer, M., and Covello, V. 1986. *Risk Assessment and Risk Assessment Methods: The State of the Art*, Plenum, New York.

National Academy of Sciences (NAS). 1983. *Risk Assessment in the Federal Government: Managing the Process*, National Academy Press, Washington, D.C.

Rossman, L. A. 1986. The use of knowledge-based systems in RCRA and Superfund programs. Paper presented at NCASI Meeting, October 1986, USEPA, Cincinnati, Ohio.
Russell, M., and Gruber, M. 1987. Risk assessment in environmental policy-making, *Science*, 236:286–290.

Chapter

16

Describing Risk

Robert Wachbroit

Center for Public Issues in Biotechnology
Maryland Biotechnology Institute
University of Maryland, Baltimore County
Catonsville, Maryland 21228

Consider two familiar ways of describing an eight-ounce cup containing four ounces of liquid. We could say that the cup was half *full* or that the cup was half *empty*. Although both descriptions are true, each of them plainly conveys quite different impressions. They color the situation differently, and so, depending upon the context, they could be seen as offering different assessments and suggesting different courses of action. Nevertheless, standing behind these true but tendentious descriptions is a neutral and objective description: the eight-ounce cup contains four ounces of liquid.

This "standing behind" is important. Decisions based on tendentious descriptions naturally invite controversy. It is difficult to imagine a serious controversy fueled by the difference between the "half-empty" and "half-full" descriptions because the neutral description is so clearly available. Neutral descriptions help resolve controversy if in no other way than by distilling the information from the bias, thereby allowing the decision to have a neutral basis.

This example illustrates by analogy a fundamental problem with describing risk: there are counterparts to the "half-empty" and "half-full" descriptions—e.g., we could describe risks in terms of mortality rates or in terms of survival rates.[1] However, there does not appear to be any counterpart to the straightforwardly objective description of four ounces of liquid in an eight-ounce cup. Several studies have shown that the unavailability of neutral risk descriptions is a fact of human psychology (Tversky and Kahneman 1981). Indeed, this lack may also reflect deeper issues concerning the often acknowledged

[1]For an illustration of how these alternative descriptions affect judgment, see McNeil et al. (1982).

difficulty of separating facts from values. Be that as it may, it is at least a fact of social psychology: people do not respond equally to alternative but logically equivalent descriptions of risk.

The lack of neutral descriptions raises a problem for one of the primary ambitions of risk analysis, which is to aid technological decision making by impartially evaluating technological risks. A problem in describing risks translates directly into a problem in perceiving and communicating risks since how we perceive or communicate risks depends upon how we describe those risks to ourselves or to others.

In this chapter I shall survey briefly some of the conceptual problems in describing risks. These are not problems about how we happen to talk about risk; these are problems about our understanding of risk. They not only fuel some of the debates over risk analysis, they also reflect the role risk analysis should play in policy debates. I will suggest that some of these problems are pseudoproblems: they rest on a mistaken view of what the role of risk analysis should be. That is to say, these problems are problems only if we require risk analysis to play a certain role in policy decisions, one that I suggest cannot be maintained once we take seriously the lack of neutral risk descriptions.

A good place to begin is by defining risk. A simple characterization of "risk" is probability of harm. There are other characterizations of risk, including risk as the magnitude of harm and risk as the expectation value of the harm (i.e., the product of the probability of harm and the magnitude of harm) (Vlek 1987, p. 174). We can find support in everyday speech for these various characterizations of "risk." Nevertheless, insofar as an understanding of risk builds on a prior understanding of probability and of harm, the choice of definition is not important for this discussion.[2] This understanding of risk suggests that we should proceed first by examining issues raised in identifying harms and then by considering the probability claims involved in risk judgments.

Understanding Harm

Differences in risk assessments can often be traced to differences in what harms are to be recognized. We can distinguish "thin" from "thick" conceptions of harm. On a "thin" conception, harms are physical harms, understood in terms of mortality and morbidity. Thus, the thin harms that could occur in an intentional release of a genetically engineered organism consist of the possible physical harms—that is to say, death or disease. On a "thicker" conception of harm, we look at a much broader range of losses and damages. Many of these losses and damages might be considered "social harms." They might include economic losses, social disruption, the abandonment of certain conventions and values, and the undermining of political and social institutions.

Although technical risk analysis focuses on thin harms, thick harms are by no means insignificant or any less real. We can get a dramatic illustration of

---

[2]For an example of an understanding of risk that does not rest on probability and harm, see Rayner and Cantor (1987).

this by considering the accident at the Three Mile Island (TMI) nuclear reactor in 1979. As one commentator has noted,

> Despite the fact that not a single person died at TMI, and few if any latent cancer fatalities are expected, no other accident in our history has produced such costly societal impacts. The accident at TMI devastated the utility that owned and operated the plant. It also imposed enormous costs (estimated at 500 billion dollars by one source) on the nuclear industry and on society, through stricter regulations, reduced operations of reactors worldwide, greater public opposition to nuclear power, reliance on more expensive energy sources, and increased costs of reactor construction and operation. (Slovic 1989, p. 10)

Risk perceptions or analyses based on a thin conception of harm will tend to underestimate greatly the costs associated with mishaps.

This point is particularly important to keep in mind when discussing biotechnology because thick conceptions of harm are so often raised with regard to biotechnology risks. This is why public concerns over biotechnology have persisted despite growing scientific assurances that genetic engineering can be safely performed. [See, for example, Schneider (1988).] It is not necessarily that the public distrusts the scientists; rather, the public and the experts may be using quite different conceptions of harm. In ways that are perhaps more dramatic than other technologies, biotechnology vividly raises issues regarding thick harms. Medical biotechnology doesn't simply promise new ways of promoting or maintaining health. Medical biotechnology, especially in the form of genetic testing, also raises questions about our understanding of what health is, and what our reproductive responsibilities are. It has also begun to raise issues regarding the impact of genetic testing on determinations of workplace safety and on insurance classifications. Agricultural biotechnology doesn't simply promise new ways of efficiently producing food. It also raises questions about the value of the family farm, the centralization of food production, and the meaning of the "all natural ingredients" label. Environmental biotechnology doesn't simply promise new ways of cleaning and reclaiming polluted areas. It also points to previously unthinkable ways that we might be able to control the environment or control wildlife to suit our tastes, and so calls into question many of our environmental values.

In short, the development of biotechnology will have an impact that extends far beyond the benefits of its products or the thin harms that might result from biotechnology's failure in certain cases. Indeed, many potential thick harms are not the results of such failures at all; they are the possible results of biotechnology's *success*. In some ways, the public is more worried about biotechnology's succeeding than about its failing.

Thick conceptions of harm are notoriously difficult to measure or even to compare. There are so many different kinds of harm, and different kinds of harms do not often translate into different degrees of harm without distortion. For example, losing a family heirloom may be a much greater harm than the loss of anything else with the same market value. Treating the loss as an economic harm would misrepresent the harm the family suffers (MacLean 1983).

It is equally important to note that thick conceptions of harm are controversial. No one disputes that death or disease is a harm. But consider the effects of biotechnology on the relations between universities and industries. Private biotech companies and public universities are developing joint enterprises, and an increasing number of university professors are pursuing research beneficial to companies in which they have a financial stake. Are the closer ties that appear to be developing between public universities and private industries a harm? Plainly, reasonable people can disagree on this matter, as they sometimes do over many of the candidates for social harms.

The problem of what a harm is becomes even more complicated when we broaden our perspective to an international point of view. Different nations and cultures can have different conceptions of harm because in many cases thick conceptions of harm are tied to ethical and cultural values. For example, one country might value a particular species of animal more highly than its neighboring countries do because of the role that animal plays in that country's history or folklore. That country would therefore see threats to that species as a greater harm than its neighbors would. This could easily lead to sharp disagreements between that country and its neighbors concerning the appropriate trade-offs and levels of safety if these animals are at some significant risk.

A natural response to some of these problems would be to relativize harms: harms should be described in terms of who is being harmed. This response may appear to settle some of the problems of determining whether something is a harm. Instead of wondering whether the loss of a particular species of plant is a harm, we note that the loss would be a harm for country A (e.g., a loss of income) but not a harm for country B. Unfortunately, this response's easy identification of harms renders the comparison of harms much more difficult. A loss of two million dollars is plainly worse than a loss of one million dollars, but is a loss of two million dollars by country A worse than a loss of one million dollars by country B? Worse for whom? A difficulty in comparing harms directly translates into a difficulty in comparing risks.

Understanding Probability

Let us now turn to some issues surrounding the probability claims that are part of risk judgments. Several studies of social psychology have noted the fallacies about probabilities to which the public seems particularly vulnerable. For example, people tend to overestimate the probabilities of conjunctive events (i.e., the probability of event A *and* event B occurring) while underestimating the probabilities of disjunctive events (i.e., the probability of event A *or* event B occurring), and they appear to believe that deviations from a random process will get "corrected" (the so-called "gambler's fallacy") (Tversky and Kahneman 1974).

In addition to these problems in social psychology there are important *conceptual* problems in probability claims, which affect both expert and lay judgments. Unfortunately, there has not been as much attention paid recently to these conceptual problems.

A standard issue in the foundations of probability theory is understanding what the "probability" of a single event could mean. Suppose the probability of some harmful occurrence resulting from an intentional release of a genetically engineered organism were extremely small, say 10^{-54}. Most decision makers would find this probability small enough to ignore. A probability of 10^{-54} shouldn't be a reason to block the release of the organisms. Nevertheless, that low probability is perfectly compatible with the actual occurrence of a harmful event. Indeed, two, three, or more occurrences would not by themselves show that there was anything wrong with the original probability claim. And yet, as we all know, the public would reject the original low probability claim in the face of actual occurrences.

Is this irrational? That depends upon how we interpret probability claims.

On one standard interpretation of probability, one used by statisticians, a probability claim is a claim concerning the long-run frequency of the *type* of event in question.[3] Thus, a probability of 10^{-54} means that, in the long run, the frequency of that type of event occurring is 10^{-54}; that type of event will occur once every 10^{54} times. This interpretation says nothing at all about what the frequency would be in the short run. As we all know, it is quite possible for a coin to be a fair coin and yet come up heads ten times in a row. And what is true for the short run is also true for single events, since a single event is just a very short run. There is simply no sense to be attached according to this interpretation to the probability of a single event.

Thus, someone working with a frequency interpretation of probability could acknowledge accident after biotechnological accident and still retain a rational belief in the long-run safety of an experimental procedure or release. But long-run frequencies are not a sound basis for public policy. As the famous economist John Maynard Keynes once remarked, "In the long run, we are all dead."

These thoughts might move us toward another standard interpretation of probability—the so-called "subjective interpretation."[4] Someone making a probability claim on this account is simply stating a degree of belief or level of confidence that the event in question will occur. The numerical value reflects the odds that person would accept were he or she to bet on the event's occurring. Where do these "levels of confidence" come from? The subjective interpretation is completely silent on this point, although it does specify precisely how these levels of confidence should be revised in the light of new evidence.[5] Nevertheless, this interpretation of probability does give some content to the probability of a single event, but it does so at the cost of making the probability claim subjective.

Because they express a degree of belief, subjective probability claims can easily vary from person to person. Thus, if risk claims are to be understood according to this interpretation of probability, we are then faced with the question "On whose subjective probability claims ought we to base public policy?" We cannot simply say, "Those people who are likely to be correct," since that would lead us into an infinite regress of probabilities, starting with the probability of someone's probability claims being correct.

---

[3]A standard exposition of this interpretation of probability is Reichenbach (1949).

[4]The classic exposition of this interpretation is Ramsey (1931).

[5]This revision proceeds via Bayes' Rule. For a lucid exposition of this, see Horwich (1982).

A more responsible approach would be to look to the probability judgments of those who have the right expertise and experience. If I am trying to make a decision about whether to undergo surgery, rationality would seem to require that I be more interested in the physician's subjective probability claims of my survival than in my banker's subjective probability claims.

One of the problems with using this approach in the case of biotechnology risks is that the experts disagree. For example, in the March 1987 issue of *Science*, Dr. Frances Sharples and Dr. Bernard Davis each wrote an article examining the risks involved in the environmental release of genetically engineered organisms (Sharples 1987, Davis 1987). As one commentator has remarked, "Printed side-by-side, the pieces were implicitly accorded equal scientific weight" (Jasanoff 1989). Each author invoked quite different models and analogies. Dr. Sharples, an ecologist, considered a release comparable to the introduction of exotic organisms into a new environment, noting how often in the past such introductions quickly proliferated. Dr. Davis, a geneticist, considered a release comparable to playing "evolutionary roulette": organisms do not survive, much less proliferate, if they are not well adapted to the natural environment, which is unlikely for laboratory-engineered organisms. In the discussion, each author criticized the other author's conceptual framework.

My interest here is not in taking sides in this debate. My point is that such disagreements between experts can easily leave the public not knowing whom to believe. If the experts cannot agree on the probabilities, the public cannot be blamed if it is apprehensive. Disagreements over the probabilities can easily turn into a dispute over authorities. The result, unfortunately, can be that the public believes no one or that the public believes what it wants to believe.

In a well-received article, Stephen Stich (1978) proposes a solution to this problem of determining probabilities. The case he considers is that of an *accidental* release of genetically engineered organisms into the environment. He notes that such an event can be broken down into a sequence of smaller events:

1. A pathogenic bacterium must be synthesized.
2. The chimerical bacteria must escape from the laboratory.
3. The strain must be viable in nature.
4. The strain must compete successfully with other microorganisms which are themselves the product of intense natural seclection.

Since the probability of the potential epidemic is the product of the probabilities of each of these four events, it must be less than the probability of any one of these events. According to Stich, some of these probabilities, for example, the second one, "are amenable to reasonably straightforward empirical assessment" (Stich 1978, p. 195). Given a low assessment for escape, the overall probability would have to be low, regardless of the subjective estimates for the other probabilities.

Unfortunately, Stich does not explain what "empirical assessment" he has in mind. Empirical assessments of probabilities are typically assessments of long-range frequencies; they apply to aggregates, not to single cases. Thus, if

Stich's empirical assessments are these long-run frequencies, we are left in the dark about the probability of a single case. And the probability of a single case may matter.

The Importance of Neutrality

The problems surveyed so far point to a general problem about describing risk to others—we might call it the problem of "risk communication." Given that risk descriptions are never neutral, that the range of harms identified is often controversial, and that the estimated probabilities are always subjective, how can risks be conveyed without prejudice?

This is not the usual formulation of the problem in risk communication. The problem of risk communication is often expressed as a problem about how informed experts should deal with an uninformed public: Given the public's ignorance of science and probabilities, how can the experts inform the public about risks without provoking uninformed or irrational responses? For example, as one writer remarks,

> ...many people are not comfortable with mathematical probability as a guide for living. If you inform people that their risk of developing cancer from a 70-year exposure to a certain carcinogen at ambient level ranges between 10^{-5} and 10^{-7}, their response may be, "Okay, but can I drink the water?" (Daggett 1989, p. 31)

The predictable response to this situation is to say that the public needs to be better educated. Nevertheless, we should realize how tendentious both this formulation and the response are. We have noted that the range of harms identified in risk assessments can be controversial. If the difference between the experts and the public is over what possible harms ought to be considered in the assessment, then it is by no means obvious that the public is always wrong on this matter. As far as understanding probabilities is concerned, it is worth rereading the above passage. Although the writer intended it to illustrate how people fail to understand small numbers, the same passage can be read to illustrate the problem of applying probabilities to single cases: if the probability expresses a long-run frequency, then it says nothing about the particular case of my drinking the water. Better education is indeed important—for everyone.

Let us return to the less tendentious formulation of the problem of risk communication: Given the problems in describing risks, how can risks be communicated responsibly? In particular, if risk descriptions are not neutral, how can risk communication be a matter of *informing* rather than a matter of *handling* the public?

It is tempting to believe that neutrality can be achieved through balance. If we tell all sides of the story, then the biases will cancel each other out. Unfortunately, even if we put aside the worry that this approach might more likely confuse than inform, the attempt at balancing will not work. First of all, telling all sides is an unrealistic demand and a dubious ideal. We don't know how many sides there ever are to a story. Surely there are more possible sides than those that get expressed. Furthermore, some possible sides are just too crazy and irresponsible to be told. Second, implicit in talk of balance is the idea of prejudicial distance: Two differing views do not balance each other if one is far more extreme than the other. But we don't know whether one risk description is more extreme than another without an idea of what a center position consists of. The unavailability of neutral risk descriptions means that we don't know what a center position would be. In short, if we knew what was necessary in order to balance risk descriptions, we wouldn't need to appeal to balancing in order to provide neutral risk descriptions.

Rather than consider other suggestions for approximating neutral risk descriptions, I want instead to take seriously the idea that there are no neutral risk descriptions. And so I want to question an underlying assumption in this discussion. Are neutral risk descriptions necessary for responsible risk communication?

An affirmative answer appears to rest on the following picture of communication: There is a clear distinction between the informational content of a message and its presentation. What is being communicated is the information, but whether the communication is successful (effective, clear, etc.) depends upon the presentation. The theoretical distinction between information and presentation can of course be obscured in practice. A message can be presented so that it appears to say more or less than it really does. That is the genius of some advertisements and propaganda.

On this picture, a nonneutral risk description obscures the informational content of the risk message because risk information is considered to be itself neutral. Informing consists of conveying just the information. Intending to convey anything more or less would be deceptive and manipulative.

This picture of communication fits with a picture of the role of risk analysis in policy debates. Most commentators agree that the management of risks is a political or social decision because it turns on the goals of the society and what trade-offs are deemed acceptable. Risk management ought therefore to be as controversial as any political matter can be. The role of risk analysis, as it is often assumed, is to help rationalize the public debate over risk management. First we determine what harms we are likely to face, and then we debate what actions we ought to take. How could it be otherwise? If we disagree about the harms we are likely to face, then a debate about actions would be at cross-purposes. It would hardly be open to a rational resolution. Thus, the requirement that risk descriptions be neutral is part of a picture in which the information about the risks is above the fray. Although there can be controversies over risk information, these are controversies over technical matters—they are not political or social controversies.

These pictures cannot be maintained once we take seriously that risk descriptions cannot be neutral. The unavailability of neutral risk descriptions suggests that risk information itself is nonneutral. Instead of obscuring or distorting, a nonneutral description may in fact accurately convey the risk information. Responsible risk communication does not demand neutrality.

These theoretical discussions of risk communication can be illustrated with mundane examples. Many of the descriptions of the risks associated with drunk driving, drugs, or certain sexual activities are hardly neutral. They are often full of graphic details couched in alarming language. Not only is providing such descriptions sometimes not seen as irresponsible, providing anything less alarming could be seen as irresponsible. The point can be stated generally: Suppose you have two alternative risk descriptions, each equally supported by the available evidence, but one more alarming than the other. Which description you use to communicate the risks should depend upon whether you believe the risks are alarming. Acting otherwise would be irresponsible.

Of course, accepting this responsibility does not confer a license to frame misleading descriptions. If you know that the audience will draw what you believe are false or unsupported conclusions from the risk description, then the audience is being misled—manipulated rather than informed. In that case, the risk description is hardly a proper alternative to one that doesn't mislead.

Nevertheless, risk analysis is a controversial matter. Even if a communication doesn't mislead, it will still be controversial. But that is what we should expect—and foster (cf., Cannell and Otway 1988). Once we acknowledge the nonneutrality of risk information, we begin to see describing risks—and so, communicating risks—for what it is: it is not something that occurs prior to the public debate, it is part of that debate. Risk communication is part of risk management.

References

Cannell, W., and Otway, H. 1988. Audience perspectives in the communication of technological risks. *Futures* 20:519–531.

Daggett, C. J. 1989. The role of risk communication in environmental gridlock. In *Effective Risk Communication: The Role and Responsibility of Government and Nongovernment Organizations*, V. T. Covello, D. B. McCallum, and M. T. Pavlov (eds.), Plenum, New York, pp. 31–36.

Davis, B. 1987. Bacterial domestication: underlying assumptions. *Science* 235:1329–1335.

Horwich, P. 1982. *Probability and Evidence*, Cambridge University Press, Cambridge.

Jasanoff, S. 1989. The problems of public perceptions of biotechnological risk. Conference presentation: Patents, Regulation and Public Issues, June 14, 1989, Instructional Television System, College Park, Maryland.

MacLean, D. 1983. Valuing human life. In *Uncertain Power: The Struggle for a National Energy Policy*, D. S. Zinberg (ed.), Pergamon, New York, pp. 93–111.

McNeil, B. J., Pauker, S. G., Sox, H. C., Jr., and Tversky, A. 1982. On the elicitation of preferences for alternative therapies. *N. Engl. J. Med.* 306:1259–1262.

Ramsey, F. P. 1931. Truth and probability. In *Foundations of Mathematics*, Routledge & Kegan Paul, London, pp. 156–198.

Rayner, S., and Cantor, R. 1987. How fair is safe enough? The cultural approach to societal technology choice. *Risk Anal.* 7:3–9.

Reichenbach, H. 1949. *Theory of Probability, an Inquiry Into the Logical and Mathematical*, 2d ed. Univ. of California Press, Berkeley.

Schneider, K. 1988. Biotechnology lags despite success. *The New York Times*, 18 January, p. A10.

Sharples, F. S. 1987. Regulation of products from biotechnology. *Science* 235:1329–1335.

Slovic, P. 1989. The perception and management of therapeutic risk. Unpublished paper presented to Royal College of Physicians.

Stich, S. P. 1978. The recombinant DNA debate. *Philosoph. Publ. Affairs* 7:187–205.

Tversky, A., and Kahneman, D. 1974. Judgment under uncertainty: heuristics and biases. *Science* 185:1124–1131.

Tversky, A., and Kahneman, D. 1981. The framing of decisions and the rationality of choice. *Science* 211:453–458.

Vlek, C. 1987. Risk assessment, risk perception and decision making about courses of action involving genetic risk: an overview of concepts and methods. In *Genetic Risk, Risk Perception, and Decision Making*, G. Evers-Kiebooms, J.-J. Cassiman, H. Van den Berghe, and G. d'Ydewalle (eds.), March of Dimes Birth Defects Foundation Original Article Ser., vol. 23, no. 2, Liss, New York, pp. 171–207.

Chapter

17

Biotechnology and Biosafety: Perspective of an International Donor Agency

Joel I. Cohen

American Association for the Advancement of Science
U.S. Agency for International Development
Washington, D.C. 20523-1809

Judith A. Chambers

American Association for the Advancement of Science
U.S. Agency for International Development
Washington, D.C. 20523-1809

Introduction

Research programs generating new agricultural technologies, germ plasm, and animal health care products constitute a major portion of international development assistance. These research initiatives for the benefit of the developing world have recently included biotechnology and have been supported primarily by the Rockefeller Foundation and the U.S. Agency for International Development (AID) (Directorate-General for International Cooperation 1989).

Other development agencies have begun to consider such initiatives in a cautious manner, recognizing that recent advances in cellular and molecular biology are not shortcuts circumventing conventional processes. More precisely, agricultural biotechnology constitutes alternative tools which, when applied at the cellular and molecular level, must be integrated with conventional research (Cohen et al. 1988*a*).

Integrating biotechnology, and specifically genetic engineering, into international agricultural research raises new concerns regarding acquisition and integration of technologies, biosafety, and allocation of scarce research funds. Development strategies are needed to ensure the most rational use of finite resources while also ensuring that products of biotechnology can be safely delivered to farmers and consumers. This chapter describes support for agricultural biotechnology provided by international agencies and focuses on the need for such research to incorporate appropriate biosafety considerations.

Biosafety and the Developing World: Points to Consider

Complications surrounding the establishment of biotechnology regulations in the developing world are due to a variety of influences. Foremost among these is the diverse array of players involved in agricultural biotechnology research and testing (Cohen et al. 1988*b*). These include the International Agricultural Research Centers (IARCs); donor agencies which provide support for the IARCS and national programs; governments and research institutions of developing countries; governments, research institutions, and private industry of industrialized nations; and various environmental and public advocacy concerns. Each of these groups has its own agenda and its own definition of what constitutes safe and appropriate use of biotechnology products in the environment. As such, one may presume that for each of these players, biosafety policy formation for biotechnology might reflect an individualized risk-benefit analysis.

In addition, major constraints on the development of responsible regulatory policy for recombinant DNA research exist in the developing world. These include:

- Lack of national regulatory structure and finances to support it
- Lack of confidence in decision-making expertise
- Lack of coordination with international organizations
- Requests received from many sectors: public, private, donor, IARCs
- Institutional biosafety committees not yet implemented
- Lack of funds and technical expertise for risk assessment analysis and modeling
- Apprehension that too many regulations will stifle scientific innovation

As noted above, regulatory infrastructure in a particular country may be nonexistent or underdeveloped and unable to cope with the monetary and enforcement requirements necessary to establish and implement policy (Umminger 1989). Establishment of informed public consent, although desirable, may not presently be feasible. Therefore, host-country approval will, in most cases, be granted without the presence of a national regulatory structure.

It should also be recognized that sophisticated systems providing for risk-benefit analysis, expert systems, and extensive data on approved testing will not generally be available in developing countries. The provision of such services is quite beyond the financial and technical capabilities of most developing countries eager to receive relevant biotechnology products. Lack of personnel trained in risk assessment, environmental impact analysis, and modeling reflects the overall need to provide comprehensive training in biotechnology.

Concern that imposition of standards will thwart the ability of developing countries to legitimize biotechnological research is also a pervasive theme. For example, in Thailand the regulatory system which is currently operative is decentralized and informal for fear that formalization of the regulatory process will impede further developments in biotechnology research (Umminger 1989).

In addition, a universal approach to formulating regulatory policy, despite the similarities of the concerns and problems facing less-developed nations, is nonexistent among these countries. Policies of industrialized nations, such as those in Western Europe, also lack uniformity. Nonetheless, the enhancement of host-country capabilities in biosafety, including national abilities to evaluate, approve, and monitor specific requests for testing, is essential if these countries are to fully participate in and benefit from the exchange of this technology with public and private sector elements in more-developed countries (Cohen 1989).

AID would be able to provide support to developing countries as they attempt to establish responsible biosafety policies for recombinant DNA research. Such support would be firmly based in AID's proven history of cooperation with host countries regarding technical aspects of research and in well-established networks for coordinating activities between national programs, other donor agencies, and the IARCs. However, other opportunities must be explored to assist the countries no longer eligible for AID assistance.

In the interim, AID and other donors must work cooperatively with each host country to obtain appropriate approval for testing. It has been suggested that assurance should be provided that research involving genetic engineering that is sponsored by the U.S. government and tested abroad be conducted under conditions that meet U.S. standards for the protection of health and the environment (Mellon 1988). Such a measure was recently recommended regarding research in cooperation with developing and European countries to ensure appropriate safety standards (German Foundation for International Development 1989). Thus, prior to obtaining host-country approval, AID federally funded research must be in agreement with, and approved by, the appropriate U.S. regulatory agency.

Once approval for domestic testing has been obtained, and the product shown to be safe and efficacious in the United States, obtaining approval from a host country should be simplified. However, it is not possible to predict (1) complications which may arise as national biosafety programs are developed and (2) costs required for domestic approval, testing, export, and finally, host-country testing. Further complications will no doubt arise for those products designed for use in the developing world which may not receive the best possible test in the United States.

As an example, cassava or potato genetically engineered for protection against viral infection endemic in Asia or Africa will be difficult to test in the United States. However, this can be rectified by testing plants in contained facilities, such as the Foreign Disease-Weed Science Research Unit of the U.S. Department of Agriculture (USDA) located at Fort Detrick, Maryland. This type of domestic testing prior to subsequent tests in the developing world has been carried out in the case of recombinant vaccines, but not yet in the case of transgenic plants.

This is not to say that testing in the United States can substitute for actual host-country testing. Each country has a unique set of environmental conditions, many of these tropical in nature, which cannot be successfully emulated in the United States. For example, concerns have been raised regarding complex crop-weed interactions which may exist between transformed crops and related weed species. In the case of animal vaccines, many diseases are endemic in Africa which do not occur in the United States. Each of these parameters will have to be carefully considered as confined field testing is initiated.

Range of Safety Parameters

Experience with recombinant vaccine projects has increased our understanding of the range of safety parameters which warrant consideration before host-country testing can be approved. As new projects develop, more thorough consideration must be given to the range of relevant biosafety considerations and the cost associated with providing data which addresses both safety and efficacy parameters (Fig. 17.1). These factors can include the following: risk-benefit, efficacy, characterization, safety, spread or dispersal, human health parameters, and environmental fate (Tiedje et al. 1989).

As indicated in Fig. 17.1, public opinion has a direct effect on the processes involved in product development and can be a powerful force in ensuring that sound developmental practices are initiated for the products of scientific research. This means providing data to satisfy questions regarding both efficacy and safety for each product being considered for testing. Public education and awareness should thus be a critical component in the transfer of technology from the laboratory to the field. This is particularly true of new technologies, such as those employing recombinant DNA techniques, since initial public perceptions about product safety and efficacy may have far-reaching implications for further technological advancement. The role of public opinion and response, and its impact on product development, should not be underestimated in either developed or developing countries. While commercial firms understand this necessity, most research sponsored through donor agencies or the IARCs has not previously incorporated such considerations, as will be necessary for products derived from biotechnology.

Thus, maximum coordination will be required to ensure that requisite parameters are understood and established for both efficacy and safety. Gaining such information will require direct communication between regulators, principal scientists, and donor agencies. Without establishing such communication links, research projects run the risk of "spinning" the

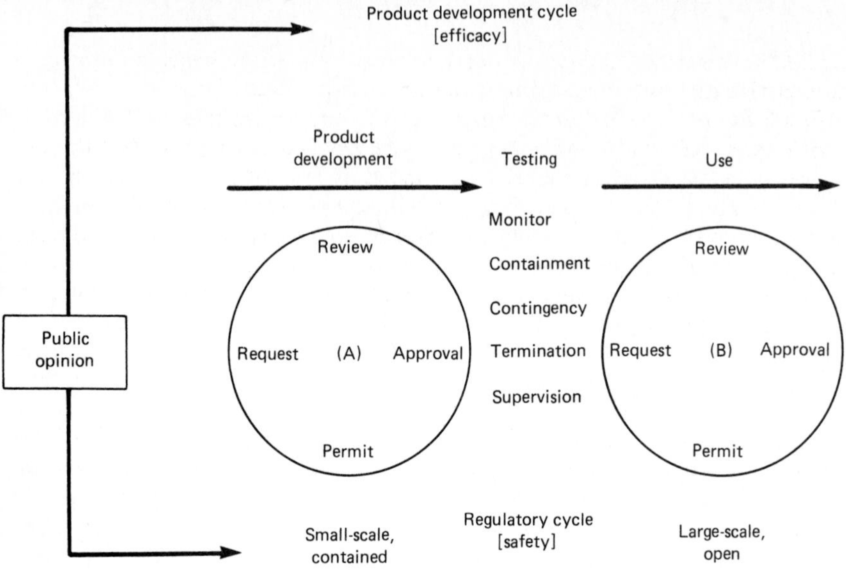

Product development cycle
[efficacy]

Product development
Testing
Use

Monitor
Containment
Contingency
Termination
Supervision

Review
Request (A) Approval
Permit

Review
Request (B) Approval
Permit

Public opinion

Small-scale, contained
Regulatory cycle
[safety]
Large-scale, open

Figure 17.1 Product development and regulatory cycle with emphasis on efficacy, safety, and approval for testing.

request-review-approval-permit cycle multiple times until appropriate questions have been answered. This will be costly, as each time a request is submitted, costs are entailed for the subsequent testing and documentation. Obviously, these costs must be minimized so that development costs do not increase exponentially.

Once contained testing is warranted, consideration must be given to each host country's ability to monitor the test, the test's degree of containment, contingency plans if problems occur, termination procedures, and responsibility for supervision. Here also, costs associated with such experimental modifications will need to be estimated in research proposals so that more realistic funding levels can be determined.

While various international organizations, such as the World Health Organization (WHO), may assist countries in their approval process, the decision should finally be made at the national level. Such approval becomes more complicated in the case of the IARCs, which have a regional or global mandate for their research and retain special international status that provides for greater autonomy.

In addition, many environmental and safety parameters will vary: those in the United States may be different from those in any one developing country. While some of these differences may be obvious, special attention will need to be paid to ensure that an appropriate assessment of such differences is obtained well in advance of host-country testing. It would be of great value for regulatory agencies in the industrialized countries to make their personnel available for site inspections and recommendations regarding each proposed test. In the case of AID, such inspections will help ensure that problems are corrected prior to authorization of testing.

However, to do this, the domestic regulatory mechanism would have to make staff available for some international travel and allow staff time for provision of recommendations and reports to both agency and host-country biosafety committees. In addition, special training would be required to sensitize domestic regulatory officials to conditions and facilities available in the developing world. Initially, this would put some additional burden on domestic regulatory agencies but would also assist them in safely moving the products of biotechnology to those most in need.

Institutional Biosafety Committees

An urgent need persists to develop some mechanism whereby biotechnological advances can be safely and responsibly assimilated into a given country's overall agenda for scientific advancement and economic progress. One possible solution may be offered in the establishment of Institutional Biosafety Committees (IBCs), as has been recommended in the Philippines, Kenya, and Thailand. Specific objectives of national IBCs have been summarized: (1) demonstrate that a responsible, objective approval and recommendation process has been instigated, (2) provide a technically based forum to address major biosafety concerns, (3) produce final recommendations for approval and present these to national decision maker, (4) ensure compliance with essential legislative regulations, and, (5) issue formal host-country approval (Cohen 1990).

One important feature of these committees is that the cost for such review should not be an added burden to the publicly funded principal investigator; rather, it should be supported by the institutions concerned.

In more practical terms, an IBC could verify that small-scale field trials or contained preliminary tests of recombinant products are first properly conducted in an industrialized country with established biosafety mechanisms in place. For example, in the case of a recombinant vaccine developed in an industrialized country but targeted for the developing world, the IBCs of targeted countries could require and monitor a safety assessment program whereby the vaccine is (1) tested for purity; (2) evaluated for stability by backpassage studies in tissue culture and the host animal; (3) evaluated for host-animal safety using the recommended route of administration at elevated dosages; (4) evaluated with respect to vector integrity; (5) evaluated for the shed, spread, and fate of virus, especially in the host animal; and (6) clinically tested for safety in laboratory animals, and, if necessary, in human subjects (Chambers and Cohen 1990).

After certain preliminary tests have been performed and approved by an IBC and other appropriate international agencies, further confined testing could be initiated at an international center or host-country national facility (Fig. 17.2). Data from these tests would lead to final decisions by the developing country itself, although certain international agencies might be requested by a host country for consultation. Such a system has recently been recom-

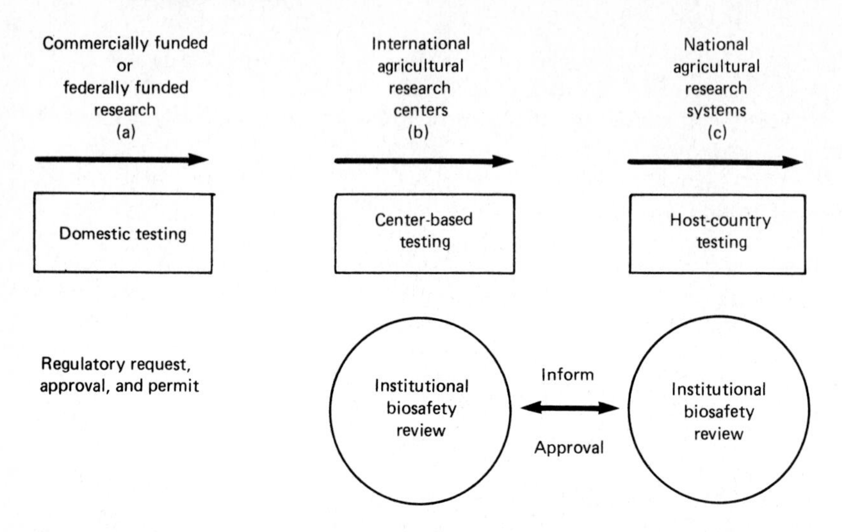

Figure 17.2 Interrelation between biotechnology research sponsored by federal agencies involving the international centers and eventual host-country testing.

mended for adoption by the Consultative Group on International Agricultural Research (CGIAR) (Plucknett et al., in press).

Biosafety and the International Agricultural Research Centers

The CGIAR is made up of donors who provide support to its thirteen International Agricultural Research Centers (IARCs). The centers themselves operate as nearly autonomous institutions, and it is at the center level that decisions concerning research programs are made. The Board of Trustees, the Director General, and the staff of each center plan and carry out research programs according to a mandate agreed upon with an external Technical Advisory Group and the CGIAR. In this framework, biotechnology is considered as an approach to research that brings forth new information or techniques to help solve difficult problems. Each of the thirteen centers now has an active program in biotechnology or is involved in policy implications of the new technologies (Plucknett 1988).

Few IARCs will be able to develop new basic knowledge in support of biotechnology; rather, they will become involved in the application of new scientific advances. The centers will thus become dependent upon scientific advances and the resultant technologies in their effort to adapt technologies to developing-country problems. A comprehensive survey and discussion of current biotechnology activities at the CGIAR centers has recently been prepared. This indicates a strong reliance upon collaborative research as the IARCs work to make advances in biotechnology (Cohen 1989, Plucknett 1988).

Some of this research now includes transformation and genetic engineering. Such collaborative research efforts of industrialized countries and CGIAR centers will pose new issues once products arise requiring field testing. Both donor countries and the CGIAR center will have to be fully cognizant of biosafety protocol and have in place an approval system which is rapid, yet thorough.

Increased activities by the IARCs in adapting the tools of biotechnology to the demands of developing-country agriculture, will cause a shift in allocation of resources and personnel. While it will be necessary to continue conventional plant improvement programs essential for integrating new technologies and completing product development, centers may also be required to add personnel able to address biosafety concerns, implement biosafety protocols, and meet the novel demands of biotechnology-based research. This alteration in center-based research toward a more strategic direction reflects but one of many challenges the IARCs must face because of recent advances in technology development. Some of these changes represent a departure from the way research was managed during the Green Revolution, but such a shift is essential if the CGIAR centers are to successfully and safely provide products of biotechnology for farmers in the developing world (Plucknett et al., in press).

Insistence by federal agencies on including institutional biosafety review of selected biotechnology-based research indicates the appeal of such an approval process. For example, in addition to the regulatory role of its Animal and Plant Health Inspection Service (APHIS), the USDA has also established an Agricultural Biotechnology Research Advisory Committee to review research sponsored by USDA. International centers constituting the CGIAR system should take action to ensure that their research undergoes a similar type of formalized review and is coordinated with host-country approval (Fig. 17.3).

Biosafety Initiatives within AID

Most international donor agencies are intentionally decentralized in order to increase effectiveness of host-country operations. This is especially true of AID. In most cases, project development requires clearances and authorization from one of the agency's central bureaus, respective in-country mission, host-country officials, and, finally, the U.S. Congress (Fig. 17.4). Such a complex chain of approval makes it difficult to monitor biotechnology, as projects may begin at any of the above starting points. This is especially the case for biotechnology projects requiring biosafety approval.

Federal agencies have met a similar challenge before by providing environmental assessments for all bilateral assistance projects to assure compliance with the National Environmental Protection Act (NEPA). Such an assessment must be prepared by each office initiating and funding a bilateral project. An environmental analysis is also required for biotechnology projects, although special concerns need to be addressed in these assessments regarding approval prior to field testing an organism created using recombinant DNA techniques. Thus, special expertise in biotechnology will be required by AID,

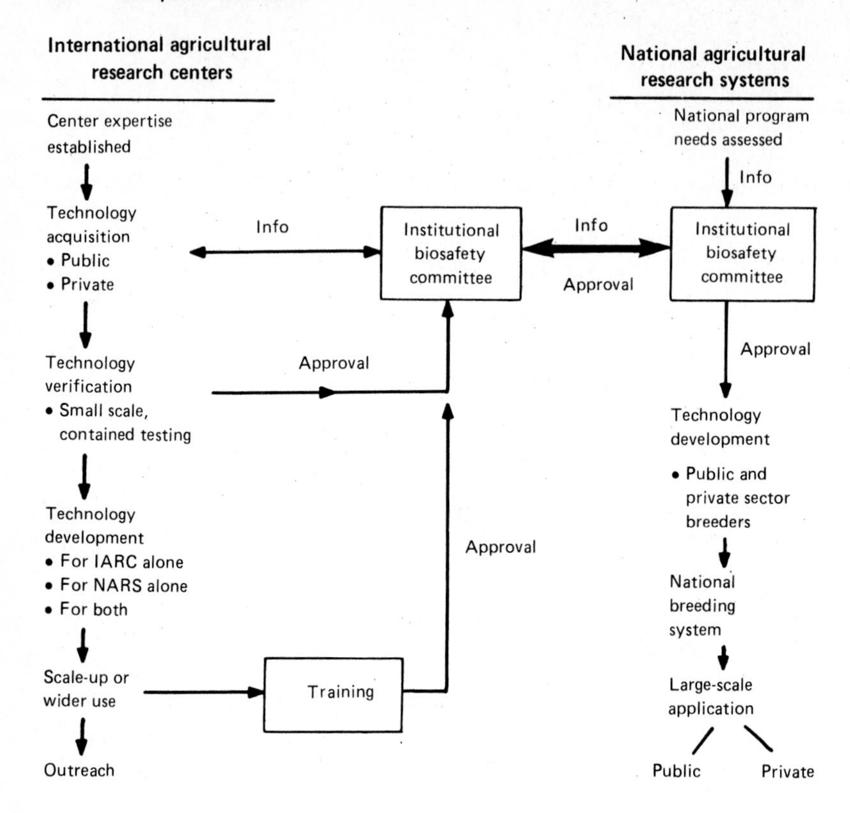

Figure 17.3 Proposed institutional biosafety committees established at international agricultural research centers and coordinating with host-country national programs.

as has been previously needed regarding environmental science. Such agency expertise will be invaluable in establishing links between NEPA activities and those required for biosafety approval of biotechnology ventures.

With these concerns in mind, AID established a Standing Committee on Biotechnology in 1987 to serve as an intra-agency mechanism to provide (1) an agency forum for advice on issues surrounding biotechnology and on the development of mechanisms to address those issues which are technical, regulatory, and programmatic in nature, and (2) an entity within AID which will provide liaison with U.S. government scientific and regulatory agencies principally concerned with supporting and regulating biotechnology (AID 1987).

By establishing this committee, AID recognized the role which biotechnology research occupied in its development activities and also addressed the agency's need to safely and responsibly monitor and implement the products of its research. The Standing Committee serves a central purpose in the review and dissemination of information and the coordination of domestic and host-country regulations affecting biotechnology research and offers specific procedural guidance for AID projects.

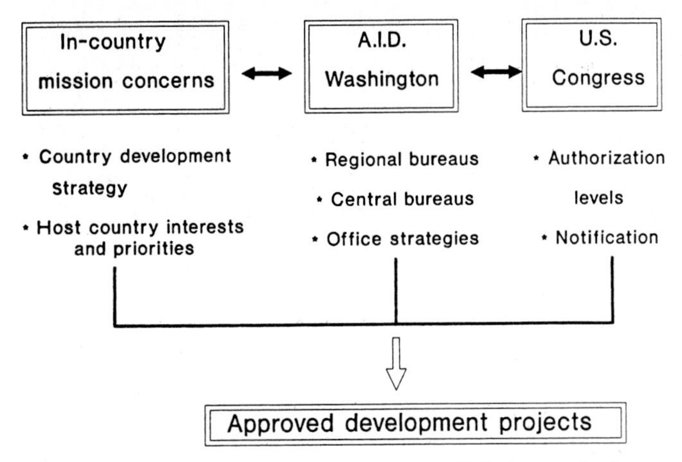

Figure 17.4 Program development within the U.S. Agency for International Development illustrating complexities derived from a decentralized organization.

Presently, the agency lacks a precise mechanism for ensuring that all projects with possible biosafety implications are appropriately routed through the Standing Committee. Individual projects are brought to the attention of the Standing Committee at the discretion of individual program officers. Depending on the nature of the research and its perceived biosafety impact, the Standing Committee may then convene a Subcommittee on Biosafety to address specific regulatory concerns posed by this research. Clearly, the success of the entire process depends on individual recognition of the possible biosafety concerns associated with projects being handled by the various offices, bureaus, and missions of AID worldwide. As more projects incorporate elements of recombinant DNA research, the present structure may be inadequate to effectively monitor all projects for potential biosafety implications.

Even at present, the extent to which such activities are implemented varies throughout the agency. For example, the Science Advisor's Office, which was established in 1981 to support basic experimental research, provides funds for research activity under the Program in Science and Technology Cooperation (PSTC). Prior to obtaining funds under this program, the applicant must, in conjunction with submission of a preproposal, complete a Special Concerns Review Sheet. This form requires the applicant to specify the nature of the research with respect to recombinant DNA activities and field release of organisms, and to assess the proposed research in the context of biosafety or environmental impact concerns. In addition, for certain projects which may potentially involve field testing of organisms, contractual language has been included in the grant which requires compliance with federal guidelines and regulatory procedures, compliance with AID's environmental procedures, and notification to and written approval from AID and the host-country government.

The AID Office of Agriculture includes, as a standard provision of grants supporting biotechnology research, specific language regarding regulatory compliance and approval from U.S. regulatory agencies (see Appendix to this chapter, and Jones 1987). Thus, there is a formal, written requirement by the office that any risks posed by the research be fully addressed in the context of U.S. government regulations prior to obtaining host-country approval. In addition, this approval must be obtained in writing as must that of that of the AID in-country mission. Provision for consideration of biosafety is consistent with a recent National Research Council report, "Plant Biotechnology Research for Developing Countries," which recommends that AID assume an active role in assisting developing countries in their implementation and monitoring of biosafety regulations (National Research Council, 1990). The establishment of a hierarchical structure which links activities within individual offices and missions to those of the Standing Committee may eventually be required.

Other International Initiatives in Biotechnology

Other development agencies exploring potential support for biotechnology include the World Bank, the United Nations Food and Agriculture Organization (FAO), and the United Nations Industrial Development Organization (UNIDO). UNIDO recently established (UNIDO 1989) the International Center for Genetic Engineering and Biotechnology (ICGEB) with component laboratories in Italy and India. The ICGEB provides advisory services, at the request of member countries, on matters relating to biosafety policy. Furthermore, UNIDO, in conjunction with WHO and UNEP (United Nations Environment Program), established, in June 1985, a Working Group (WG) on biotechnology safety whose purpose is "to establish a process through which the potential risks arising from this rapidly evolving technology can be assessed and appropriate safety measures designed" (UNIDO 1985).

Three points of consideration with specific implications for international agencies were identified: (1) Does biotechnology pose any special problems for developing countries distinct from those confronting developed countries? (2) How can the WG's recommendations be legitimized and given wide-ranging support? (3) What should be the role of the ICGEB? To date, the working group has met on four occasions. A consultative report, which addressed biosafety practices and regulations from an international perspective, has been prepared (Karny 1986). This report outlines specific roles for the WG and ICGEB in providing a forum for information exchange, in the development of risk assessment methodology and safety guidelines, in conducting training in safe laboratory practices, and in assisting individual countries to adapt the guidelines to their own needs.

The World Bank has conducted an exhaustive review of options and implications of biotechnology, the complete report to be published in 1990 (Persley 1989). This study recognized regulatory requirements to derive from the legitimate need to ensure environmental safety and public health. The study went on to concur with the advantages to be derived from institutional biosafety committees at the IARCs and within the developing-country national programs.

A conference convened by the Technical Center for Agricultural and Rural

TABLE 17.1 Newsletters Providing Information on Biotechnology Activities Relevant to the Developing World

Newsletter	Sponsor
Anaplasmosis Babesiosis Network Newsletter	AID
Genetic Engineering and Biotechnology Monitor	UNIDO
Rice Biotechnology Quarterly	Rockefeller Foundation
Biotechnology and Development Monitor	Director General International of the Ministry of Foreign Affairs, The Hague, and the University of Amsterdam
The International Plant Biotechnology Network Newsletter	AID

Cooperation (CTA) and the FAO in Luxembourg developed a list of priority recommendations for plant biotechnology to be considered for future support by FAO (CTA/FAO 1989). This meeting included a recommendation calling for establishment of appropriate biosafety and ecological assessment guidelines. It was suggested that various donor agencies could provide details of existing guidelines. Finally, an international information system on standards for field testing and the introduction of genetically engineered organisms would be extremely useful.

International newsletters and networks focusing on biotechnology activities in the developing world may also have an instrumental role in the dissemination of information on biosafety issues and policy. A list of some of the more relevant newsletters and their sponsors is presented in Table 17.1. A recent article on biosafety considerations and recombinant animal vaccines (Chambers and Cohen 1990) indicates how international publications may provide an effective forum for raising biosafety awareness in the development community at large. This article addressed complications regarding recombinant vaccine development and addressed them in the context of international development assistance.

Recombinant Vaccine Development: A Case Example

To date, AID's Office of Agriculture has supported two ongoing vaccine development projects which have served as examples for the necessary incorporation of major points discussed above. These projects demonstrate the appropriate use of biotechnology in AID-sponsored programs and illustrate the need for (1) established biosafety procedures, (2) possible private sector involvement for product development, and (3) support from international organizations in the implementation of recombinant technology. Coordinated activities in each of these areas are essential to successfully move newly developed vaccines from the laboratory bench to the field. These projects and the associated issues are described in some detail below.

Recent developments in the production of recombinant animal vaccines serve to illustrate the potential of biotechnology to solve old problems by innovative, and often more effective, means. Conventional vaccine production has been historically hampered by problems relating to efficacy, stability, safety, and production costs. Such problems are frequently compounded in the developing world, where countries are plagued by inadequate transportation, storage, and delivery systems and often lack the essential personnel and resources required to produce and administer the vaccine.

Recombinant vaccines, which are comprised of known antigenic determinants linked to infection-competent viral vectors, offer significant advantages over conventionally produced vaccines. They are less costly and easier to produce since standard fermentation techniques, rather than costly tissue-culture techniques, are utilized (Bookrack 1981). More importantly, since genetically engineered vaccines generally consist of only one or a small number of antigenic components, as opposed to an entire infectious organism, they do not generally pose the hazards associated with adverse side effects or actual transmission of disease that are typically posed by the conventional vaccines (Board on Science and Technology for International Development 1982).

In support of these claims, the recombinant vaccine recently developed as a result of a cooperative research agreement between the AID Office of Agriculture and the University of California, Davis, offers great promise for the future eradication of rinderpest, an acute, highly contagious viral disease of cattle. Although the Plowright vaccine is currently being used to control rinderpest, it requires a chain of refrigeration for its dissemination, and sophisticated methods of production, both of which are difficult to provide for many developing countries (Faulkner 1983).

The new recombinant vaccine, which has been engineered in a vaccinia virus vector, is thermostable and, therefore, should be easier to disseminate (Yilma et al. 1988). Furthermore, vaccinia-based recombinant vaccines, in general, offer safety advantages since attenuation of the virus, such as that demonstrated for the Wyeth strain of vaccinia used in the smallpox vaccine program, may be accomplished by insertional inactivation or deletion of the viral genes responsible for virulence (Buller et al. 1985).

Inherent biosafety concerns arising from the use of genetically engineered organisms are presently being addressed on a case-by-case basis. In the case of the recombinant rinderpest vaccine, contained testing was conducted in the United States at the USDA's Plum Island Animal Disease Center. Containment was necessary because rinderpest is not found in the United States and must be carefully supervised to ensure that it is not spread to domestic cattle.

Such testing is expensive, but mandatory to assure all parties concerned that the vaccine is safe, efficacious, and characterized so as to warrant approval for export. Receipt of this approval from APHIS assures AID and the University of California that all appropriate regulatory concerns have been addressed. Thus, approval to export the vaccine has been treated essentially the same as a vaccine licensed domestically when in fact its intended use is for Africa (Glosser 1988).

As of this time, approval to export the recombinant rinderpest vaccine for confined testing in Africa has not been granted, pending further testing and written approval resulting from host-country biosafety review. Such approval is consistent with AID provisions for regulatory compliance. A similar mechanism for approval has more recently been proposed by the National Institutes of Health (NIH) for testing of NIH-sponsored research which warrants testing abroad (NIH 1989). Preliminary discussions concerning testing of a recombinant vaccinia-vectored vaccine in Africa have attracted the interest of a number of international organizations such as WHO, the Organization of International Epizootics (OIE), the Organization of African Unity, and the Pan African Rinderpest Campaign.

Each of these groups has recently convened special meetings to discuss the benefits and risks of live, vaccinia-vectored vaccines. The recommendations of these deliberations, and particularly the OIE's report on its meeting (OIE 1989) are playing a significant role in establishing the requirements for the final form of the recombinant rinderpest vaccine which will be approved for confined testing in Africa. In addition, these reports will be consulted during host-country review of the proposed small-scale, confined trial. The involvement and interest of these organizations attests to the fact that those involved in development-oriented projects employing recombinant technologies must be prepared to devote significant time and resources to meeting biosafety concerns.

As many of these parameters are still evolving, both domestically and internationally, neither AID nor the University of California could have anticipated them. However, those initiating new projects must plan in greater detail for expenditures necessary to meet obligations related to product safety. Total costs associated only with safety parameters of the recombinant rinderpest vaccine have still not been determined. Such unexpected costs place an extreme financial burden on development agencies and project budgets.

A separate concern relative to these vaccine programs may derive from inability on the part of the university cooperators to deliver the commercially fruitful products of their laboratory research. The eventual characterization, scale-up, and manufacture of a recombinant vaccine product might be most efficiently handled by the private sector, which has had a successful history in vaccine development. At this point in time, the AID office of Agriculture has no standard operative policy in effect for pursuing private sector cooperation. Again, such activities, when they have existed, have been pursued on a case-by-case basis and are time-intensive.

It has been recognized by AID that such private sector involvement will potentially require sole rights to federally funded research. This means that international testing could involve what would become proprietary products and thus regulatory approval would be requested by the individual companies concerned. Nevertheless, AID would continue to have jurisdictional responsibilities to ensure that those private sector initiatives in biotechnology which receive federal funding and are targeted for developing countries are in compliance with U.S. regulations and agency recommendations.

The potential for private sector involvement in donor-supported biotechnology research further emphasizes the need to assist developing countries in the establishment of biosafety guidelines regulating such research. The necessity derives from two points of consideration. First, such capability training would ensure that countries which are perceived to have a "relaxed" regulatory environment are not exploited as testing grounds for controversial research. Alternatively, the presence of responsible regulatory infrastructure may make these countries more desirable partners for private sector collaboration. The establishment of formal regulatory mechanisms governing the testing of genetically engineered products in a developing country would reduce the risks of liability to outside investors in this research.

Transgenic Potatoes: Where to Go from Here?

As mentioned above, research in the genetic manipulation of crop plants is being carried out in numerous countries through the CGIAR. Current activities include plant tissue and cell culture, pathogen-free plant production, in vitro germ plasm conservation, molecular diagnostics, embryo rescue, and genetic engineering. One of these centers, the International Potato Center (CIP) located in Lima, Peru, has recently begun to apply transformation technologies to potatoes. Transgenic plants have been produced which express amino acids derived from an inserted synthetic gene, referred to as "high essential amino acid encoding" (Yung et al. 1988).

This research has involved Louisiana State University (LSU) and CIP in collaborative research supported by AID. Transformed plants have been produced which express a high content of those essential amino acids most deficient in plant-derived proteins. These transformants include improved cultivars from CIP, a local Peruvian cultivar (Mariva), and Russet Burbank. These transgenic plants must remain in greenhouses at LSU and CIP pending regulatory approval for field testing.

This research thus involves a federal agency (AID), a university (LSU), and an international center (CIP). The project has yet to form a technical advisory group similar to those established for recombinant vaccine projects. This advisory group would directly involve appropriate representatives from APHIS so that approval could be obtained for domestic field testing. However, future plans will have to include testing of transgenic plants in Peru.

APPENDIX:

Compliance with Federal Guidelines and Regulatory Procedures[1]

The recipient will implement this research activity in accordance with: a) the National Institutes of Health Guidelines for Research Involving Recombinant DNA Molecules; b) procedures issued by the USDA, EPA, or other appropriate Federal agency, regarding testing of genetically engineered organisms; c) A.I.D.'s environmental procedures; and d) such other Federal guidelines and procedures as may apply during the course of research.

Additionally, the cooperator cannot commence testing in any foreign location until written approval for such testing is obtained from A.I.D. and the government of the country where testing is planned. Testing shall be conducted in accord with all applicable regulations of that country.

In addition, however, and prior to commencement of any such testing, the cooperator shall make a judgement and communicate the same to A.I.D. as to whether the regulations, procedures or facilities of the country in question are adequate to ensure testing in an environmentally sound manner. In the event such judgement is that they are not, the cooperator and A.I.D. will consult and agree on the conditions to be applied to the testing which will have such environmental effect. Reports submitted under this activity to A.I.D. will address regulatory issues as above related to the activity.

References

Agency for International Development (AID). 1987. Charter for agency standing committee on biotechnology. Agency for International Development, Washington, D.C.

Board on Science and Technology for International Development. 1982. *Priorities in Biotechnology Research for International Development-Proceedings of a Workshop.* National Academy Press, Washington, D.C.

Bookrack, H. L. 1981. Genetic engineering in plants and animals. Paper prepared for a U.S. Department of Agriculture workshop on emerging technologies, Chicago.

Buller, R. M. L., Smith, G. L., Cremer, K., Notkins, A. L., and Moss, B. 1985. Decreased virulence of recombinant vaccinia virus expression vectors is associated with a thymidine kinase-negative phenotype. *Nature* 317:813.

Chambers, J. A., and Cohen, J. I. 1990. Recombinant animal vaccines for the developing world: biosafety considerations. *Anaplasmosis Babesiosis Newsl.* 5:1–2.

Cohen, J. I., Plucknett, D. L., Smith, N. J. H., and Jones, K. A. 1988*a*. Models for integrating biotech into crop improvement programs. *Bio/Technology* 6:387–392.

Cohen, J. I., Jones, K. A., Plucknett, D. L., and Smith, N. J. 1988*b*. Regulatory concerns affecting developing nations. *Bio/Technology* 6:744.

Cohen, J. I. 1989. Biotechnology research for the developing world. *Trends Biotechnol.* 7:295–303.

Cohen, J. I. 1990. International donor support for agricultural biotechnology. *Food Policy.* 15(1):57–66.

CTA/FAO. 1989. Plant biotechnologies for developing countries. Conclusions and recommendations. Technical Center for Agricultural and Rural Cooperation, Wageningen, Netherlands.

[1]Section exerpted from the AID Office of Agriculture standard form for grants supporting biotechnology research.

Directorate-General for International Cooperation. 1989. Biotechnology and development cooperation. The Hague, Section for Research and Technology, New Governmental Organizations, The Netherlands.

Faulkner, D. E. 1983. *World Animal Review*, special supplement. United Nations Food and Agriculture Organization, pp. 5–9.

German Foundation for International Development. 1989. Biotechnology in plant breeding and plant nutrition for the benefit of the third world. Feldafing, 8133, Germany.

Glosser, J. W. 1988. Regulation and application of biotechnology products for use in veterinary medicine. *Rev. Sci. Tech. Offi. Inti. Epiz.* 7:223–237.

Jones, K. A. 1987. A.I.D.'s activities in biotechnology: regulatory considerations. Contract DAN-1406-0-00-7008-00, Agency for International Development, Washington, D.C.

Karny, G. M. 1986 .An international approach to biotechnology safety. UNIDO/IS 627, Austria.

Mellon, M. 1988. *Biotechnology and the Environment. A Primer on the Environmental Implications of Genetic Engineering*. National Wildlife Federation, Washington D.C.

National Institutes of Health (NIH). 1989. Recombinant DNA research; actions under guidelines. *Fed. Reg.* 54(47):10508–10510, March 13.

National Research Council. 1990. *Plant Biotechnology Research for Developing Countries*. National Academy Press, Washington, D.C., 1990

Organization of International Epizootics (OIE) 1989. Report of the expert consultation on requirements for vaccinia-rinderpest recombinant vaccines. Paris, France 75017.

Persley, G. (ed). 1989. Biotechnology study project papers: synthesis report. World Bank, Washington, D.C.

Plucknett, D. L. 1988. An overview of biotechnology research in the CGIAR. Original paper presented at the CGIAR/Japanese Forum on Biotechnology in Tokyo, Japan. CGIAR Secretariat, World Bank, Washington, D.C.

Plucknett, D. P., Cohen, J. I., and Horne, M. *Future Role of the IARCs in the Application of Biotechnology in Developing Countries*. In Agricultural Biotechnology: Opportunities for International Development. Biotechnology in Agriculture Series No. 2, Oxon: CAB International, U.K. In press.

Tiedje, J. M., Colwell, R. I., Grossman, Y. L., Hodson, R. E., Leuski, R. E., Mack, R. N., and Regal, P. J. 1989. The planned introduction of genetically engineered organisms: ecological considerations and recommendations. *Ecology* 70(2):298–315.

Umminger, B. 1989. Biosafety in agricultural research. In J. I. Cohen (ed.), *Strengthening Collaboration in Biotechnology: International Agricultural Research and the Private Sector*. Agency for International Development, Washington, D.C., pp. 437–442.

United Nations Industrial Development Organization (UNIDO). 1985. Safety guidelines and procedures for bioscience-based industry and other applied microbiology. UNIDO Secretariat, Austria.

United Nations Industrial Development Organization (UNIDO). 1989. Biotechnology for development. UNIDO, Vienna.

Yang, M. S., Espinoza, N. O., Nagpala, P. G., Dodds, J. H., White, F. F., Schnoor, K. L., and Jaynes, J. M. 1988. Expression of a synthetic gene for improved protein quality in transformed potato plants. *Plant Sci.* 64:2724–2736.

Yilma, T., Hsu, D., Jones, L., Owens, S., Grubman, M., Mebus, C., Yamanaka, M., and Dale, B. 1988. Protection of cattle against rinderpest with vaccinia virus recombinants expressing the ha and f gene. *Science* 242:1058–1061.

Index

ABOUT THE EDITORS

MORRIS LEVIN is a senior research scientist at the University of Maryland's Maryland Biotechnical Institute. He previously served as program coordinator for biotechnology risk assessment with the U.S. Environmental Protection Agency.

HARLEE STRAUSS is president of H. Strauss Associates, Inc., and a consultant to government and industry on environmental biotechnology. She has research appointments at MIT and at Tufts University.